ADVANCES IN

Applied Microbiology

VOLUME 19

CONTRIBUTORS TO THIS VOLUME

S. G. Bradley

A. C. Chang

I. Chet

J. J. Ellis

Melvin S. Finstein

D. D. Focht

Y. Henis

C. W. Hesseltine

D. J. D. Hockenhull

Hidehiko Kumagai

Hubert A. Lechevalier

Erwin F. Lessel

Irving Marcus

Merry L. Morris

Koichi Ogata

T. G. Pridham

Donald V. Richmond

Hideaki Yamada

ADVANCES IN

Applied Microbiology

Edited by D. PERLMAN

School of Pharmacy
The University of Wisconsin
Madison, Wisconsin

VOLUME 19

 1975

ACADEMIC PRESS, New York San Francisco London
A Subsidiary of Harcourt Brace Jovanovich, Publishers

COPYRIGHT © 1975, BY ACADEMIC PRESS, INC.
ALL RIGHTS RESERVED.
NO PART OF THIS PUBLICATION MAY BE REPRODUCED OR
TRANSMITTED IN ANY FORM OR BY ANY MEANS, ELECTRONIC
OR MECHANICAL, INCLUDING PHOTOCOPY, RECORDING, OR ANY
INFORMATION STORAGE AND RETRIEVAL SYSTEM, WITHOUT
PERMISSION IN WRITING FROM THE PUBLISHER.

ACADEMIC PRESS, INC.
111 Fifth Avenue, New York, New York 10003

United Kingdom Edition published by
ACADEMIC PRESS, INC. (LONDON) LTD.
24/28 Oval Road, London NW1

LIBRARY OF CONGRESS CATALOG CARD NUMBER: 59-13823

ISBN 0–12–002619–8

PRINTED IN THE UNITED STATES OF AMERICA

CONTENTS

List of Contributors ix

Culture Collections and Patent Depositions

T. G. Pridham and C. W. Hesseltine

I.	Introduction	1
II.	History of Patent Culture Depositions	3
III.	ARS Culture Collection Policies	4
IV.	The Deposit	7
V.	Preparation of Materials for Preservation and Distribution	12
VI.	Preservation and Storage	14
VII.	Records	18
VIII.	Availability and Distribution	18
IX.	The Future	19
X.	Concluding Remarks	21
	References	21

Production of the Same Antibiotics by Members of Different Genera of Microorganisms

Hubert A. Lechevalier

I.	Introduction	25
II.	Same Antibiotics Produced by Different Organisms	27
III.	Conclusion	42
	References	43

Antibiotic-Producing Fungi: Current Status of Nomenclature

C. W. Hesseltine and J. J. Ellis

I.	Introduction	47
II.	Antibiotic-Producing Fungi	47
III.	Rules of Botanical Nomenclature	48
IV.	Purpose of the Botanical Code	48
V.	Numbers of Fungi	49
VI.	Botanical Rules Specifically for Fungi	49
VII.	Comparison of the Botanical and Bacterial Codes	50
VIII.	Descriptions of Fungi	53

IX.	Type Cultures in Culture Collections	54
X.	Publications Dealing with Fungal Nomenclature	56
	References	57

Significance of Nucleic Acid Hybridization to Systematics of Actinomycetes

S. G. Bradley

I.	Introduction	59
II.	DNA Nucleotide Composition	60
III.	DNA:DNA Association	60
IV.	The Cot Concept	63
V.	Optical Reassociation	64
VI.	Phylogenetic Implications	66
VII.	Neutral Mutations	69
	References	70

Current Status of Nomenclature of Antibiotic-Producing Bacteria

Erwin F. Lessel

I.	Introduction	71
II.	Antibiotic-Producing Bacteria as Problems to Nomenclature	71
III.	Two Major Nomenclatural Problems	72
IV.	Major Changes in International Code of Nomenclature	73
V.	Requirement for Deposition of Type Strains	74
VI.	Effect of Changes in International Code	75
VII.	Speciation of *Streptomyces*	75
	References	76

Microorganisms in Patent Disclosures

Irving Marcus

I.	Introduction	77
II.	Responsibilities and Requirements	78
	Appendix	83

Microbiological Control of Plant Pathogens

Y. Henis and I. Chet

I.	Introduction	85
II.	Naturally Occurring Microbiological Control	86

III.	Induced Microbiological Control of Plant Pathogens	87
IV.	Mechanisms of Microbiological Control	98
V.	Integrated Control	105
	References	107

Microbiology of Municipal Solid Waste Composting

MELVIN S. FINSTEIN AND MERRY L. MORRIS

I.	Introduction	113
II.	Self-Heating Variations	116
III.	The Temperature Ascent	121
IV.	The Temperature Descent	134
V.	Batch and Continuous Composting	138
VI.	Operational Factors	141
VII.	Conclusion	148
	References	148

Nitrification and Denitrification Processes Related to Waste Water Treatment

D. D. FOCHT AND A. C. CHANG

I.	Introduction	153
II.	Biochemistry of Nitrification and Denitrification	155
III.	Environmental Factors Affecting Nitrification and Denitrification	161
IV.	Comparative Waste Treatment Methods for Nitrification and Denitrification	173
V.	Summary and Conclusions	181
	References	182

The Fermentation Pilot Plant and Its Aims

D. J. D. HOCKENHULL

I.	Why a Pilot Plant?	187
II.	The Pilot Plant as an Introduction to Production Management	197
	References	208

The Microbial Production of Nucleic Acid-Related Compounds

KOICHI OGATA

| I. | Introduction | 209 |
| II. | Production of 5'-IMP and 5'-GMP by the Enzymic Hydrolysis of RNA | 210 |

III.	Production of 3′-, 2′-Nucleotides and 5′-Deoxynucleotides by the Enzymic Hydrolysis of RNA and DNA	213
IV.	Excretion of RNA Derivatives	214
V.	Fermentative Production of Nucleosides, Nucleotides, Ribose, Orotic Acid, and DNA	216
VI.	Salvage Synthesis of Nucleosides and Nucleotides	225
VII.	Conversion of Nucleosides	226
VIII.	Formation of Nucleoside Derivatives	230
IX.	Production of Coenzyme A	236
X.	Conclusion	240
	References	241

Synthesis of L-Tyrosine-Related Amino Acids by β-Tyrosinase

Hideaki Yamada and Hidehiko Kumagai

I.	Introduction	249
II.	Physicochemical Properties of β-Tyrosinase	250
III.	Reaction Mechanism	258
IV.	Immobilization of β-Tyrosinase on Sepharose	273
V.	Enzymic Preparation of L-Tyrosine and L-Dopa	274
VI.	Conclusions	285
	References	285

Effects of Toxicants on the Morphology and Fine Structure of Fungi

Donald V. Richmond

I.	Introduction	289
II.	Morphological Changes Induced by Toxicants	291
III.	Effects on Dimorphic Fungi	301
IV.	Effects of Toxicants on Fine Structure	306
	References	316

Subject Index	321
Contents of Previous Volumes	325

LIST OF CONTRIBUTORS

Numbers in parentheses indicate the pages on which the authors' contributions begin.

S. G. BRADLEY, *Department of Microbiology, Virginia Commonwealth University, Richmond, Virginia* (59)

A. C. CHANG, *Department of Soil Science and Agricultural Engineering, University of California, Riverside, California* (153)

I. CHET, *Department of Plant Pathology and Microbiology, The Hebrew University of Jerusalem, Faculty of Agriculture, Rehovot, Israel* (85)

J. J. ELLIS, *ARS Culture Collection Research, Fermentation Laboratory, Northern Regional Research Laboratory, Peoria, Illinois* (47)

MELVIN S. FINSTEIN, *Department of Environmental Science, Cook College, Rutgers University, New Brunswick, New Jersey* (113)

D. D. FOCHT, *Department of Soil Science and Agricultural Engineering, University of California, Riverside, California* (153)

Y. HENIS, *Department of Plant Pathology and Microbiology, The Hebrew University of Jerusalem, Faculty of Agriculture, Rehovot, Israel* (85)

C. W. HESSELTINE, *Chief, Fermentation Laboratory, Northern Regional Research Laboratory, Peoria, Illinois* (1, 47)

D. J. D. HOCKENHULL, *Glaxo Laboratories Ltd., Ulverston, Cumbria, England* (187)

HIDEHIKO KUMAGAI, *The Research Institute for Food Science, Kyoto University, Uji, Kyoto, Japan* (249)

HUBERT A. LECHEVALIER, *Waksman Institute of Microbiology, Rutgers University, The State University of New Jersey, New Brunswick, New Jersey* (25)

ERWIN F. LESSEL,* *American Type Culture Collection, Rockville, Maryland* (71)

IRVING MARCUS,† *U.S. Patent Office, Washington, D.C.* (77)

MERRY L. MORRIS, *Department of Environmental Science, Cook College, Rutgers University, New Brunswick, New Jersey* (113)

---

* Present address: Lederle Laboratories, a Division of American Cyanamid Co., Pearl River, New York.

† Present address: 8411 Spencer Court, Chevy Chase, Maryland.

KOICHI OGATA, *Department of Agricultural Chemistry, Kyoto Univerversity, Kyoto, Japan* (209)

T. G. PRIDHAM, *Research Leader, ARS Culture Collection Research, Fermentation Laboratory, Northern Regional Research Laboratory, Peoria, Illinois* (1)

DONALD V. RICHMOND, *Long Ashton Research Station, University of Bristol, Bristol, England* (289)

HIDEAKI YAMADA, *The Research Institute for Food Science, Kyoto University, Uji, Kyoto, Japan* (249)

ADVANCES IN

Applied Microbiology

VOLUME 19

Culture Collections and Patent Depositions[1]

T. G. PRIDHAM[2] AND C. W. HESSELTINE[3]

Northern Regional Research Laboratory, Peoria, Illinois

I.	Introduction	1
II.	History of Patent Culture Depositions	3
III.	ARS Culture Collection Policies	4
IV.	The Deposit	7
	A. Time of Deposit	7
	B. Nature of Deposit	8
	C. Acceptability of Deposits	9
	D. Description of Material	9
	E. Statement of Availability by Depositor	10
V.	Preparation of Materials for Preservation and Distribution	12
	A. Retention of Original Submissions	12
	B. Cultivation	12
	C. Characterization	13
	D. Identifications	13
	E. Nomenclature	13
	F. Verification of Claims for Deposited Cultures	14
VI.	Preservation and Storage	14
	A. Maintenance on Agar Slants	14
	B. Storage of Cultures under Oil	15
	C. Storage of Cultures in Deep Freeze	15
	D. Storage of Cultures in Soil	15
	E. Storage of Cultures in Lyophile	16
	F. Storage of Cultures in or over Liquid Nitrogen	16
	G. Viability and Characterization Checks	17
	H. Maintenance in Perpetuity	17
	I. Reserve Collections	18
VII.	Records	18
VIII.	Availability and Distribution	18
IX.	The Future	19
	A. Official Designation of Patent Collections	19
	B. Extra Demands on Collections	20
	C. National and International Patent Legislation	20
X.	Concluding Remarks	21
	References	21

I. Introduction

Man's exploitation of microorganisms over the centuries is a well-known fact. Within the last three decades, this exploitation has had considerable

[1] Presented in part at the symposium on "Antibiotics and Antibiotic-Producing Organisms. Classification and Patent Issues" at the 13th Interscience Conference on Antimicrobial Agents and Chemotherapy, Washington, D.C., September 19, 1973.

[2] Research Leader, ARS Culture Collection Research, Fermentation Laboratory, Northern Regional Research Laboratory, Peoria, Illinois.

[3] Chief, Fermentation Laboratory, Northern Regional Research Laboratory, Peoria, Illinois.

impact not only on the progress of microbiology and chemistry and their many subdisciplines but also on the legal profession.

Literally millions of strains of microorganisms have been studied, in varying degree, by researchers in industry, government, and academe in efforts to discover new products; to develop new processes for the production of microbial metabolites useful in medical, agricultural, and industrial applications; and to control food spoilage. As a result of these intensive studies, scientific contributions have been made in chemistry, biochemistry, microbial physiology, microbial taxonomy, and in both chemical and biological nomenclature.

Patent laws and regulations, allowing some degree of protection to the inventors or discoverers of new processes and metabolites, have led to the buildup of a large body of patent literature along with the scientific literature. Many microbiologists and chemists today must consider both kinds of literature to keep abreast of developments in their fields of specialization. The scientific merits of the patent system have been questioned from time to time; however, in our opinion, the system overall has led to considerable scientific progress. Nevertheless, much valuable information probably still lies hidden and unreported because of the patent system.

The steadily growing use of microorganisms has imposed new problems on the microbiologists, chemists, and patent examiners and attorneys involved. Microbiologists and chemists must consider both scientific law and juridical law in the conduct of their work. Whether he likes it or not, the microbiologist curator of a collection of "patent cultures" becomes involved with legal aspects of satisfying requirements of patent offices throughout the world so far as preservation, maintenance, and distribution of these cultures is concerned. Requests for cultures and questions about their availability make it necessary to examine records on particular strains to determine whether a U.S. patent has been issued or, if not yet issued, whether the depositor has indicated that the culture could be distributed prior to issuance of the patent. The very tone of questions that one receives regarding patent cultures requires the assessment of each case on an individual basis. Clearly, some of the questions that have been posed to us have had legal implications.

Although a number of papers addressed to problems concerning patent cultures have appeared, many deal with legal aspects rather than microbiological or chemical ones: Casida (2), Daus (5,6), Edelblute (7), Hayhurst (15,16), Kent (21), Kurylovich (22), Levy and Wendt (23), Neshatayeva and Kiselyov (25), and Stoy (34).

This paper describes how we have handled our "patent culture collection" and points out some problems from the viewpoint of the microbiologist, who must be aware not only of new developments in his own discipline, but also of possible national and international legal implications of his activities.

II. History of Patent Culture Depositions

From the beginning, cultures deposited in connection with patent applications occupied a unique position because of various requirements for confidentiality, special records, and special handling. Today, most major collections have a "patent culture collection."

The practice of depositing microorganisms in culture collections other than those of the patent applicants apparently existed nowhere in the world prior to 1949. So far as we know, the first such deposition was made by the American Cyanamid Company, in August 1949, when company representatives brought cultures of strain Lederle A-377 of *Streptomyces aureofaciens* Duggar to Peoria. Arrangements were made to deposit this strain in our Collection (NRRL), now known as the ARS Culture Collection, and it was accessioned as strain NRRL 2209. Later, in May 1950, representatives of Charles Pfizer & Company, Incorporated, deposited cultures of *Streptomyces rimosus* Finlay *et al.*, accessioned as strain NRRL 2234. *Streptomyces venezuelae* Ehrlich *et al.* also was one of the first antibiotic-producing streptomycetes deposited in a major collection, the American Type Culture Collection (ATCC), apparently in response to patent office requirements. It may well have been the first culture so deposited. Although we do not know the circumstances of its deposit in the ATCC, the type strain A65 = P.D. 04745 is listed in the second known U.S. patent on chloramphenicol production as strain ATCC 10712. The application date for this patent was March 16, 1948, and the issuance date October 4, 1949.

Other companies then began depositing cultures of microorganisms in connection with patent applications in the ATCC; the Collection of the Institute of Microbiology, Rutgers University; and our ARS Culture Collection. Most of these depositions, if not all, were made in confidence. Our arrangements were made with the understanding that progeny of the cultures would be made available and distributed to bona fide requestors from the time of issuance of the U.S. patent(s).

Aside from the three collections mentioned, and although there is no legislation for the practice (29,41), other collections have become involved. Based on statements made in granted patents, the U.S. Patent Office recognizes deposition of cultures in a number of foreign culture collections, e.g., The National Collection of Industrial Bacteria (NCIB), the Commonwealth Mycological Institute (CMI or IMI), and the Forest Products Research Laboratory (FPRL) in the United Kingdom; the Centraalbureau voor Schimmelcultures (CBS) in the Netherlands; the Research Laboratories of Hindustan Antibiotics, Ltd. (HACC) in India; and the Institute for Applied Microbiology (IAM), the Fermentation Research Institute (IFO), the National Institute of Animal Health (NIAH), and the Faculty of Agriculture Hokkaido University (AHU) in Japan, among others. These collections have been cited in U.S. patents with application dates in the last several years.

These depositions apparently were made to satisfy demands for complete disclosure of specifications in applications and, in retrospect, would seem to have solved the problems that (i) it is difficult to set down on paper how to isolate a particular microorganism from a sample (generally soil) so that one "skilled in the art," i.e., a competent microbiologist or perhaps a chemist, could *readily* obtain the microorganism from nature; (ii) there being no precise definition for many species of microorganisms, it is difficult to write a description that will permit a competent microbiologist to *readily and accurately* recognize a particular taxon; and (iii) without the proper microorganism the processes would be inoperative, and no amount of words ever could allow them to be operable.

The practice of patent culture deposition has continued over the years; the ARS Culture Collection now maintains a "patent culture collection" of more than 1000 strains, the great majority of which are Actinomycetales. U.S. patents covering the use of about one-half of these strains have been granted.

III. ARS Culture Collection Policies

Because there are limited guidelines governing deposition of cultures of microorganisms in connection with patent applications (4,14,21,46), each collection has developed its own particular policies. Our practices are based on some legal advice, on guidelines published from time to time in various journals including the *Official Gazette* of the U.S. Patent Office, and on our own in-house experiences and policies. Such policies, of course, are subject to change from time to time. Within recent years, a number of new companies and organizations have entered the picture. Because some of these were unfamiliar with culture deposition practices, we have prepared a procedures and policies statement as a guide for prospective depositors who request this information. The latest revision is given below.

Procedures and Policies for Deposition of Cultures for
Patent Purposes in the ARS Culture Collection

The ARS Culture Collection serves as a depository for cultures which are involved in fermentation patents and, therefore, will be glad to receive such materials in connection with patent applications. When such a culture is received, it is assigned a number in the collection and is maintained thereafter in a living state. Immediately after receipt, a letter is written to the depositor advising of the number assigned and including one of the following statements:

Furthermore, insofar as is practicable in carrying out the business of the Department of Agriculture, we shall refrain from distributing this culture pending the issuance of the U.S. Patent to your company, with the exception, however, that access to this culture by other parties will be granted upon receipt of written authorization from your company specifying the name and the ARS Culture Collection designation (NRRL number) of the culture and identifying the party who is to receive it. (Restricted distribution.)

OR

As of this date, the subject culture(s) will be made available to anyone who requests the same. (Nonrestricted distribution.)

OR

With reference to 886.O.G. 638, progeny of this (these) strain(s) will be avilable during pendency of the patent application to one determined by the Commissioner of patents to be entitled thereto under Rule 14 of the Rules of Practice in Patent Cases and 35 U.S.C. 122. All restrictions on the availability of progeny of the strain(s) to the public will be irrevocably removed upon the granting of the patent(s) of which the strain(s) is (are) the subject.

Deposition of strains of microorganisms in the ARS Culture Collection in connection with patent applications affords reasonable permanency of the deposit and ready accessibility thereto by the public if a patent is granted.

Pertinent references concerning deposition of strains for patent purposes are: USPQ 157: 437–444 (1967); OG 848: 863–867 (1968); OG 849: 5–11 (1968); USPQ 168: 99–104 (1971); USPQ 169, No. 6: II–III (1971); and OG 886, No. 4: 638 (1971).
It is suggested that you seek advice from your attorney as to which type of statement you should use. The ARS Culture Collection letter then can be attached to the patent application for the Patent Examiner.
There is no charge for the deposit or maintenance of cultures.
The ARS Culture Collection is unable to accept for deposit strains of viruses and would have to carefully consider any request to deposit strains of bacteria, yeasts, molds, *Actinomycetales*, and parasitic agents listed in classes 2 and 3 of the U.S. Department of Health, Education, and Welfare's "Classification of Etiological Agents on the Basis of Hazard" and the U.S. Department of Agriculture's publica-

tions PA-873 and PA-967. Also we are unable to accept microorganisms that would be considered fastidious or mixtures of microorganisms which cannot be lyophilized. Potential depositors also should be familiar with the various U.S. laws and regulations regarding shipment and import of microorganisms.

The ARS Culture Collection does not issue a catalog or list. It has no regulations imposing restrictions on the use of such cultures deposited for patent purposes. Such materials are distributed according to the depositor's wishes which, in turn, generally are based on his interpretation of patent office requirements. Use of such materials, once distributed, is the responsibility of the requestor. Cultures are automatically removed from any restriction category, once a U.S. patent issues wherein the particular microorganism is involved.

Curators in the ARS Culture Collection do not attempt to make an identification or to name any organism which has been deposited in connection with a patent application, nor do they carry out research work with such deposits until a U.S. Patent issues or cultures are otherwise released. It is not necessary, of course, to provide a precise identification but the depositor should at least state to what genus the microorganism belongs. Also, if special media are required for its maintenance, the curators need to know this. Ordinarily, one or two agar slant cultures and one or two lyophilized preparations are received from depositors. Depositors also are responsible for resupplying material should the need ever arise and this responsibility extends beyond the life of the patent.

The depositor has the option of sending cultures for deposit in the ARS Culture Collection in three ways:

1. Thirty lyophilized preparations, clearly labeled with the depositor's original strain designation and preferably in tubes no longer than indicated in the drawing below:

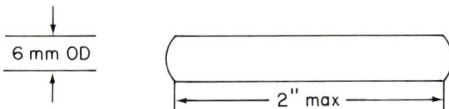

One of these is checked for viability, the NRRL number placed on each tube, and the supply of tubes stored at 3 to 5°C. *Bona fide* letter requests for the culture would be shipped from this stock.

The ARS Culture Collection will no longer accept materials for deposit under option 1 unless they meet the specifications cited above. Larger sized tubes greatly complicate storage.

2. One lyophilized preparation, clearly labeled with the depositor's original strain designation. On receipt, the microorganism is cultivated on appropriate agar media and thirty lyophilized preparations made. One of these is checked for viability, the remainder handled as in option 1. This option, and option 3 below, is acceptable provided cultures submitted are not fastidious and do not require more than usual normal operating procedures.

3. One, or preferably two, agar slant cultures of the microorganisms growing on an appropriate medium. Sufficient material is prepared by our curators to make thirty lyophilized preparations, check one for viability and handle the remainder as in options 1 and 2. When the initial agar slant cultures deposited appear suitable, lyophilizations often are made from that material.

Cultures deposited in the ARS Culture Collection are considered as public property and the property of the ARS Culture Collection. Therefore, no strains are removed, returned to depositors *in toto,* or completely destroyed except for very good reason, e.g., the receipt and accessioning of a virulent pathogen by mistake.

Progeny (agar slant cultures or lyophilized preparations) of strains of microorganisms deposited in connection with patent applications may be obtained (when restrictions, if any, are removed), free of charge by letter request stating the name of the microorganism and its strain number (either the depositor's number or our NRRL number) or by providing an otherwise satisfactory reference to the strain(s) in question.

We do not provide depositors with the names of requestors of microorganisms.

IV. The Deposit

A. TIME OF DEPOSIT

Generally, cultures are accessioned in our patent culture collection on the day the deposit material is received and the depositor is so informed of the date. This procedure is based on the premise that the deposited material is, in fact, viable and authentic. In most instances the materials sent for deposit are viable. We have no knowledge of the relationship between the time of deposit of a particular strain and the date of the patent application. There has been no need for such information. The date of deposit, however, is of particular importance to depositors as exemplified in the latest guideline appearing in the *Official Gazette* of the U.S. Patent Office (46).

We do not authenticate cultures beyond simple generic or group (bacteria, yeasts, molds, Actinomycetales) placement. Sometimes, however, it may take as long as 1 month to grow cultures satisfactorily. Also, other responsibilities of our curators could delay preparation of materials for distribution stocks. These facts should be taken into consideration by the depositors in the event there might be a request for cultures within a short time of deposit. Such a circumstance conceivably could occur. It probably would be difficult for us to honor any request for material 1 or 2 days after receipt of the cultures for deposit. One could inoculate an agar slant of appropriate medium and ship it before any growth has occurred, but we look with disfavor on such a practice. To help resolve the problem of immediate requests, we will accept lyophilized preparations (30 such preparations is a reasonable and practical number) provided they meet certain physical requirements as indicated in our procedures and policies statement.

The practice of having the depositor lyophilize his own material in sufficient quantity for subsequent use in honoring requests also has other advantages. In the past, we have been accused of occupying a privileged position with regard to patent depositions because we, too, are engaged in fermentation research. Access to active cultures would, of course, provide our curators with certain kinds of information. However, the so-called privileged position never can be escaped entirely, because records must be kept. The scientific integrity of our curators gives assurance that advantage is not taken of this position.

B. Nature of Deposit

The kinds of materials we have received in connection with patent applications have ranged from cultures preserved in soil to lyophilized preparations. Generally, agar slant cultures are provided. In some instances the deposit of a single strain consists of an agar slant culture and a soil culture or an agar slant culture and one lyophilized preparation. In the letter accompanying the deposit, all materials should be clearly identified with the acronym (abbreviation, sigla) of the depositor's collection, a number designation in his collection, and the name of the organism. This has not always been the case. Occasionally, in the past, only the names of certain microorganisms were provided and, therefore, release of the cultures was obstructed. The same problem of appropriate delineation of microorganisms occurs in publications in scientific journals. When attempts to obtain release from the depositor for distribution of the culture fails, one other recourse is to check with the patent office regarding these cultures. Certain information in the applications may allow resolution. Otherwise, the particular cultures involved can be effectively restricted for years.

Occasionally, either obviously mixed or contaminated cultures are received. Receipt of such materials requires extra correspondence and work. Conceivably, legal questions could be raised. Because it often is difficult to pinpoint the origin of contamination, the competency of the depositor or of the curator can be questioned. In the case of accidentally contaminated cultures, legal questions as to whether the depositor or the curator was responsible could be raised. The 30-lyophilized preparation option would place the responsibility for culture purity with the depositor.

The use of deliberately mixed cultures for certain processes poses a different kind of problem. We are not prepared to accept such mixtures because the ratio of the components may be difficult to maintain. We can accept the individual components. If the components of such mixtures can be lyophilized and deposit of the mixture is preferred, the depositor is encouraged to prepare his own mixtures in proper ratio and exercise the 30-lyophilized preparation option.

The future may well hold more problems for curators insofar as the nature of deposits is concerned. New directions of microbiological and chemical research have resulted in the appearance of more exotic and fastidious kinds of microorganisms as well as mixtures of microorganisms in processes. Further exploitation of the more commonly occurring, less fastidious, and easily isolatable microorganisms is becoming increasingly difficult. One might expect, then, that human, animal, and plant pathogens; extreme thermophiles or psychrophiles; diatoms; protozoa; nematodes; viruses; and cell lines of all kinds will appear among patent deposit strains in the years to come. Because of other responsibilities and because handling of these microorganisms falls outside their expertise or limitations, our present curators cannot work with such materials.

C. Acceptability of Deposits

Because of the nature of the ARS Culture Collection and the missions and goals of the U.S. Department of Agriculture, the Agricultural Research Service, and our laboratory (the Northern Regional Research Laboratory), we currently are unable to accept certain microorganisms, particularly the viruses. Phages (bacterial or fungal viruses) are a possible exception. Processing of phages might be difficult, and we would prefer deposit of lyophilized phage and host preparations. Also, any request to deposit strains of bacteria, yeasts, molds, Actinomycetales, and parasitic agents listed in classes 2 and 3 of the U.S. Department of Health, Education, and Welfare's "Classification of Etiologic Agents on the Basis of Hazard" (40) and the U.S. Department of Agriculture's publications PA-873 and PA-967 (38,39) would have to be very carefully considered. We maintain liaison with the appropriate agencies in this regard. We do not plan to expand our operations in this direction.

Moreover, potential depositors should be aware of the packaging standards and permits for importation, exportation, and shipping of cultures required by the U.S. Public Health Service, the U.S. Department of Agriculture's Animal and Plant Health Division and Plant Quarantine Division, the U.S. Department of Commerce's Export Division, the U.S. Bureau of Custom's Import Division, the Department of the Army's Industrial Health and Safety Directorate, and other Federal and State agencies.

D. Description of Material

When we receive only one agar slant culture or one lyophilized culture, it is our policy to prepare a number of lyophilized preparations of patent culture deposits as close, in number of generations to the original material as possible. Often, sufficient material is not available from the original slant culture alone. Therefore, it is necessary that our curators have access to a reasonable description of the microorganism. Sometimes, only a generic name is sufficient. Occasionally, we receive lengthy taxonomic descriptions. We need only enough information to cultivate the microorganism appropriately and to recognize whether the culture is contaminated or mixed. We do not authenticate identifications of the organisms. Authentication would require a great deal of effort, particularly where yeasts, bacteria, and Actinomycetales are concerned. Many physiological tests would be required. Our curators accept the name given to the deposited microorganism until the U.S. patent has been granted, or until a nonrestricted culture has been sent to at least three requestors. After that time, the name of the microorganism is in the same scientific status as many others, i.e., subject to continuing changes in taxonomic concepts and in nomenclature. Thus, the original acronym and number (not the name) supplied by the depositor assume major importance as the only real fixed denominator so far as the history of the strain is concerned. Hopefully, most microbiologists are interested in describing their cultures and in naming them. One must remember, however, that the names are subject to change with the advance of science.

E. Statement of Availability by Depositor

The availability and confidentiality of patent deposition cultures has been a matter of considerable concern to both foreign and domestic depositors and to the U.S. Patent Office. Some letters accompanying deposits are quite specific with regard to availability statements. Others say nothing in this regard. For the latter, we automatically assume that the culture is to be held in what we call "restricted status," i.e., progeny of the strain will not be sent to anyone other than the depositor or persons designated by him or the U.S. Patent Office, until the patent issues. Appearance of the name and our acronym (NRRL) and strain number in non-U.S. patents or other publications does not remove this restriction. We have received a number of letters requesting information on availability and confidentiality of deposits.

As a result of the controversy about availability of cultures and ensuing decisions by the U.S. Patent Board of Appeals and the Courts (*3,6,29,41–44*), we have had to recategorize our records on patent cultures. We now maintain three separate files: those on strains which are the subject of issued U.S. patents; those on strains deposited and accompanied with instructions that cultures be made available prior to grant of patent; and those on strains which are to be maintained in "restricted status" until grant of the U.S. patent.

An interesting series of letters concerning availability of cultures of microorganisms (not necessarily directed to the patent culture problem) appeared in the American Society for Microbiology's News Letters in 1971 and 1972 (8). It is obvious from many of the letters that the special restrictions on economically important microorganisms apparently were either unknown or considered unimportant by some of the correspondents. Also of interest is the proposal (13) currently under consideration which may allow publication of U.S. patent applications prior to issuance of patents. This proposal, if adopted, could result in requests for patent cultures prior to issuance of the U.S. patent, or before the depositor has released the microorganism for distribution.

It should be remembered that patent culture availability during the pendency of a patent application is determined by the Commissioner of Patents and the depositor (46). Because of the confidential nature of applications, the only persons privy to the name, acronym, and strain number of microorganisms are the depositors and their attorneys, certain personnel in the U.S. Patent Office, and the curators of the collection wherein the culture is deposited. Therefore, except for actions taken by these individuals, no requests for the culture could be initiated until the patent issues. Prior publication of the name, acronym, and strain number does sometimes occur when foreign patents issue or foreign applications are published before the U.S. patent is granted. However, in such cases we continue to restrict distribution until the depositor or the U.S. Patent Office communicates with us. Notification by depositors that the U.S. patent has been granted has been minimal. Justification for distribution of many strains which are the subjects of U.S. patents have been based on our routine scanning of the *Official Gazette* of the U.S. Patent Office. Unusual titles to inventions and the omission of acronyms and strain numbers has complicated this procedure. Inclusion of the acronyms and strain numbers in the *Official Gazette* abstracts would help resolve this problem.

From a microbiological viewpoint, if information on the nature of a particular process were known, a microbiologist would like to have a culture or its description so that available files of information could be checked, or a selected number of cultures in his own collection could be screened to learn whether any produced the metabolites or carried out the process concerned. Except for interferences, such information and cultures are unknown to others in the field prior to appearing in print.

The U.S. Patent Office, with its complete knowledge of all U.S. applications and the legal requirements for making cultures available, is in the best position to publish the names, acronyms, and strain numbers of patent microorganisms and of the metabolites or processes involved. Such publication would provide a measure of relief because the information now is provided in haphazard fashion as a result of the involvement of more than one major collection of microorganisms.

V. Preparation of Materials for Preservation and Distribution

A. RETENTION OF ORIGINAL SUBMISSIONS

We have not yet had call to provide samples of original material deposited. In some cases, none is available because the only submission was one lyophilized preparation. Depositors in the ARS Culture Collection are obligated to supply material should ours die out or be expended. So far as possible, we have tried to store remnants of original depositions, e.g., original agar slant cultures. Many of these agar slant cultures, of course, have dried down and may or may not be viable. Replicates of original material, extra lyophilized preparations, cultures in soil, and used original agar slant cultures are stored at 3–5°C, constituting a separate collection in itself. Information on the original labels has been of value in one or two instances.

It would seem wise for depositors of patent cultures to include an additional lyophilized preparation (or soil culture) with their deposits for their own protection. Curators are not infallible, and the availability of the microorganism depends upon having viable material to distribute. Provision of a second lyophilized preparation, where possible, would ensure accurate resolution of questions arising at some later date.

B. CULTIVATION

One receipt, submitted material is assigned an NRRL strain number, and original records are prepared. Depending upon the nature of the microorganism it is turned over to the appropriate curator to prepare material and make 30 lyophilized preparations if possible. It is essential for us to have some information relative to the genus or group of microorganisms involved so that the proper medium for cultivation can be used. In general, many of the media developed by our curators for cultivation of bacteria, yeasts, molds, and Actinomycetales have been sufficient. Occasionally, additional study is required and other media and conditions must be used. Comments from depositors concerning recommended cultivation media are welcome and useful.

Strains destined for lyophilization are cultivated on solid media in petri dishes or as agar slants. Each curator has developed his own techniques in order to obtain sufficient material to lyophilize. For strains that must be maintained as active agar slant cultures, the appropriate maintenance media are used. Most of these strains are preserved under sterilized mineral oil and require periodic transfer.

C. Characterization

Because patent cultures occupy a unique position so far as confidentiality is concerned, our curators limit their observations to those necessary to assure them that the microorganism is pure and meets the general characteristics outlined in the deposit letter. Such characterizations are minimal and are limited to the appearance of the culture growing either in inoculum broth or on the cultivation media used for preservation and maintenance purposes and in viability checks immediately after lyophilization. Such information is recorded only when difficulties are encountered or questions arise. Consequently, our checks on such microorganisms concern taxa primarily above the species level.

Characterizations of microorganisms by the depositors are necessarily more detailed and complete. Some guidelines have been established, including those by foreign patent offices and by taxonomic committees (*4,14,19,20,28,32,33,49*). Criteria for characterizations are continually changing and need continuous updating in accord with contemporary advances in science.

D. Identifications

Again, because of the unique position of patent cultures, our curators perform no identification work on such microorganisms and do not study them in depth until the U.S. patent has issued or cultures have been supplied to at least three bona fide requestors. Many deposits carry no specific epithet. Some do not even carry generic names. Depending upon the interests of a particular curator, he or she may engage in characterization of particular cultures once they are available, one objective being the identification and naming of the microorganism. We believe that all depositors should provide at least a genus name for their strains.

E. Nomenclature

Our curators do not authenticate the names of strains of patent microorganisms, or assign names to those which have none until after issuance of the U.S. patent or until the strain has otherwise been released from any restrictions by the depositor. At that point our curators as well as other interested microbiologists may study the microorganism. As a result, the name may be changed and additional information on the strain be made available. Name changes should be proposed only after careful study of the microorganism, using the best and most contemporary criteria available. Treatment of some aspects of this problem has been discussed (26,49). Needless to say, depositors should make every effort to follow the appropriate code of nomenclature (Bacteriological or Botanical) in naming their microorganisms. They should also realize that the Bacteriological Code does not recognize names of new taxa published in patents.

F. Verification of Claims for Deposited Cultures

Over the years, very few questions have been raised regarding the reliability of patent cultures distributed by the ARS Culture Collection insofar as the claimed productivity is concerned. At least we have not been so informed, except very occasionally, that a particular culture does not perform in the way it should. Consequently, we can only assume that the strains and their progeny are doing what they are supposed to do. We would be interested in learning whether any of the cultures do not perform as they should, along with any other pertinent information. Further, we assume that progeny of the strains are performing adequately in a qualitative fashion. Whether they perform, or need to perform, adquately from a quantitative standpoint is another question.

The Grain Division of the U.S. Department of Agriculture's Agriculture Marketing Service has established a plant Variety Protection Office to certify protection for 17 years to breeders of new plant varieties reproduced through seeds (36,37). This office includes a staff of experts who verify that a variety submitted in connection with an application for certification of a new plant variety actually is new and different and fulfills other requirements under the Plant Variety Protection Act of 1970 (36). Such a procedure presumably could be instituted for patent cultures. The scope of such a function for patent cultures would indeed be broad; it could become prohibitive in terms of manpower and cost.

We have considered having depositors check materials we prepare to verify their activities. The history of the ARS Culture Collection has not suggested that such a procedure is necessary. If adopted, it would involve additional effort on the part of depositors and would delay the date of accessioning strains.

VI. Preservation and Storage

A. Maintenance on Agar Slants

Probably very few patent strains are maintained for long periods of time by routine, periodic transfer on appropriate agar media. Other more reliable methods of preservation are available. Some cultures have not yet responded to efforts to lyophilize them. They must be preserved by maintenance on agar slants stored in the cold or under sterilized mineral oil. These preparations require periodic transfer to fresh agar media. Possibilities for contamination or death are greater, and the cultures require more attention and concern on the part of curators.

B. Storage of Cultures under Oil

A method we routinely use for preservation of cultures not amenable to lyophilization is based on the addition of sterilized mineral oil to well grown agar slant cultures. These cultures must be transferred periodically to fresh agar slants and overlaid with sterilized mineral oil. Care must be taken in the overlaying technique so that every portion of the agar and culture is covered with the oil. If this is not done, dehydration of the medium occurs and the culture dies or becomes very difficult to retrieve. The transferring process with oil-overlaid cultures is messy.

Although we maintain only a small number of cultures in this fashion, including some patent cultures, they do constitute a problem and require consideration of alternative preservation methods. The surge of development in hydrocarbon microbiology could lead to further problems should one attempt to preserve strains by this method. Utilization of the oil overlay by particular microorganisms which attack hydrocarbons is a possibility. No definitive data indicating qualitative productivity loss of cultures maintained under oil is known to us. To provide requestors, new agar slants must be prepared and grown up from the oil cultures. These cultures are subject to all the problems concerned with routine transfer of cultures, including run-down or loss of productivity.

C. Storage of Cultures in Deep Freeze

Storage of agar slant cultures or other preparations of microorganisms in deep freeze is another method by which patent cultures could be preserved for long periods of time. We do not use this method, although it might be useful for cultures not amenable to lyophilization. As with agar slants and cultures overlaid with oil, transfer to fresh media would be necessary at appropriate intervals.

D. Storage of Cultures in Soil

We occasionally receive strains of microorganisms in previously sterilized soil in connection with patent deposits. We have used the method to only slight extent and have abandoned it for the Actinomycetales because of disappointing viability results after extended storage at room temperature (27). Any soil cultures received in connection with patent deposits are retained in our separate collection of original deposition material stored in the cold.

E. Storage of Cultures in Lyophile

One of the most significant advances in culture collection technology was the development of the lyophile process for freeze-drying microorganisms to preserve them with minimal change for long periods of time. This process, pioneered here by Drs. L. J. Wickerham and K. B. Raper, has been used for the great majority of cultures maintained at the ARS Culture Collection for many years (18). Preparations made in 1941 still are viable. Details of the process using relatively simple and inexpensive equipment are given in Wickerham and Andreasen (48) and Haynes et al. (17). We use bovine blood serum exclusively as a suspending agent. Thirty preparations of each patent strain are made initially. Before that material is completely exhausted, additional preparations are made, generally starting with material from the existing lot. We have seen no reason to change the methods or the simple equipment originally devised for the purpose.

Bacteriophages recently reported to be present in some commercially produced bovine serum (12) may present a potential problem. Milk or any other natural material used as a suspending agent also may contain such phages depending upon how it is processed. Whether the presence of these phages might exert effects on microorganisms during the process of suspending cells, spores, conidia, or hyphae; during the lyophilization process itself; or during the period that these materials are stored *in vacuo* in the cold is unknown. The possibility of transduction or transformation presents itself. Research in this area is needed, and a reexamination of nonnatural suspending agents would be desirable. Also, when considering patent depositions in some other collections, one must bear in mind that certain countries may have scruples against using animal serum or milk as suspending agents.

F. Storage of Cultures in or over Liquid Nitrogen

Within the past decade, liquid nitrogen preservation of microorganisms, whether in or above the liquid phase, has achieved recognition as a suitable means for culture preservation, particularly for those cultures that are not amenable to lyophilization (35). Disadvantages lie in the need for constant monitoring of the liquid nitrogen supply, some inconvenience in removing culture preparations from storage tanks, and the cost of nitrogen. We do not presently use this method of preservation for any of our cultures. However, the slow but continuing accession of nonlyophilizable microorganisms has prompted us to consider this method for future use. We are aware of one report of qualitative changes in productivity of a strain possibly related to its preservation with a liquid nitrogen process (24). Understandably, there is more information on viability of cultures after extended storage with lyophilized preparations than with liquid nitrogen preservation.

G. Viability and Characterization Checks

Over the years we have received a number of complaints that some of our cultures, generally distributed as lyophilized preparations, are not viable. Routinely, a viability check is made on each lot of lyophilized preparations made by our curators and records are carefully kept. No lot is acceptable unless viability can be demonstrated shortly after the materials are prepared. On receipt of complaint we check another preparation from the same lot. Only rarely have we been unable to get the strain to grow. Thus, we attribute most reports of nonviability of our preparations to lack of experience in handling lyophilized preparations or failure to follow the instructions accompanying each shipment. Our curators estimate that substantially less than 1% of the thousands of lyophilized preparations tested over the past 30 years have proved to be nonviable. The viability checks also provide some characterization information as indicated before.

H. Maintenance in Perpetuity

Depositors must provide assurance to the Patent Office that patent cultures will be permanently available to the public (46). Permanent availability has caused some discussion and elicited questions from potential depositors. Projection of the continued availability of progeny of microorganism strains deposited in culture collections throughout the world is dependent on many factors. From a reasonable and practical standpoint, the deposit of strains in the permanent collection of the ARS Culture Collection would constitute maintenance in perpetuity. Likewise, deposit of strains in some of the other major collections throughout the world would constitute maintenance in perpetuity. No culture collection in the world, however, can absolutely guarantee such maintenance. Should a reputable collection be disbanded or cease to operate, the responsible parties undoubtedly would take steps to assure continued maintenance of their holdings by passing them on to other collections or making other arrangements. We have salvaged several important collections of microorganisms in this manner. No strains are removed from the ARS Culture Collection, or destroyed, except for very good reason. Once a strain becomes available, other collections may receive it and accession it. Progeny of the original deposit are available from a number of collections for many years to come, and continued availability is assured.

I. Reserve Collections

Because we maintain a considerable number of valuable nonpatent and patent strains, we established some time ago additional duplicate, separately stored collections of microorganisms to preclude losses should any natural disasters or other calamities occur. Thus, we have duplicate collections maintained in other parts of our building complex and selected materials in other strategic locations. Thankfully, we have not yet had to rely on these collections. Other major collections probably have taken similar steps to protect their holdings. Although the development of the lyophile and liquid nitrogen preservation processes guarantees that certain microorganisms will be available for many years to come, adequate safeguards should be instituted if not already in use.

VII. Records

Maintaining records on ARS Culture Collection holdings takes up considerable time and space, as it does with other collections. Often, we must refer to records and correspondence of some 30 years ago to answer various kinds of questions that arise. A separate file is maintained for each patent strain so that we can readily retrieve information on these microorganisms. The files are carefully checked with each culture request to determine whether the strain is available at that time. The system has worked quite well so far. In retrospect, adoption of computerized record keeping some 20 or 30 years ago would have made all of the major collections more useful. There is a prodigious amount of information available from these collections, and curators provide much assistance to other researchers in providing such information. Computerization of the information today would be a formidable and full-time task but may become necessary.

VIII. Availability and Distribution

With regard to availability and distribution of patent cultures, one wonders why such cultures are requested. It appears to us that the prime reasons for requests are: for comparative purposes to determine whether similar cultures have already turned up in the requestors screening and evaluation procedures; for pure taxonomic research; or for use in research of a related nature—e.g., a researcher studying hydroxylation of terpenoids might request a culture strain patented for use as a hydroxylating agent for steroids and alkaloids; for use in connection with interferences or infringement cases; for evaluation with respect to possible licensing arrangements for production of particular metabolites; or for teaching and demonstration purposes.

As touched on before, a prime issue with regard to patent cultures is their availability before and during application procedures and after granting of the patent. Distribution of these cultures is contingent on their availability. There is little question in our minds that precisely designated strains should be deposited in reputable culture collections and that they should be available to bona fide requestors when the U.S. patent issues. Therefore, before distributing patent cultures, careful checks of the files must be made. We would be hard put to distribute some cultures shortly after their receipt. Some culture preparations might take several weeks before any shipments could be made. Hayhurst (16), in a discussion on legal aspects of industrial microorganisms in the British Commonwealth, has pointed out the desirability of depositing certain strains at the time of filing applications with no restrictions on availability save normal terms as to price, safety precautions, or the like. These limitations, in addition to those cited above, all affect the time at which a culture could be shipped. Moreover, he indicates that not only the type strain but also the best strain or strains known to the inventor should be deposited.

IX. The Future

A. Official Designation of Patent Collections

Thus far, no culture collection in the United States has been officially *designated* as a suitable repository for deposition of patent strains. The U.S. Patent Office has *recognized* several collections, including the ARS Culture Collection, as being suitable. Legislation proposed within the past several years has included elements on the official designation of certain collections by the Commissioner of Patents. Such proposed legislation is not yet law. Both microbiological and legal problems would accompany such designations. For example, we are unable to accept certain kinds of microorganisms. Other collections may be in somewhat similar circumstances because of Federal and State regulations regarding entry and shipping of certain kinds of microorganisms. Some collections charge for their materials. We do not.

There are hundreds of culture collections throughout the world. The designation of any particular collections would require considerable study of facilities, personnel, continuity, and policies to establish worthiness and reliability for serving the best interests of the depositors, the patent office, and the general public.

B. Extra Demands on Collections

What began as an interesting and exciting episode in the story of industrial microbiology, i.e., patent depositions, has exacted some payment as the years have gone by. The several patent collections in this country now have reached sizes where much service work is involved in maintaining and distributing these materials. The problems would have been of greater magnitude were it not for lyophilization and, more recently, liquid nitrogen preservation. A look into the future suggests that greater demands may be made on patent collections. Changes in the direction of research efforts offer prospects for deposition of more exotic and fastidious microorganisms. Microorganisms more difficult to cultivate, and mixtures of microorganisms may become involved. All these will create new demands upon the time and expertise of curators.

In retrospect, our personal belief is that the U.S. Patent Office might have established its own patent deposition arm staffed with competent culture collection microbiologists and with chemists as soon as it became apparent that deposition of cultures in connection with patent applications was to be a continuing thing Such an arm, we believe, could more effectively have handled the microbiological, chemical, and legal problems which have arisen in connection with these microorganisms. Close liaison with legal counsel and Federal and State agencies undoubtedly would have been advantageous.

C. National and International Patent Legislation

A number of bills have been introduced in the U.S. Congress directed toward reform of U.S. Patent Laws in recent years. Similarly, efforts to improve international patent legislation have been under way (1,9–11,16,30,31,34,47) Patents concerned with processes involving microorganisms make up only a small, but important, part of the total endeavor. Hopefully, the legislators and attorneys involved will treat the microbiological aspects as fully as possible. Some of the international problems concern designation and recognition of particular culture collections as suitable repositories, descriptions of microorganisms in patents, availability of patent strains, reciprocal agreements between countries so far as distribution is concerned, and financial support for collections maintaining patent strains and distributing them. Not all countries have culture collections that could be considered national ones. Some of the problems appear to be insurmountable. Until there is appropriate liaison between the interested legal and scientific personnel, we shall continue to operate in a maze of semantics and uncertainties.

X. Concluding Remarks

We have attempted to point out some of the problems encountered in maintaining a special collection of patent deposition strains over the past 24 years. Some, particularly those involving legal aspects, the microbiologist cannot disregard. Additionally, the direction of microbiological and chemical research suggests that, in the future, curators of culture collections will be confronted with new and difficult questions related to the cultivation, maintenance, and distribution of new, more exotic, and fastidious kinds of microorganisms. Hopefully, our comments will lead to added dialogue so that some of the problems can be seen in better perspective and efforts made to resolve them.

We believe that we have a threefold responsibility in our patent-related culture collection operations: to the depositors of patent strains; to the general public, not excluding the scientific one; and to ourselves, as such operations affect our missions, goals, and program of research. In meeting these responsibilities, we have tried to approach each problem with reasonableness, common sense, and sound judgment.

REFERENCES

1. Anonymous. (1964). *Chem. Eng. New* **42**, 86–106.
2. Casida, L. E., Jr. (1968). In "Industrial Microbiology" (L. E. Casida, Jr., ed.), Chapter 15, pp. 191–207. Wiley, New York.
3. Court of Customs and Appeals (CCPA), Northern District of Iowa. (1971). *Patent, Trademark Copyright Weekly Rep. (USPQ)* **168**(20); 99–104 = 168 USPQ 99.
4. Danish Patent Office. (1962). Saertryk kan leveres. Andre meddelelser. Saerlige retningslinier vedrørende behandlingen af patentansøgninger, der angar mikrobiologiske fremgangsmader. Dansk Patenttidende No. 6, pp. 47–48.
5. Daus, D. G. (1967). *Econ. Bot.* **21**, 388–394.
6. Daus, D. G. (1972). *J. Pat. Off. Soc.* **54**, 187–208.
7. Edelblute, H. W. (1964). In "The Encyclopedia of Patent Practice and Invention Management" (R. Calvert, ed.), pp. 567–587. Van Nostrand-Reinhold, Princeton, New Jersey.
8. Editor, American Society for Microbiology News. (1971–1972). *ASM (Amer. Soc. Microbiol.) News* **37**(4), 17; **38**(1), 54; **38**(2), 110–112; (4), 221–223; (5), 264–265; (6), 321; (7), 389.
9. Federico, P. J. (1972). *J. Pat. Off. Soc.* **54**, 102–125.
10. Federico, P. J. (1972). *J. Pat. Off. Soc.* **54**, 147–174.
11. Fergusson, J. D. (1973). *Chem. Ind. (London)* **11**, 504–505.
12. Gardner, S. (1973). *Fed. Regist.* **38**, 11080–11081.

13. Gottschalk, R. (1973). *Off. Gaz.* **911**(3), 760–761 = 911 O.G. 760.
14. Governmental Patent Agency (Japan). (1969). "Standard for Examination by Industrial Categories," Parts 1 through 5 (Part 3.12: Description of microorganism necessary for patentable invention; Part 3.14: Deposition of microbial organisms to be used; Part 3.2: Judgment of its identity). GPA, Japan.
15. Hayhurst, W. L. (1971). *Ind. Property* **10**, 189–198.
16. Hayhurst, W. L. (1973). In "Genetics of Industrial Microorganisms. Bacteria." (Z. Vaněk, Z. Hoštálek, and J. Cudlin, eds.), pp. 377–396. Publ. House Czech. Acad. Sci., Prague.
17. Haynes, W. C., L. J. Wickerham, and C. W. Hesseltine. (1955). *Appl. Microbiol.* **3**, 361–368.
18. Hesseltine, C. W., W. C. Haynes, L. J. Wickerham, and J. J. Ellis. (1970). In "Culture Collections of Microorganisms" (H. Iizuka and T. Hasegawa, eds.), pp. 21–38. Univ. of Tokyo Press, Tokyo.
19. International Subcommittee on Taxonomy of the Actinomycetes. (1963). *Int. Bull. Bacteriol. Nomencl. Taxon.* **13**, 169–170.
20. International Subcommittee on Taxonomy of the Actinomycetes. (1964). *ASM (Amer. Soc. Microbiol.) News* **30**, 13–14.
21. Kent, A. P. (1972). *Chem. Technol.* 599–605.
22. Kurylovich, V. (1971). *Antibiotiki (Moscow)* **16**, 395–400.
23. Levy, D., and L. B. Wendt. (1955). *J. Pat. Off. Soc.* **37**, 855–872.
24. Martin, J. F., and L. E. McDaniel. (1973). *Bacteriol. Abstr. Annu. Meet. Amer. Soc. Microbiol.* p. 108, Abstract M207.
25. Neshatayeva, E. V., and O. M. Kiselyov. (1972). *Prikl. Biokhim. Mikrobiol.* **8**, 65–74; *Microbiol Abstr.* **7A**, 186 (Abstract 7A5353) (1972).
26. Pridham, T. G. (1971). *Int. J. Syst. Bacteriol.* **21**, 197–206.
27. Pridham, T. G., and A. J. Lyons. (1973). *Abstr. Annu. Meet. Amer. Soc. Microbiol.* p. 4, Abstract E18.
28. Rehacek, Z. (1973). In "Genetics of Industrial Microorganisms. Bacteria" (Z. Vaněk, Z. Hoštálek, and J. Cudlin, eds.), pp. 423–433. Publ. House Czech. Acad. Sci., Prague.
29. Reynolds, E. L. (1968). *Off. Gaz.* **855**(1), 16–24 = 855 O.G. 16.
30. Robbins, L. J. (1960). *J. Pat. Off. Soc.* **42**, 830–848.
31. Schuyler, W. E., Jr. (1970). *Off. Gaz.* **876**(2), 341–388 = O.G. 341.
32. Silvestri, L. G., and D. Gottlieb. (1964). In "Global Aspects of Applied Microbiology" (M. P. Starr, ed.), pp. 109–112. Wiley, New York.
33. Sneath, P. H. A. (1970). In "The *Actinomycetales*. The Jena Symposium on Taxonomy" (H. Prauser, ed.), pp. 371–377. Fischer, Jena.
34. Stoy, A. (1973). In "Genetics of Industrial Microorganisms. Bacteria." (Z. Vaněk, Z. Hoštálek, and J. Cudlin, eds.), pp. 397–401. Publ. House Czech. Acad. Sci., Prague.
35. Swoager, W. C. (1972). *Amer. Lab.* **4**, 45–52.
36. U.S. Congress. (1970). Public Law 91–577. United States Statutes at Large. 84 (Part 2), pp. 1542–1559.
37. U.S. Department of Agriculture. Agricultural Marketing Service. Grain Division, (1973). *Off. J. Plant Var. Protect. Off.* **1**(1).
38. U.S. Department of Agriculture, Agricultural Research Service. (1968). *U.S., Dep. Agr., Agr. Res. Serv. [Publ.]* **PA-873**.
39. U.S. Department of Agriculture, Agricultural Research Service. (1971). *U.S. Dep. Agr., Agr. Res. Serv. [Publ.]* **PA-967**.
40. U.S. Department of Health, Education, and Welfare. Health Services and Mental

Health Administration. (1972). "Classification of Etiologic Agents on the Basis of Hazard," 3rd ed. USDHEW, Washington, D.C.
41. U.S. Patent Office Board of Appeals. (1965). *Patent, Trademark Copyright Weekly Rep.* (*USPQ*) **153**, 473–474 = 153 USPQ 473.
42. U.S. Patent Office Board of Appeals. (1967). *Patent, Trademark Copyright Weekly Rep.* (*USPQ*) **157**, 437–444 = 157 USPQ 437.
43. U.S. Patent Office Board of Appeals. (1968). *Off. Gaz.* **848**(4), 863–867 = 848 O.G. 863.
44. U.S. Patent Office Board of Appeals. (1968). *Off. Gaz.* **849**(1), 5–11 = 849 O.G. 5.
45. U.S. Subcommittee on Taxonomy of Actinomycetes. (1964). *AMS* (*Amer. Soc. Microbiol.*) *News* **30**(2), 13.
46. Wahl, R. A. (1971). *Patent, Trademark Copyright Weekly Rep.* (*USPQ*) **169**(6), II–III; *Off. Gaz.* **886**(4), 638 = 886 O.G. 638.
47. Whittenburg, J. V. (1970). *Advan. Appl. Microbiol.* **13**, 383–398.
48. Wickerham, L. J., and A. A. Andreasen. (1942). *Wallerstein Lab. Commun.* **5**, 165–169.
49. Woodruff, H. B., S. A. Currie, T. C. Hallada, and I. Putter. (1973). *In* "Genetics of Industrial Microorganisms. Bacteria." (Z. Vaněk, Z. Hošťálek, and J. Cudlin, eds.), pp. 403–422. Publ. House Czeck. Acad. Sci., Prague.

Production of the Same Antibiotics by Members of Different Genera of Microorganisms

HUBERT A. LECHEVALIER

Waksman Institute of Microbiology, Rutgers University,
The State University of New Jersey,
New Brunswick, New Jersey

I.	Introduction	25
II.	Same Antibiotics Produced by Different Organisms	27
	A. Nucleosides	27
	B. Pyrimidotriazines	31
	C. Prodiginines	32
	D. Pyrrolidones	33
	E. Phenazines	33
	F. Phenoxazinones	35
	G. Amino Acids	36
	H. Aminoglycosides	36
	I. Macrolides	38
	J. Sulfur-Containing Antibiotics	38
	K. Steroids	40
	L. Quinones	40
	M. Pyrones	42
	N. Phenols	42
III.	Conclusion	42
	References	43

I. Introduction

In 1959, Krassilnikov stated that, based on his own experience, the same antibiotics were always produced by the same species of microorganisms and, inversely, the same species of microorganisms always produced the same antibiotics. He added that authors who published data to the contrary were confused about either the identity of the producing organisms or of the antibiotics being elaborated. Krassilnikov's point of view, if true, would certainly simplify litigations in the field of patented antibiotics: if you have my antibiotic, you have my organism, thus you have my process.

In contrast to Krassilnikov, we have recognized that the same antibiotics may be produced by different organisms and that inversely the same organism may produce different antibiotics, usually in the form of mixtures (Waksman and Lechevalier, 1962).

A survey of the literature points out that the same antibiotic may be formed by organisms that are so different that they are placed in different genera. These different genera may even belong to different families, to different orders, and in some cases to different classes and even to different divisions.

Chemically, an antibiotic is nothing very special—it is simply a secondary microbial metabolite that happens to have some antimicrobial activity.

Secondary metabolites play no obvious role in the economy of the producing organism. They are derived from some of the intermediates of primary metabolism, and their formation represents an expenditure of energy (Turner, 1971). In the case of antibiotics it has been argued that their production might confer an ecological advantage; thus two very different organisms producing the same antibiotic might share the same ecological benefits.

However, the ecological advantages associated with the production of antibiotics are mainly obvious in petri dishes. Because it is probable that the production of an antibiotic is a mere chance event, any advantage that might be derived from it are fortuitous. Antibiotics are derived biosynthetically from primary metabolites that are found in most living cells; thus it is not too surprising that in the giant biochemical lottery that controls the chemism of microorganisms, the same combination of numbers may be drawn more than once.

Discussed hereafter are antibiotics known to be produced by organisms that are very different one from another. The production of antibiotics belonging to the same chemical families by different groups of microorganisms are not included, but only the production of the *very same* substances by organisms different enough to be placed in separate genera. No issue is made of the fact that an antibiotic may be produced by members of different genera that are not well separated one from another. For example, the genera *Streptomyces, Streptoverticillium, Chainia, Elytrosporangium,* and *Kitasatoa* are too closely related to be considered in the following survey. From the point of view of taxonomy, in the absence of producing cultures for evaluation by modern techniques, reports of the production of the same antibiotic by streptomycetes and nocardiae are considered worthless. Also, one must keep in mind that the same fungus may be placed in different genera by different authors often because different names are given to the perfect and imperfect forms. In addition, production of the same substances by strains of *Penicillium* and *Aspergillus* is so common that such events have not been considered in the preparation of this review.

What follows is thus not a complete list of antibiotics that have been reported to be produced by strains belonging to various genera, but rather a selected compilation of antibiotics whose producing organisms have a good chance, in my estimation, of having been properly identified as very different organisms.

II. Same Antibiotics Produced by Different Organisms

In the following list (summarized in Table I), which does not include lichen products, the antibiotics are grouped according to chemical similarity.

A. NUCLEOSIDES (Fig. 1)

3′-Amino-3′-deoxyadenosine is an antitumor agent with slight activity against some yeasts. It has been found to be produced by a species of *Helminthosporium* (Gerber and Lechevalier, 1962), *Cordyceps militaris* (Guarino and Kredich, 1963), and *Aspergillus nidulans* (Suhadolnik, 1970).

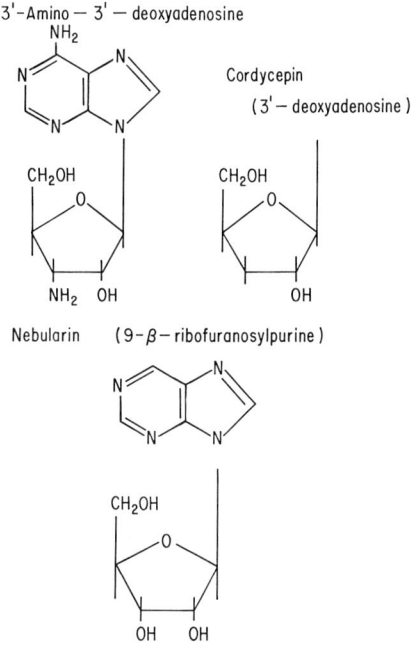

FIG. 1. Nucleosides

3′-Deoxyadenosine (cordycepin) is an antitumor agent which also is active against some gram-positive bacteria. It is produced by *Cordyceps militaris* (Frederiksen et al., 1965) and *Aspergillus nidulans* (Kaczka et al., 1964).

9-β-Ribofuranosylpurine (nebularin) is an antitumor agent which also is active against mycobacteria and fungi. It has been found to be pro-

TABLE I
Antibiotics Known To Be Produced by Very Different Organisms

Chemical family	Antibiotic	Type of activity[a]	Produced by Genus	Types of organisms
Nucleosides	3′-Amino-3′-deoxyadenosine	T, F	*Helminthosporium* *Aspergillus*	Imperfect fungi
	Cordycepin (3′-deoxyadenosine)	T, B+	*Cordyceps* *Aspergillus*	Ascomycetes Imperfect fungi
	Nebularin (9-β-ribofuranosylpurine)	T, B+, F	*Cordyceps* *Clitocybe*	Ascomycetes Basidiomycetes
Pyrimidotriazines	Toxoflavin (xanthothricin)	B	*Streptomyces* *Pseudomonas* *Streptomyces*	Actinomycetes Gram − bacteria Actinomycetes
Prodiginines	Prodigiosin	B+, F, P	*Serratia* *Vibrio*	Gram − bacteria
			Streptomyces	Actinomycetes
	Undecylprodigiosin	B+	*Streptomyces* *Actinomadura*	Actinomycetes
Pyrrolidones	Tenuazonic acid	V, P, T	*Alternaria* *Aspergillus* *Sphaeropsidales*	Imperfect fungi
Phenazines	Iodinin	B+, F	*Brevibacterium*	Gram + bacteria
			Microbispora *Actinomadura* *Streptomyces*	Actinomycetes
	1,6-Phenazinediol 5-oxide	Same, but less active	Same	Same
	1,6-Phenazinediol	Same, but even less active	Same	Same
Phenoxazinones	Questiomycin A (2-aminophenoxazin-3-one)	B+, F	*Streptomyces* *Microbispora* *Actinomadura*	Actinomycetes

PRODUCTION OF SAME ANTIBIOTICS BY DIFFERENT GENERA 29

	2-Acetamidophenoxazin-3-one	B+, F	Brevibacterium Penicillium	Gram + bacteria Imperfect fungi
			Microbispora Actinomadura Brevibacterium Penicillium	Same
Amino acids	Actinomycins	B+, T	Streptomyces Micromonospora	Actinomycetes
	Anticapsin	B−	Streptomyces Bacillus	Actinomycetes Gram + bacteria
	Cycloserine	B	Streptomyces Pseudomonas	Actinomycetes Bacteria
Aminoglycosides	Mannosidostreptomycin	B	Streptomyces Micromonospora	Actinomycetes
	Neomycin B	B	Streptomyces Micromonospora	Actinomycetes
Macrolides	Erythromycin A	B+	Streptomyces Arthrobacter	Actinomycetes Bacteria
Sulfur-containing substances	6-Aminopenicillanic acid	B	Penicillium Cephalosporium Aspergillus Trichophyton	Imperfect fungi
	Benzylpenicillin (penicillin G)	B	Penicillium Aspergillus Trichophyton Malbranchea	Imperfect fungi
	Penicillin N (cephalosporin N, synnematin B)	B	Cephalosporium Paecilomyces Streptomyces	Imperfect fungi Actinomycetes
	Gliotoxin (aspergillin)	B, F	Penicillium Gliocladium Aspergillus Trichoderma	Imperfect fungi

(Continued)

TABLE I (*Continued*)

Chemical family	Antibiotic	Type of activity[a]	Produced by Genus	Types of organisms
Steroids	Fumigacin (helvolic acid)	B	*Aspergillus* *Cephalosporium*	Imperfect fungi
			Streptomyces	Actinomycetes
	Fusidic acid (ramycin)	B+	*Fusidium* *Cephalosporium*	Imperfect fungi
			Mucor	Phycomycetes
Quinones	Oosporein (chaetomidin)	B, F	*Oospora* *Acremonium* *Beauveria*	Imperfect fungi
			Chaetomium	Ascomycetes
			Phlebia	Basidiomycetes
	Citrinin	B, F	*Aspergillus* *Penicillium* *Clavariopsis*	Imperfect fungi
			Crotalaria	Higher plant
Pyrones	Kojic acid	B, F	*Aspergillus* *Penicillium*	Imperfect fungi
			Acetic acid bacteria	Gram − bacteria
Phenols	Flavipin	F	*Aspergillus* *Epicoccum*	Imperfect fungi

[a] T = antitumoral; F = antifungal; B = antibacterial; B+ = mainly active against gram-positive bacteria; B− = mainly active against gram-negative bacteria; V = antiviral; P = antiprotozoal.

duced by *Streptomyces yokosukaensis* (Isono and Suzuki, 1960; Nakamura, 1961) and *Clitocybe nebularis* (Löfgren et al., 1954).

B. PYRIMIDOTRIAZINES (Fig. 2)

Bongkrek is a coconut food product popular with the natives of Mid-Java. It is produced much like tempeh (Hesseltine, 1965), which is made from soybean, but unlike tempeh it may be very toxic. Bongkrek is prepared by the action of *Rhizopus oligosporus* on finely divided copra from which most of the oil has been removed. Occasionally, this product is toxic owing to the growth of *Pseudomonas cocovenenans*,

Toxoflavin (xanthothricin)

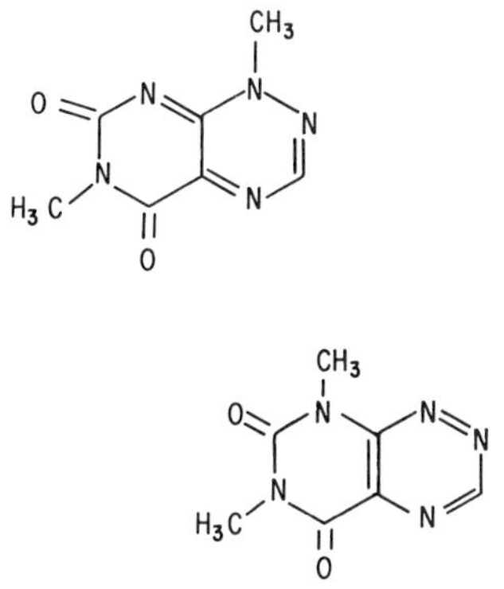

Fervenulin

FIG. 2. Pyrimidotriazines.

which can produce two toxic principles from precursors present in the coconut. Both these toxins have antimicrobial activity. One of these, bongkrekic acid, is a highly unsaturated and very labile fatty acid (Nugteren and Berends, 1957). The other toxic principle is a yellow pigment, called *toxoflavin*, which is active against gram-positive and gram-negative bacteria (Van Damme et al., 1960). The toxicity of this compound, which has no antibiotic activity under anerobic conditions, is probably due to the production of hydrogen peroxide, a strong poison for catalase-deficient organisms. Yeasts, which are rich in catalase, are not affected by toxoflavin.

Xanthothricin, an antibiotic produced by a species of *Streptomyces*, closely related to *S. albus* (Machlowitz et al., 1954), is identical to toxoflavin (Daves et al., 1961). In addition, toxoflavin is similar to fervenulin (Daves et al., 1962), a weak antibiotic produced by *Streptomyces fervens* (DeBoer et al., 1960).

C. Prodiginines (Fig. 3)

Prodigiosin is a tripyrrolic red antibiotic pigment characteristic of *Serratia marcescens* (Williams and Hearn, 1967; Gaughran, 1969). It is active against bacteria, mainly gram-positive, and has some antifungal (Gerber, 1971) and antiprotozoal activity (Williams and Hearn, 1967).

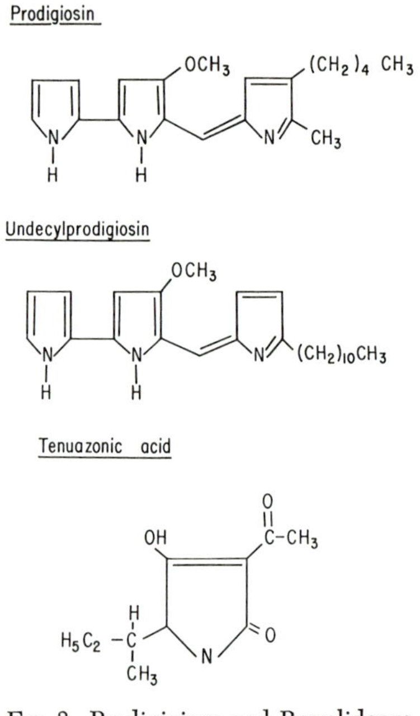

Fig. 3. Prodiginines and Pyrrolidones.

This antibiotic pigment was also found to be produced by bacteria with properties quite different from those of the members of the genus *Serratia*. Lewis and Corpe (1964) isolated two marine mesophilic monotrichously flagellated prodigiosin-producing bacteria, and D'Aoust and Gerber (1974) isolated and identified prodigiosin from the marine psychrophile *Vibrio psychroerythrus*. In addition, Perry (1961) reported the production of prodigiosin by a *Streptomyces* species.

Undecylprodigiosin is one of the natural analogs of prodigiosin. It is active against gram-positive bacteria (Harashima et al., 1967) and has been found to be produced by a number of unidentified strains of *Streptomyces* (N. N. Gerber, personal communication), by *Streptomyces longisporus ruber* (Wasserman et al., 1966), and by *Actinomadura pelletieri* (Gerber, 1971).

D. Pyrrolidones (Fig. 3)

Tenuazonic acid was described by Rosett et al. in 1957 as one of the metabolites of Alternaria tenuis. It has no activity against bacteria and yeasts but is active against certain viruses and protozoa (Miller et al., 1963). In addition, it has antitumor activity (Kaczka et al., 1963; Shigeura, 1963). Tenuazonic acid has been found to be produced not only by A. tenuis, but also by a strain of Aspergillus (Miller et al., 1963) and by a member of the Sphaeropsidales (Kaczka et al., 1963).

E. Phenazines (Fig. 4)

Iodinin (1,6-phenazinediol 5,10-dioxide) was first reported as an antibiotic pigment from a bacterium isolated from milk. The surface of

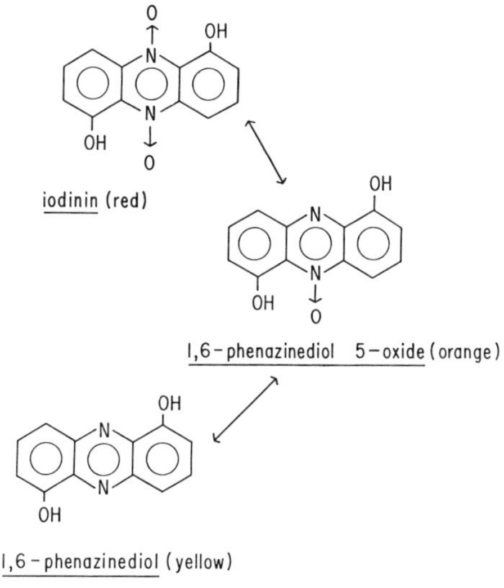

Fig. 4. Phenazines.

its growth was covered with crystals of a pigment that resembled ordinary iodine. The bacterium was described as gram-negative and non-motile and was named *Chromobacterium iodinum* (Davis, 1939). The systematic position of this organism, which later was placed in the genus *Pseudomonas,* has been questioned, especially because the culture, in its present-day form, is a gram-positive rod (Sneath, 1956). Present assignment is to the genus *Brevibacterium* (Irie et al., 1960; Lechevalier, 1965). Iodinin is also produced by many different bacteria: *Microbispora aerata, M. amethystogenes, M. parva* (Gerber and Lechevalier, 1964), *Actinomadura dassonvillei* (Gerber, 1966), *Streptomyces thioluteus*

(Gerber and Lechevalier, 1965), *Streptosporangium amethystogenes* (Prauser and Eckard, 1967) and an unidentified gram-negative nonmotile rod-shaped bacterium (Gerber, 1969).

The production of the purple-red iodinin is accompanied by that of 1,6-phenazinediol-5-oxide, which is orange, and of 1,6-phenazinediol, which is yellow. All these compounds have activity against gram-positive bacteria and fungi, but, as can be seen in Table II, the diol is the

TABLE II
ANTIMICROBIAL ACTIVITY OF 1,6-PHENAZINEDIOL, 1,6-PHENAZINEDIOL 5-OXIDE, AND IODININ[a]

Organism	Activity (µg/ml for total inhibition)		
	Phenazinediol	1,6-Phenazinediol 5-oxide	Iodinin
Bacteria			
Sarcina lutea 14	5	5–6	0.08
Corynebacterium fimi 22	5	5–6	0.08
Escherichia coli 54	>75	>9.0	>2.0
Proteus vulgaris 73	>75	>9.0	>2.0
Actinomycetes			
Mycobacterium smegmatis 607	25	5.0	1.5
Nocardia rhodochrous 271	37.5	8.0	2.0
Nocardia coeliaca 3520	15	5[b]	0.4
Micropolyspora brevicatena 1086 W/F	37.5	>9	>2.0
Microellobosporia cinerea 3855	12	5	0.5
Actinoplanes sp. W13	10	5	0.1
Fungi			
Saccharomyces cerevisiae 216	50	5[b]	0.4
Hansenula anomala 25	25	5[b]	0.5
Trichophyton mentagrophytes 171	20	4	0.4
Ceratostomella ulmi 185	15	3	0.5

[a] From Gerber and Lechevalier (1965).
[b] Static activity.

least active and the dioxide the most potent. Increased activity is accompanied by a decrease in solubility. This family of three compounds represents a biosynthetic pathway. Disrupted cells of *Brevibacterium iodinum* can carry out the synthesis of the mono- then the dioxide from the diol. Going in the reverse direction, some fungi and actinomycetes can solubilize and detoxify iodinin by producing first the monooxide and then the diol (Gerber and Lechevalier, 1965).

F. PHENOXAZINONES (Fig. 5)

Questiomycin A (2-aminophenazin-3-one) was isolated from the culture filtrates of a species of *Streptomyces* by Anzai and co-workers (1960). They noted that the compound had antibiotic activity against some gram-positive bacteria, especially mycobacteria, and some fungi. Subsequently this brown antibiotic pigment has been demonstrated

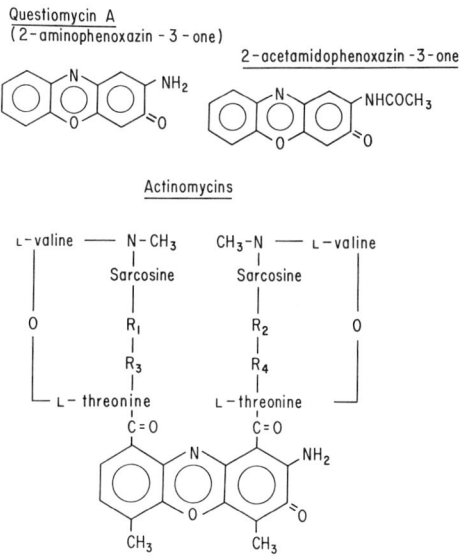

FIG. 5. Phenoxazinones.

among the metabolic products of several bacteria and fungi: *Microbispora aerata* and *Brevibacterium iodinum* (Gerber and Lechevalier, 1964); *Streptomyces thioluteus* and *Actinomadura dassonvillei* (Gerber, 1966) and some strains of *Penicillium notatum* (Pfeifer et al., 1972).

2-Acetamidophenoxazine-3-one is a yellow pigment with mild antibiotic activity against some gram-positive bacteria, including some *Actinomycetales*, and some dermatophytes. So far, it has been found among the metabolic products of *Microbispora aerata, Brevibacterium iodinum* (Gerber and Lechevalier, 1964), *Actinomadura dassonvillei* (Gerber, 1966) and some strains of *Penicillium notatum* (Pfeifer et al., 1972).

The actinomycins constitute a family of red antibiotics which share a phenoxazinone nucleus to which are attached two cyclic polypeptides. The structure of the polypeptides differs from one actinomycin to another. In Fig. 5, R_1, R_2, R_3, and R_4 stand for various amino acid substituents which differentiate one actinomycin from another. In general, actinomycins are mainly active against gram-positive bacteria, but it was the discovery in 1952 that actinomycins had antitumor activity that spurred renewed interest in these toxic, though brightly colored substances (Waksman and Lechevalier, 1962). Various mixtures of actinomycins are known to be produced by scores of species of *Streptomyces*. Most interesting was the report by Fisher *et al.* (1951) of the production of an actinomycin by a strain of *Micromonospora*. Their description of the amino acids found in the hydrolyzates of their actinomycin is compatible with the possibility that their product was actinomycin IV (presence of proline and valine in addition to threonine, sarcosine, and N-methylvaline).

The description of the producing organism by Fisher *et al.* (1951) leaves little doubt that they were not dealing with a strain of *Streptomyces*, and assignment to the genus *Micromonospora* seems reasonable: "The organism, *Micromonspora* sp. 608, was isolated on solid media (from a sample of Costa Rican soil) as a waxy, confluent, irregular-edged, yellow to orange colony segmented by spokelike striations toward the center. Microscopic examination showed slender mycelia 0.7 microns in diameter with spherical conidia 1.4 microns in diameter borne singly on short conidiophores in grapelike clusters typical of the genus *Micromonspora*. Growth on various media indicated a similarity to *Micromonospora globosa*."

Unfortunately this organism was not kept in the collection of Merck, Sharp and Dohme and is thus not available for examination (H. B. Woodruff, personal communication).

G. Amino Acids (Fig. 6)

Anticapsin is an amino acid with antibiotic activity against *Salmonella gallinarum*. It also inhibits the synthesis of the hyaluronic acid capsule of *Steptococcus pyogenes*. It is produced by *Streptomyces griseoplanus* (Neuss *et al.*, 1970) and by *Bacillus subtilis* (Walker and Abraham, 1970). In addition, when linked to L-alanine, anticapsin becomes the antibiotic bacilysin (tetaine) (Kaminski and Sokolowska, 1973), produced by strains of *B. subtilis* and *B. pumilus*.

Although cycloserine has limited activity against bacteria it is mainly active against mycobacteria. It is produced by several species of *Streptomyces* (Waksman and Lechevalier, 1962). Cycloserine has also been found to be produced by species of *Pseudomonas* (Stapley *et al.*, 1969).

H. Aminoglycosides (Fig. 7)

Mannosidostreptomycin, like streptomycin, is active against gram-positive and gram-negative bacteria, but it is less active on a weight basis.

PRODUCTION OF SAME ANTIBIOTICS BY DIFFERENT GENERA 37

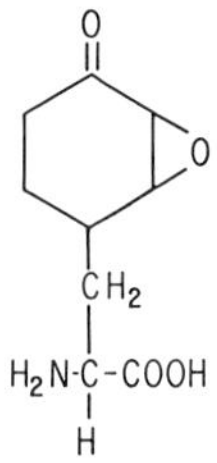

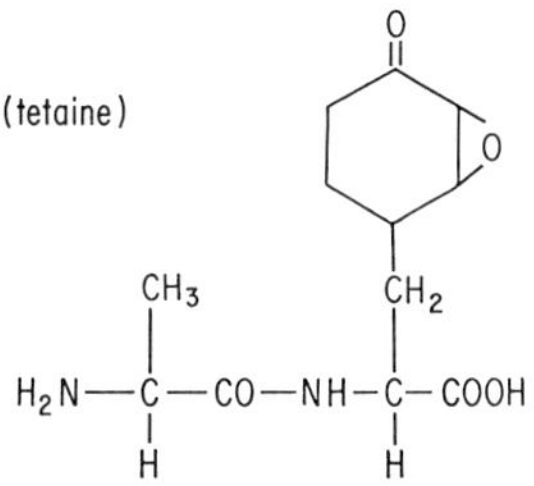

FIG. 6. Amino Acids.

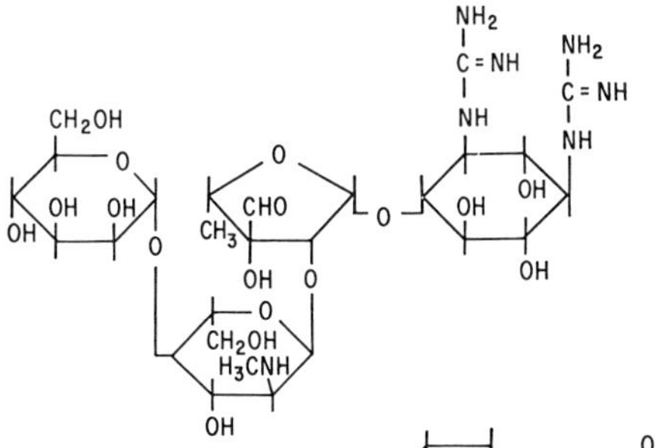

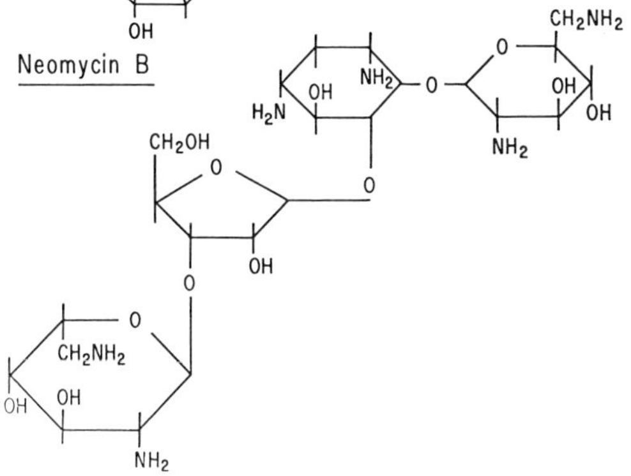

FIG. 7. Aminoglycosides.

It is cleaved to streptomycin by the action of the enzyme mannosidostreptomycinase. In addition to being produced by *Streptomyces griseus* (Waksman and Lechevalier, 1962), it has been isolated from the culture fluids of *Micromonospora pallida* (Gause et al., 1970).

Neomycin B is the most active component of the neomycin complex, a group of white, water-soluble aminoglycosidic antibiotics which are active against gram-positive and gram-negative bacteria (Rinehart, 1964). It is produced by several strains of *Streptomyces* (Waksman and Lechevalier, 1962), including one described as *Elytrosporangium brasiliense* (Sukapure et al., 1969). In addition, it has been found to be produced by a strain of *Micromonospora* (Wagman et al., 1972).

I. Macrolides (Fig. 8)

Erythromycin A is a basic macrolide which is mainly active against gram-positive bacteria. It is produced not only by *Streptomyces eryth-*

Fig. 8. Macrolide.

reus, but also by a species of *Arthrobacter* (French et al., 1970). The latter bacterium produces erythromycin A essentially free from the closely related but less active erythromycins B and C.

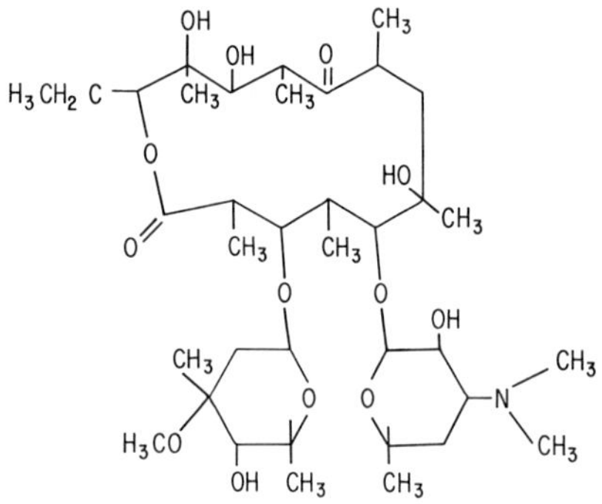

J. Sulfur-Containing Antiobiotics (Fig. 9)

1. Penicillins

The penicillins form a large group of antibiotics which are mainly active against gram-positive bacteria and which share a 6-aminopenicillanic acid nucleus.

6-Aminopenicillanic acid is frequently, but not invariably, produced with penicillins (Cole, 1966). For example, it can be obtained from cultures of *Penicillium chrysogenum* especially in the absence of precursors for the side chains of penicillins. 6-Aminopenicillanic acid has antibacterial activity of its own and is an intermediate in the manufacture of semisynthetic penicillins. It is known to be produced by several species of *Penicillium*, by strains of *Cephalosporium* and *Emericellopsis* (Cole and Rolinson, 1961), various dermatophytes, including *Trichophyton mentagrophytes* (Uri et al., 1963) and by *Aspergillus ocharaceus* (Cole, 1966).

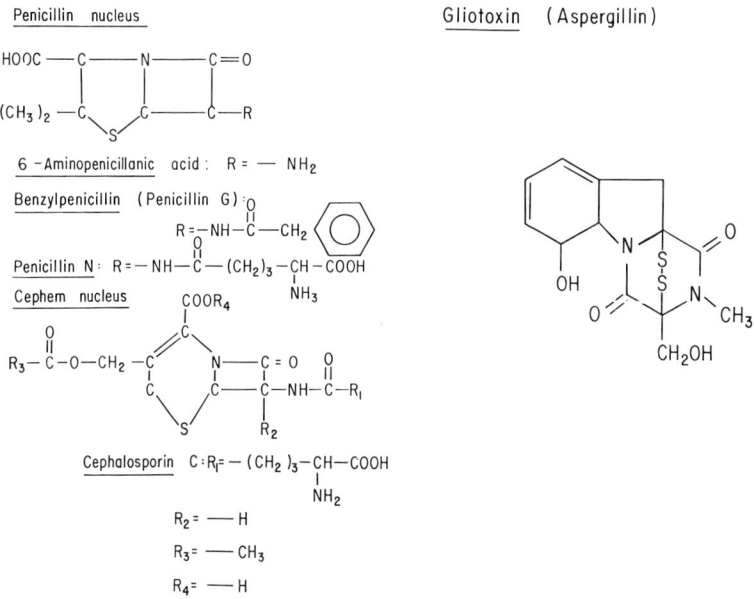

FIG. 9. Sulfur-containing antibiotics.

Benzylpenicillin is produced by numerous strains of *Penicillium* and *Aspergillus* when phenylacetic acid or α-phenylacetamine is present as a precursor in the medium. This antibiotic is also produced under the same conditions by *Trichophyton mentagrophytes* (Uri et al., 1955) and *Malbranchea pulchella* (Bai and Rao, 1963). For a description of the latter fungus, see Coonery and Emerson (1964).

Penicillin N, which has a straight D-(4-amino-4-carboxybutyl) side chain, has more activity against gram-negative bacteria than most natural penicillins (Stewart, 1965). This antibiotic, which was previously called cephalosoporin N and synnematin B, is produced by numerous strains of *Cephalosporium* which fall into the perfect genus *Emericellopsis*

(Grosklags and Swift, 1957). It also is produced by a close relative of the penicillia, *Paecilomyces persicinus* (Pisano *et al.*, 1960) and from the broth filtrates of species of *Streptomyces* (Miller *et al.*, 1962; Nagarajan *et al.*, 1971).

2. *Cephem Antibiotics*

Among the many substances isolated from strains of *Cephalosporium* and called cephalosporin, one, cephalosoporin C (Abraham and Newton, 1961), has a basic nucleus which is shared by numerous other antibiotics. This nucleus is a 6-membered dihydrothiazine ring fused to a β-lactam ring as in the penicillins. It is called a cephem nucleus and is illustrated in Fig. 9 with indication of loci for substitutions. A number of *Streptomyces* have been found to form five antibiotics with a cephem nucleus (Nagarajan *et al.*, 1791; Stapley *et al.*, 1972; Albers-Schönberg *et al.*, 1972). None of these have been found to be identical with antibiotics produced by members of different genera. It would seem, however, that the announcement of such a discovery is only a question of time.

Gliotoxin is a sulfur-containing antibiotic active against bacteria and fungi. It is known to be produced by strains of *Penicillium* and the related genus *Gliocladium*. In addition it has been isolated from cultures of *Aspergillus* and *Trichoderma* (Baron, 1950; Winstead and Suhadolnik, 1960).

K. STEROIDS (Fig. 10)

Fumigacin, first isolated in 1942, is an antibacterial substance which is produced by *Aspergillus fumigatus* (Baron, 1950), *Cephalosporium mycophylum* (Tsuda *et al.*, 1965), and *Streptomyces reticuli* var. *protomycicus* (Hata, 1964).

Fusidic acid (ramycin) is active against gram-positive bacteria. It is known to be produced by several imperfect fungi belonging to the genera *Fusidium* and *Cephalosporium* and also by one of the Phycomycetes, *Mucor ramannianus* (Vanderhaeghe *et al.*, 1965). A similar substance was found to be elaborated by *Microsporum gypseum* and strains of its perfect form, *Nannizia incurvata* (Elander *et al.*, 1969).

L. QUINONES (Fig. 11)

Oosporein is a red bi-benzoquinone pigment with anbacterial and antifungal activity. It is produced by the imperfect fungi *Oospora colorans*, *Acremonium* sp., and *Beauveria bassiana* (Vining *et al.*, 1962). In addition, it has been isolated from the metabolites of an asycomycete, *Chaetomium aureum* (Lloyd *et al.*, 1955), and of basidiomycetes, *Phlebia*

Fumigacin (helvolic acid)

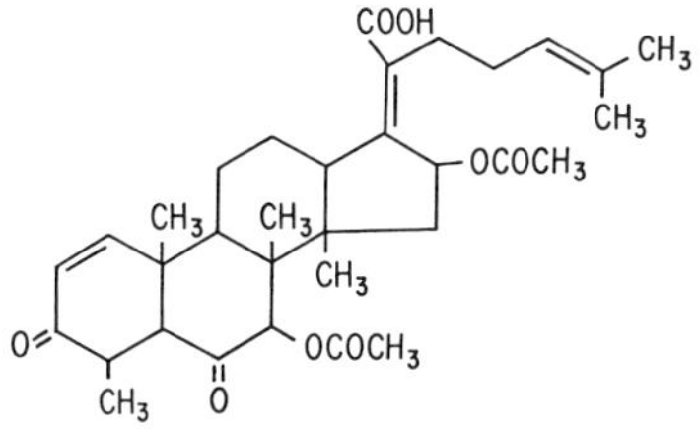

Fusidic acid (ramycin)

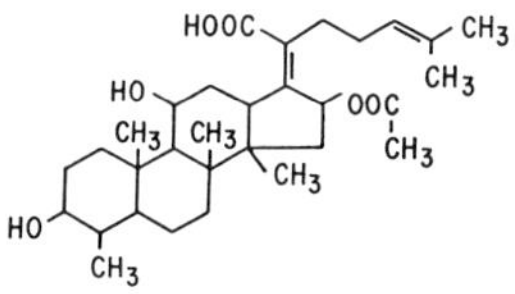

FIG. 10. Steroids.

Oosporein (chaetomidin)

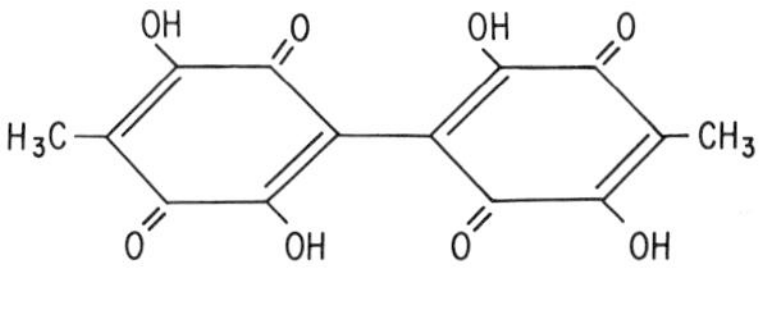

Citrinin

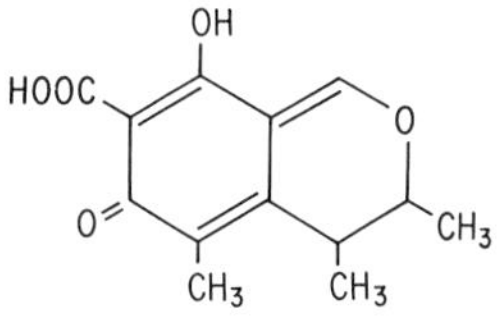

FIG. 11. Quinones.

mellea and *P. albida* (Takeshita and Anchel, 1965). The mycelia of the two cultures of *Phlebia* were septate and had clamp connections, thus ruling out the possibility of contamination with cultures of non-basidiomycetes.

Citrinin is active against bacteria and fungi. It is known to be produced by several species of *Aspergillus* and *Penicillium* (Baron, 1950) and by *Clavariopsis aquatica* (Broadbent, 1966). In addition, it is produced in the leaves, and excreted in the hairs, of *Crotalaria crispata*, a leguminous plant of the tropical belt of Australia (Ewart, 1933).

M. Pyrones (Fig. 12)

Kojic acid displays modest antibacterial and antifungal activity. It is active only at levels of milligrams per milliliter rather than micrograms per milliliter, as are most antibiotics. It is often more active against

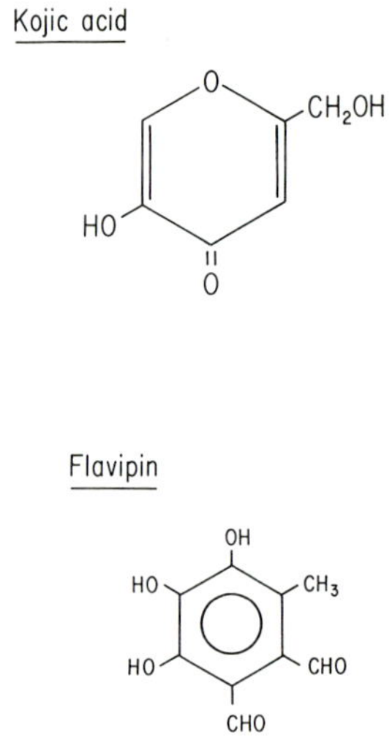

FIG. 12. Pyrones and phenols.

gram-negative organisms than against gram-positive ones (Foster and Karow, 1945). This compound has been isolated from among the metabolites of numerous species of *Aspergillus*, a *Penicillium*, and several strains of acetic acid bacteria (Beélik, 1956).

N. Phenols (Fig. 12)

Flavipin is a pale yellow antifungal substance which is stable in aqueous solutions only at low pH values. It is produced by strains of *Aspergillus* and of *Epicoccum* (Bamford et al., 1961).

III. Conclusion

The present compilation is undoubtedly very incomplete since it is based solely on published information. Unfortunately, as we all know, not all the research done in the field of antibiotics is necessarily published. Indeed, to write a paper just to say that an already known substance has been rediscovered among the products of a different organism is hardly an exciting prospect.

As far as I know, there is only the unique case of the antibiotic citrinin which is known to be obtained both from a higher plant and microorganisms. In contrast, production of the same antibiotic by different genera of imperfect fungi is without doubt the most common finding. Between the two extremes we have the production of the same substances by fungi and actinomycetes (nebularin, questiomycin A, penicillin N, and fumigacin) and the not surprisingly somewhat more common production of the same substances by actinomycetes and bacteria (toxoflavin, prodigiosin, iodinin and relatives, questiomycin A, anticapsin, and erythromycin A). Thus we arrive at the not unexpected conclusion that the more closely related organisms are, the more likely they are to produce the same secondary metabolites. The obvious corollary to this is that if one is seeking novel antibiotics, one should look at novel sources, which undoubtedly is easier said than done.

ACKNOWLEDGMENT

I wish to thank Nancy N. Gerber, Thomas G. Pridham, Edward Katz, and L. V. Kalakoutskii for drawing some of the references to my attention.

REFERENCES

Abraham, E. P., and Newton, G. G. F. (1961). *Biochem. J.* **79**, 377–393.
Albers-Schönberg, G., Arison, B. H., and Smith, J. L. (1972). *Tetrahedron Lett.* **29**, 2911–2914.
Anzai, K., Isono, K., Okuma, K., and Suzuki, S. (1960). *J. Antibiot. Ser. A.* **13**, 125–132.
Bai, M. P., and Rao, P. L. N. (1963). *Indian J. Exp. Biol.* **1**, 50–53.
Bamford, P. C., Norris, G. L. F., and Ward, G. (1961). *Trans. Brit. Mycol. Soc.* **44**, 354–356.
Baron, A. L. (1950). "Handbook of Antibiotics." Van Nostrand-Reinhold, Princeton, New Jersey.
Beélik, A. (1956). *Advan. Carbohyd. Chem.* **11**, 145–183.
Broadbent, D. (1966). *Bot. Rev.* **32**, 219–242.
Cole, M. (1966). *Appl. Microbiol.* **14**, 98–104.
Cole, M., and Rolinson, G. N. (1961). *Proc. Roy. Soc., Ser. B* **154**, 490–497.
Cooney, D. G., and Emerson, R. (1964). "Thermophilic Fungi." Freeman, San Francisco, California.
D'Aoust, J. Y., and Gerber, N. N. (1974). *J. Bacteriol.* **118**, 756–757.
Daves, G. D., Robins, R. K., and Cheng, C. C. (1961). *J. Amer. Chem. Soc.* **83**, 3904–3905.
Daves, G. D., Robins, R. K., and Cheng, C. C. (1962). *J. Amer. Chem. Soc.* **84**, 1724–1729.
Davis, J. G. (1939). *Zentralbl. Bakteriol., Parasitenk. Infektionskr., Abt. 2* **100**, 273–276.

DeBoer, C., Dietz, A., Michaels, R. M., Eble, T. E., Olson, E. C., Large, C. M., and Shell, J. W. (1960). *Antibot. Annu.* pp. 220–229.
Delcambe, L. (1967). *Inform. Bull., Int. Cent. Inform. Antibiot.* **4**, 84.
Delcambe, L. (1969). *Inform. Bull., Int. Cent. Inform. Antibiot.* **7**, 65.
Elander, R. P., Gordee, R. S., Wilgus, R. M., and Gale, R. M. (1969). *J. Antibiot.* **22**, 176–178.
Ewart, A. J. (1933). *Ann. Bot. (London)* **47**, 913–915.
Fisher, W. P., Charney, J., and Bolhofer, W. A. (1951). *Antibiot. Chemother.* **1**, 571–572.
Foster, J. W., and Karow, E. O. (1945). *J. Bacteriol.* **49**, 19–29.
Frederiksen, S., Malling, H., and Klenow, H. (1965). *Biochim. Biophys. Acta* **95**, 189–193.
French, J. C., Howells, J. D., and Anderson, L. E. (1970). U.S. Patent 3,551,294.
Gaughran, E. R. L. (1969). *Trons. N.Y. Acad. Sci.* [2] **31**, 3–24.
Gause, G. F., Brajnikova, M. G., Sveshnikova, M. A., Ukhlina, R. S., Hechaeva, N. P., Gavrilina, G. V., Lavrova, M. F., Kobcharova, I. N., Proshliakova, V. V., Kudinova, M. K., and Shanobalova, S. P. (1970). *Antibiotiki (Moscow)* **15**, 99–102.
Gerber, N. N. (1966). *Biochemistry* **5**, 3824–3829.
Gerber, N. N. (1969). *J. Heterocycl. Chem.* **6**, 297–300.
Gerber, N. N. (1971). *J. Antibiot.* **24**, 636–640.
Gerber, N. N., and Lechevalier, H. A. (1962). *J. Org. Chem.* **27**, 1731–1732.
Gerber, N. N., and Lechevalier, M. P. (1964). *Biochemistry* **3**, 598–602.
Gerber, N. N., and Lechevalier, M. P. (1965). *Biochemistry* **4**, 176–180.
Gerber, N. N., and Wieclawek, B. (1966). *J. Org. Chem.* **31**, 1496–1498.
Grosklags, J. H., and Swift, M. E. (1957). *Mycologia* **49**, 305–317.
Guarino, A. J., and Kredich, N. M. (1963). *Biochim. Biophys. Acta* **68**, 317–319.
Harashima, K., Tsuchida, N., Tanaka, T., and Nagatsu, J. (1967). *Agr. Biol. Chem.* **31**, 481–489.
Hata, T. (1964). Japanese Patent 41-12669.
Hesseltine, C. W. (1965). *Mycologia* **57**, 149–197.
Irie, T., Kurosawa, E., and Nagaoka, I. (1960). *Bull. Chem. Soc. Jap.* **33**, 1057–1059.
Isono, K., and Suzuki, S. (1960). *J. Antibiot., Ser. A* **13**, 270–272.
Kaczka, E. A., Smith, M. C., and Folkers, K. (1963). *Fed. Proc. Fed. Amer. Soc. Exp. Biol.* **22**, 306.
Kaczka, E. A., Dulaney, E. L., Gitterman, C. O., Woodruff, H. B., and Folkers, K. (1964). *Biochem. Biophys. Res. Commun.* **14**, 452–455.
Kaminski, K., and Sokolowska, T. (1973). *J. Antibiot.* **26**, 184–185.
Krassilnikov, N. A. (1959). *Ann. Inst. Pasteur, Paris* **96**, 434–447.
Latuasan, H. E., and Berends, W. (1951). *Biochim. Biophys. Acta* **52**, 502–508.
Lechevalier, H. A. (1965). In "Biogenesis of Antibiotic Substances" (Z. Vaněk and Z. Hošťálek, eds.), pp. 227–232. Publ. House Czech. Acad. Sci., Prague.
Lewis, S. M., and Corpe, W. A. (1964). *Appl. Microbiol.* **12**, 13–17.
Lloyd, G., Robertson, A., Sankey, G. B., and Whalley, W. B. (1955). *J. Chem. Soc., London* pp. 2163–2165.
Löfgren, N., Lüning, B., and Hedström, H. (1954). *Acta Chem. Scand.* **8**, 670–680.
Machlowitz, R. A., Fisher, W. P., McKay, B. S., Tytell, A. A., and Charney, J. (1954). *Antibiot. Chemother.* **4**, 259–261.
Miller, F. A., Rightsel, W. A., Sloan, B. J., Ehrlich, J., French, J. C., Bartz, Q. R., and Dixon, G. J. (1963). *Nature (London)* **200**, 1338–1339.
Miller, I. M., Stapley, E. O., and Chaiet, L. (1962). *Bacteriol. Proc.* p. A49.
Nagarajan, R., Boeck, L. D., Gorman, M., Hamill, R. L., Higgens, C. E., Hoehn,

M. M., Start, W. M., and Whitney, J. G. (1971). *J. Amer. Chem. Soc.* **93**, 2308–2310.
Nakamura, G. (1961). *J. Antibiot., Ser. A* **14**, 94–97.
Neuss, N., Molloy, B. B., Shah, R., and DeLaHiguera, N. (1970). *Biochem. J.* **118**, 571–575.
Nugteren, D. H., and Berends, W. (1957). *Rec. Trav. Chim. Pays-Bas* **76**, 13–27.
Perry, J. J. (1961). *Nature (London)* **191**, 77–78.
Pfeifer, S., Bar, H., and Zarnack, J. (1972). *Pharmazie* **27**, 536–542.
Pisano, M. A., Fleischman, A. I., Littman, M. L., Dutcher, I. D., and Pansy, F. E. (1960). *Antimicrob. Ag. Annu.* pp. 41–47.
Prauser, H., and Eckardt, K. (1967). *Z. Allg. Mikrobiol.* **7**, 409–410.
Rinehart, K. L. (1964). "The neomycins and related antibiotics." Wiley, New York.
Rosett, R., Sankhala, R. H., Stickings, C. E., Taylor, M. E. U., and Thomas, R. (1957). *Biochem. J.* **67**, 390–400.
Shigeura, H. T. (1963). *Fed. Proc. Fed. Amer. Soc. Exp. Biol.* **22**, 306.
Sneath, P. H. A. (1956). *J. Gen. Microbiol.* **15**, 70–98.
Stapley, E. O., Miller, T. W., and Jackson, M. (1969). *Antimicrob. Ag. Chemother.* **1968**, 268–273.
Stapley, E. O., Jackson, M., Hernandez, S., Zimmerman, S. B., Currie, S. A., Mochales, S., Mata, I. M., Woodruff, H. B., and Hendlin, D. (1972). *Antimicrob. Ag. Chemother.* **2**, 122–131.
Stewart, G. T. (1965). "The Penicillin Group of Drugs." Amer. Elsevier, New York.
Suhadolnik, R. J. (1970)."Nucleoside Antibiotics." Wiley (Interscience), New York.
Sukapure, R. S., Deshmukh, P. V., and Thirumalachar, M. J. (1969). *Hindustan Antibiot. Bull.* **12**, 26.
Takeshita, H., and Anchel, M. (1965). *Science* **147**, 152–153.
Tresner, H. D., Hayes, J. A., and Borders, D. B. (1971). *Appl. Microbiol.* **21**, 562–563.
Tsuda, K., Okuda, S., Hata, F., Sano, T., and Yamaguchi, H. (1965). Japanese Patent 6398.
Turner, W. B. (1971). "Fungal Metabolites." Academic Press, New York.
Uri, J., Juhász, P., and Cosban, G. (1955). *Pharmazie* **10**, 709–713.
Uri, J., Valu, G., and Bébési, I. (1963). *Nature (London)* **200**, 896–897.
Van Damme, P. A., Johannes, A. G., Cox, H. C., and Berends, W. (1960). *Rec. Trav. Chim. Pays-Bas* **79**, 255–267.
Vanderhaeghe, H., van Dijck, P., and De Somer, P. (1965). *Nature (London)* **205**, 710–711.
Vining, L. C., Kelleher, W. J., and Schwarting, A. E. (1962). *Can. J. Microbiol.* **8**, 931–933.
Wagman, G. H., Watkins, P. D., Marquez, J. A., and Weinstein, M. J. (1972). *Amer. Soc. Microbiol.* Abstract No. 10.
Waksman, S. A., and Lechevalier, H. A. (1962). "Antibiotics of Actinomycetes." Williams & Wilkins, Baltimore, Maryland.
Walker, J. E., and Abraham, E. P. (1970). *Biochem. J.* **118**, 563–570.
Wasserman, H. H., Rodgers, G. C., and Keith, D. D. (1966). *Chem. Commun* p. 825.
Williams, R. P., and Hearn, W. R. (1967). *In* "Antibiotics" D. Gottlieb and P. D. Shaw, eds.), Vol. 2, pp. 410–432. Springer-Verlag, Berlin and New York.
Winstead, J. A., and Suhadolnik, R. J. (1960). *J. Amer. Chem. Soc.* **82**, 1644–1647.

Antibiotic-Producing Fungi: Current Status of Nomenclature

C. W. HESSELTINE[1] AND J. J. ELLIS[2]

Northern Regional Research Laboratory, Peoria, Illinois

I.	Introduction	47
II.	Antibiotic-Producing Fungi	47
III.	Rules of Botanical Nomenclature	48
IV.	Purpose of the Botanical Code	48
V.	Numbers of Fungi	49
VI.	Botanical Rules Specifically for Fungi	49
VII.	Comparison of the Botanical and Bacterial Codes	50
VIII.	Descriptions of Fungi	53
IX.	Type Cultures in Culture Collections	54
X.	Publications Dealing with Fungal Nomenclature	56
	References	57

I. Introduction

This topic is perhaps too restrictive for the material we shall include. Although emphasis will be on nomenclature of antibiotic-producing fungi, we shall discuss more broadly the problem of nomenclature for fungi used in industrial fermentations. We also shall indicate some areas of mycological nomenclature that may be changed in the near future in the Botanical Code, which covers fungi.

II. Antibiotic-Producing Fungi

Although there are a large number of species of fungi, only a relative few have been found to produce antibiotics, and only seven antibiotics are produced commercially. The 1970 Information Bulletin, No. 8 (3) of the International Center of Information on Antibiotics lists 338 species of fungi that produce antibiotics. According to 1967 (11) and 1970 (13) lists assembled by Perlman, the following antibiotics are produced commercially: Fusidic acid formed by *Fusidium coccineum* Fückel, griseofulvin formed by *Penicillium griseofulvim* Dierk, penicillins formed by *Penicillium chrysogenum* Thom, variotin formed by *Paecilomyces varioti* Bainier, derivatives of cephalosporin formed by *Emericellopsis* (*Cephalosporium*), and fumagillin produced by *Aspergillus fumigatus* Fres. Since then, siccanin produced by *Helminthosporium siccans* Drechsler has been produced in Japan. According to Perlman (12), only three groups of antibiotics are produced by fungi in the United States. These are the cephalosporins, penicillins, and fumagillin.

[1] Chief, Fermentation Laboratory, Northern Regional Research Laboratory, Peoria, Illinois.

[2] ARS Culture Collection Research, Fermentation Laboratory, Northern Regional Research Laboratory, Peoria, Illinois.

III. Rules of Botanical Nomenclature

Fungi, including Myxomycetes, Phycomycetes, Ascomycetes, Basidiomycetes, Lichens, and Fungi Imperfecti, are covered by the International Code of Botanical Nomenclature (6). This code covers all plants. It originated in 1906, although a set of rules for nomenclature was published in 1867. Since then, the rules have been improved and revised in 1912, 1935, 1947, 1950, 1952, 1956, 1961, 1966, and finally 1972. References to all these codes appear at the end of the Botanical Code published in 1972. Generally, a new code is published after each Botanical Congress. The code consists of a Preamble, Principles, Rules, and Recommendations; it lists both conserved and rejected taxa; it is published in English, French, and German. In the 1972 Code, 58 genera of fungi are conserved exclusive of Lichens. The only name of importance to industrial fermentation is the genus *Candida* Berkhout, which is conserved over all its synonyms.

Initially, the rules of Botanical Nomenclature were considered at the First Botanical Congress held in Paris in 1867. These rules were primarily for vascular plants but also included bacteria and fungi. It was not until 1947 that bacteriologists broke away from the Botanical Code, primarily because the Botanical Code stated that living cultures are not acceptable as types. The Botanical Code specifies that a type specimen of a taxon of recent plants, the bacteria excepted, must be preserved permanently and cannot be a living plant or culture. The International Code of Nomenclature of Bacteria (7) incorporates many rules that apply to the Botanical Code.

IV. Purpose of the Botanical Code

The Botanical Code arose to satisfy a need for an orderly manner of naming taxa. As Shear (17) states, "the fundamental requirements of a satisfactory nomenclature are uniformity, stability, exact application and convenience."

The Botanical Code, as set forth in the Preamble:

1. Establishes a precise and simple system of nomenclature to be used in all countries.
2. Establishes a stable method of naming taxonomic groups.
3. Tries to avoid the useless creation of names.
4. Establishes rules (articles), which must be followed, and recommendations. Recommendations deal with subsidiary points; their objective is to bring about greater uniformity and clearness, especially in future nomenclature.

5. Recommends, in the absence of a relevant rule or where the consequences of rules are doubtful, following established custom.

Examples of the application of the rules and recommendations are cited after each rule and recommendation. Since fungi offer certain special problems, some rules are specifically designed for fungi.

V. Numbers of Fungi

One can ask how many fungi actually exist which are covered by the Botanical Rules. Of course, there are hazards in estimating the number of fungi. The figure most frequently quoted is 100,000 species. This number is evidently based on the total compiled in Saccardo's *Sylloge Fungorum*, Volumes 1 through 25 (*15*), plus an estimate of those proposed since. Martin (*10*) considered this figure to be "excessively conservative." He based his statement on an exercise of randomly selecting 100 vascular plants and finding from Seymour (*16*) the number of different fungi that parasitized them. The result indicated an average of three fungi per host species. Martin suggested that, if to the total number of parasitic fungi is added the number of saprobic fungi, the number of good species of fungi is at least as great as the number of Phanerogams (vascular plants). The number of Phanerogams is believed to be about 250,000. According to Ainsworth (*1*) the number of proposed new species of flowering plants is now static or on the decline. However, new species of fungi proposed averaged 700 per year between 1920 and 1950. Currently, the *Index of Fungi* (*5*) lists more than 1000 newly proposed names each year with about half of them redispositions.

VI. Botanical Rules Specifically for Fungi

Botanical rules of nomenclature that specifically deal with fungi include an article on the names of fungi with a pleomorphic life cycle. The pleomorphic life cycle refers to fungi that may have two or more spore states. The mold that produces cephalosporins exhibits such a life cycle. Consequently, the fungus that produces these compounds was first described as a species of *Cephalosporium* because only the conidial or asexual stage was known. Later the ascosporic or sexual (perfect) stage was found, which belonged to the genus *Emericellopsis* (*4*). Commonly, the fungus grows as a *Cephalosporium*, and only under certain conditions will it produce sexual spores. The valid name for a fungus having a pleomorphic life cycle is the earliest legitimate name typified by the perfect state, but the name of the imperfect state can be used for convenience. Hence, it would be correct to say that *Emericellopsis* has a *Cephalosporium* imperfect state for the species that produces the cephalosporins.

A second section of one article in the Botanical Code lists the starting dates for the nomenclature of fungi. The starting date for certain fungi belonging in the Basidiomycetes begins with Persoon's book *Synopsis Methodica Fungorum*, 1801. The starting date for Myxomycetes and Lichens begins with Linnaeus' *Species Plantarum*, 1753, and the date for the remaining fungi begins with Fries' *Systema Mycologicum*, 1821. The third starting date is the one particularly important to fermentation researchers because almost all fungi concerned started at that date or later. Incidentally, the Botanical Code states that the nomenclature of bacteria begins with Linnaeus' *Species Plantarum*, 1753. (The Bacteriological Code likewise begins with Linnaeus.) To us, 1753 is a poor starting date for bacteria because Linnaeus was unaware of bacteria and had only a fuzzy idea of what fungi were. A date after 1900 based on a treatment of all bacteria would be a much more reasonable and effective one. The proposed new Code for Bacteria (9) is just as inappropriate because it states that the starting date should be moved to January 1, 1980. One wonders why 1984 might not have been suggested. The Botanical Code can be revised only at International Congresses, and the manner in which it can be revised is clearly stated.

In the Botanical Code the type specimen is clearly defined: It states that a type specimen is a preserved specimen, not a living culture. For example, when we have described a new species of mold, a dried culture is designated the type, and it, or portions of it, are deposited in a herbarium. A type culture is not recognized in the Botanical Code because it can change in appearance with repeated transferring or it may actually be lost. For many fungi, designating a culture or type offers certain difficulties.

Of course, there are other examples of the rules dealing with fungi, but they are of little concern to us here.

VII. Comparison of the Botanical and Bacterial Codes

In Table I are given some differences between the Botanical Code and the Bacterial Code. It is appropriate to make some generalizations about the two codes as they affect applied microbiology. Since the Botanical Code has been in existence for a longer time, it is more stable and the points of controversy are fewer. For example, in the next Botanical Congress, the five issues or problem areas (8) as applied to fungi are as follows: (i) whether or not a type culture should be designated for fungi (we believe that this defect is the only serious major one in the Code that needs to be corrected); (ii) clarification of the rule that deals with pleomorphic fungi; (iii) clarification of starting dates and problems of overlap of groups with different starting

TABLE I
SOME COMPARISONS OF THE BOTANICAL AND BACTERIOLOGICAL CODES

Description	Botanical code	Bacteriological code
Starting date of nomenclature	Lichens ⎱ Linnaeus' "Species Plantarum" May 1, 1753 Myxomycetes ⎫ Uredinales (rusts) ⎬ Persoon's *Synopsis Methodica Fungorum* December 31, 1801 Ustilagenales (smuts) ⎪ Gastromycetes (puff balls) ⎭ Fungi ⎱ Fries' *Systema Mycologicum* January 1, 1821 Caeteri (all other fungi)	Linnaeus' *Species Plantarum*, May 1, 1753. It is proposed to set this date as January 1980
Language	After January 1, 1935 must be accompanied by a Latin description or diagnosis	None. Proposed code would require a description in a familiar language
Type	"Type specimen . . . must be preserved permanently and cannot be a living plant or culture." If it is impossible to preserve a specimen, then "the type may be a description or figure"	A designated type strain or in special cases a description, preserved specimen, or illustration. Proposed code: Wherever possible a strain of a living culture. If no living culture is available, a later cultured strain may be designated as neotype
Effective and valid publication	Distribution of printed material must be accompanied by a diagnosis or reference to a previously effectively published description. After January 1, 1933, no name proposed in a tradesman's catalog or in a nonscientific newspaper is valid	Printed matter for sale or distribution to the general public or bacteriological institution is valid. Inclusion of a name in a patent is not effective publication
Pleomorphic life cycle	Pleomorphic life cycles in Ascomycetes and Basidiomycetes are recognized. The valid name is the "earliest legitimate name typified by the perfect state," but names of imperfect states can be commonly used	Nothing
Provision for emendment of	Modified by act of plenary session. Provides for 11 permanent nomenclatural committees and includes a "Committee for Bacteria"	Modified by action of the International Committee on Nomenclature and approval of a general meeting

dates; (iv) provision for handling intraspecific taxa not covered by the Code; (v) registry of new names and proposals for conservation.

The starting dates of bacteria, both proposed and the current ones, seem to be unreasonable. For fungi, the starting dates have been fixed for a long time and are generally quite satisfactory. Such genera as *Penicillium* and *Aspergillus* were described in Fries, and reference is made to illustrations that clearly depict the same organisms we know as these genera today. At least some of the species recognized by Fries are *Penicillium* and *Aspergillus* in the modern sense.

Exclusion by the Bacteriological Code of patents as vehicles to publish bacterial names is most unfortunate. In the Botanical Code publication of names in patents is permitted since the article states, "Publication is effected, under this Code, only by the distribution of printed matter (through sale, exchange or gift) to the general public. . . ." It specifically excludes publication "in tradesmen's catalogs and nonscientific newspapers." Since patents are not in these categories but are part of the scientific literature, are printed, and are available for sale, descriptions of fungi in patents, provided a Latin diagnosis is given, are effectively and validly published.

Because the procedure for amending the rules is clearly defined in the Botanical Code, whenever changes are proposed they are seriously studied. The consensus of many people is required before additions or deletions can be made. In contrast, it appears that a few people have been able to control changes in the Bacteriological Code.

The requirement for a Latin description included in the Botanical Code has certain distinct advantages even though the language is no longer a common everyday one. However, with the use of a dictionary, almost everyone can read a simple Latin diagnosis. Taxonomists the world over use this procedure, and even though it is impossible for most Western people to read articles written by either Russians or Japanese, these people give Latin descriptions that can be understood everywhere. To Latinize a description requires some skill, but often help can be obtained from friends who are expert in Latin or from faculty members in classical language departments. Because a Latin description is required, persons naming new taxa must be serious about the new names they are proposing.

A special problem in industrial mycology involves the nomenclature of induced mutants that may be very different in appearance from their wild-type ancestors. The mutants should retain the species name of the wild-type material. However, the Botanical Code has a section devoted to "Names of Plants in Cultivation" in which variants produced by hybridization, mutation, or selection are of sufficient interest to receive epithets preferably in common language (i.e., fancy epithets) markedly different from the Latin epithets of species and varieties. It cites several examples such as *Primula malacoides* 'Pink Sensation.' Normally in microbiology a strain number is used, not a fancy name.

Part of the reason for differences between the Botanical and Bacteriological Codes is that species, and even varieties, of fungi are described on morphological grounds, whereas almost all the descriptions of bacteria are physiological.

VIII. Descriptions of Fungi

Some may wonder what an adequate taxonomic treatment of a new fungus should be. First, it should cite the name to be applied and the appropriate taxon designation; for example, sp. nov., which means that the authors consider it to be a new species. Following this designation should be given the synonyms or names that are not valid and the citations to where these were used first in the literature. Next comes a description that typically gives the colony appearance, including color on three or four media, which can be reproduced readily in other laboratories. Color of the colonies on each medium should be described from a color chart at various ages and for both the top and bottom of the colony. Then a description of the microscopic characters of the fungus on one or more of the media should be given. Development of structures and their mature appearance should be described. Measurements of the size of all the morphological parts should be listed, including the minimum, maximum, and average dimensions. Generally, this information should include the range and the average of at least 100 measurements. If there is a sexual state, it should be described and the conditions under which it was induced to form should be given. The description should give the data on physiological characteristics, such as the growth at different temperatures, sugars fermented or utilized, and other similar traits.

In a separate paragraph the location of the type material, who collected it, and the date it was collected should be reported, followed by a statement indicating where the type specimen was deposited. A type strain and its deposit number in an established culture collection can be named here. Because the designation of a type strain number is not prohibited by the Botanical Code, we believe a culture should be so designated, and every attempt made to have the culture preserved. The Latin diagnosis should follow the formal description in a common scientifically used language. Often, it is put first rather than after and it needs to only cover the more important characteristics of the taxon. Some prefer to repeat the complete description in Latin.

Next should appear a statement indicating the origin of the new name (its etymology). This statement should precede a description of the geographic range of the fungus, so far as is known, and the host specificity if it is a parasite. Next comes a general discussion of how many specimens or strains were studied and their sources. If there is any question as to which family it belongs, the authors should indicate the family to which they think it belongs and their reasons for placing it there.

How the new taxon differs from its most closely related species needs to be discussed, including an enumeration of the characteristics that make it distinct. Sometimes it is helpful to give a diagnostic key to the genus or portion of the genus to show how the species differs from all other known species. Information needs to be provided as to how stable the various characteristics may or may not be. Also, there needs to be information on additional strains and specimens studied and how any of these differ from the type as well as how the species differs from other species studied for comparative purposes.

The description of each new taxon requires illustrations (either photographs or line drawings) including microscopic details. Finally, the description and name must be both validly and effectively published according to the International Code of Botanical Nomenclature (6).

IX. Type Cultures in Culture Collections

Because we are dealing here primarily with fungi used in industry, and particularly in the antibiotic industry, the subject of deposit of cultures, especially those in connection with patent applications, needs to be discussed. Deposits need to be made in good permanent collections whose standards are quite high. Along with Drs. Creech and Warwick (2), one of us (CWH) has attempted to give the characteristics of superior culture collections in the report *Genetic Pools. The Conservation of the World's Genetic Resources in Plants, Animals, and Microorganisms for use in Agriculture and Industry*. These characteristics are repeated here:

1. The collection must be part of, or closely related to, a research laboratory concerned with either microbiology or fermentation, or both. For example, at the Northern Regional Research Laboratory, Peoria, Illinois, the ARS Culture Collection is one of four research units in the Fermentation Laboratory. Interactions between microbiologists and culture collection curators work to the mutual benefit of both. The microbiologist, being aware of general trends in microbiological research, is able to anticipate future areas of interest and to give guidance as to what microorganisms a culture collection should accession to meet future needs. The curator, with his knowledge of the relationships among genera

and the physiological requirements of various microorganisms, can make valuable suggestions in return.

2. A culture collection must be well funded, and this funding must be at a relatively uniform level each year. In many research operations, a program may be increased or decreased readily with changes in the amount of budgeted money. People can be shifted easily from one project to another. A culture collection, on the other hand, is a continuing operation that must be sustained without great fluctuations in either budget or people from year to year.

3. A culture collection must have adequate facilities and equipment, including transfer rooms, refrigerator space, incubators, microscopic and photographic equipment, autoclaves, and lyophilizers. Usually these facilities should be separate from those of other research groups.

4. Library facilities are necessary so that the staff may have access to the taxonomic and microbiological literature being published not only in the region or country of location, but also elsewhere in the world.

5. The collection should have an active and continuing program of isolating new strains of microorganisms from nature. Such a program will lead to the discovery of new products and reactions. New material will add to understanding the classification of special groups of microorganisms and will make it possible to discover species and genera new to science.

6. The collection must have an adequate staff to support the curators. By this is meant technical help for the preparation of media, sterilization of glassware, maintenance of supplies, and similar service duties; secretarial help to keep the voluminous records and to handle correspondence; and shops to construct special apparatus. Reliable sources of supplies are also necessary.

Optimally, each curator should have a careful, intelligent, and dedicated assistant with some microbiological training. Technicians need not be too specialized because they always must be trained in the special techniques required by each group of microorganisms. Assistants should handle periodic transfers, lyophilization and associated records, inoculation of cultures for study by the curator, seeding of flask cultures for preliminary surveys for new products, and making and recording routine observations on all cultures.

7. The curator(s) must do research as well as culture maintenance. Each must have an active research program either in taxonomy or in genetics, with preference to the former. Thus a curator will have an intimate knowledge of the strains he is maintaining and will develop a reputation as an expert in his field. Consequently, important material will be sent to him for safekeeping, for identification, and for other purposes. Other microbiologists will know from whom they may get

expert advice, cultures, or materials to isolate cultures from. Culture collection people not only should engage in research, but they should report their research in the form of papers published in scientific journals, give lectures, and occasionally take out initial patents.

8. College-trained men and women in collections must be aware of the field of basic and applied microbiology and have an appreciation of fermentation research and development. They must comprehend the problems of geneticists, fermentologists, engineers, biochemists, and organic chemists. Probably the most difficult job from the standpoint of the administration of a culture collection involves indoctrination of the curators in understanding other scientists' point of view, and bringing them to realize that they are part of a team working toward established goals whether in producing a product for sale or in the discovery of new information about microorganisms. Collection staff must be made aware of the needs of other research people.

9. Although the curators should be trained in taxonomy, the overall background of the staff should have balance. If the collection has more than one senior scientist, then the broader the interests of the group the better. It does no good to have three specialists on bacteria and yet have no mycologists, or vice versa.

10. At least in the larger collections, young people with new ideas and knowledge of new techniques should be brought into the group periodically. This means of rejuvenation may be supplemented with postdoctoral fellows and exchange of personnel from other institutions. They should not necessarily be people from other collections. In turn, the resident staff needs periodically to travel or study in other laboratories.

X. Publications Dealing with Fungal Nomenclature

Proposals for modifying the Botanical Code are published in *Taxon*, the official organ of the International Association for Plant Taxonomists. These modifications are periodically voted on by the membership and can be accepted, rejected, or referred to a committee. Conserved fungus names have also been published in *Taxon*. The names of fungi can most easily be found in two references. The first is Ainsworth's *Dictionary of the Fungi* (1), which gives all generic names, including lichens up to 1970. Synonyms are referred to their proper genera. Generally, a reference to a monographic treatment of the genus or to a recent treatment of the species is given. The second reference is the *Index of Fungi* (5), which is published periodically—twice a year lately. The *Index of Fungi* is a continuation of Petraks' Lists (14), which contained names of fungi up to 1940. These two references give the place of publication of the species of fungi, and the *Index of Fungi* includes references to new species as they are proposed.

It is to be hoped that future International Botanical Congresses will devise a logical and methodical means of correctly solving the nomenclatural problems confronting us.

REFERENCES

1. Ainsworth, G. C. (1971). "Dictionary of the Fungi." Commonwealth Mycological Institute, Kew, Surrey, England.
2. Creech, J. L., E. J. Warwick, and C. W. Hesseltine. (1972). "Genetic Pools. The Conservation of the World's Genetic Resources in Plants, Animals, and Microorganisms for use in Agriculture and Industry" (unpublished).
3. Delcambe, L. (1970). Information Bulletin No. 8. International Center of Information on Antibiotics, Etud'imprim, Liège, Belgium.
4. Grosklags, J. H., and M. E. Swift. (1957). *Mycologia* 49, 305–317.
5. "Index of Fungi." (1940–1973) (and current). Vols. 1–4. Commonwealth Mycological Institute, Kew, Surrey, England.
6. International Code of Botanical Nomenclature. (1972). Published by the International Bureau for Plant Taxonomy and Nomenclature of the International Association for Plant Taxonomy, Utrecht, Netherlands.
7. International Code of Nomenclature of Bacteria. (1966). *Int. J. Syst. Bacteriol.* 16, 459–490.
8. Korf, R. P., D. L. Hawksworth, G. L. Hennebert, Z. Pouzar, D. P. Rogers, and L. K. Weresub. (1973). *Plant Sci. Bull.* 19, 26.
9. Lapage, S. P., W. A. Clark, E. F. Lessel, H. P. R. Seeliger, and P. H. A. Sneath, (1973). *Int. J. Syst. Bacteriol.* 23, 83–108.
10. Martin, G. W. (1951). *Iowa Acad. Sci.* 58, 175–178.
11. Perlman, D. (1967). *Chem. Week* 101, 82–85, 88, 93–95, 98, and 100–101.
12. Perlman, D. (1968). *Process Biochem.* 3, 54–58.
13. Perlman, D. (1970). *Wallerstein Lab. Commun.* 33, 165–175.
14. Petraks' Lists. (1920–1939). In 8 parts. Commonwealth Mycological Institute, Kew, Surrey, England.
15. Saccardo, P. A. (1882–1931). "Sylloge Fungorum," 25 vols. Published by the author, Pavia, Italy. Vol. 26 published by Johnson Reprint Corp., New York, 1972.
16. Seymour, A. B. (1929). "Host Index of the Fungi of North America." Harvard Univ. Press, Cambridge, Massachusetts.
17. Shear, C. L. (1936). *Mycologia* 28, 337–346.

Significance of Nucleic Acid Hybridization to Systematics of Actinomycetes

S. G. BRADLEY

*Department of Microbiology,
Virginia Commonwealth University,
Richmond, Virginia*

I.	Introduction	59
II.	DNA Nucleotide Composition	60
III.	DNA:DNA Association	60
IV.	The Cot Concept	63
V.	Optical Reassociation	64
VI.	Phylogenetic Implications	66
VII.	Neutral Mutations	69
	References	70

I. Introduction

Systematics progresses through three stages: the description of cultures leading to bases for distinguishing between strains; the development of keys, classification schemes, and a system of nomenclature; and finally an assessment of evolutionary relationships. The approaches applied and the techniques used differ for each of these stages. During the descriptive phase, extensive surveys employing diverse tests and observations are needed. During the nomenclatural phase, type cultures must be established as references. During the phylogenetic phase, selected type cultures and representative strains must be examined in detail for evidence about their genetic relationships (5).

Genetic homologies can be determined by use of the various mechanisms for genetic exchange: transduction, transformation, or syncytic recombination (2). It must be noted, however, that a limited number of genetic determinants can be transferred from one organism to another by episomes or plasmids that are not necessarily homologous with the recipient's basic genome. In numerical taxonomy, an attempt is made to describe a significant sample of the genome of a group of organisms by analyzing a very large number of overt characteristics. Relationships based upon numerical analysis, therefore, are approximations of the degree of genetic homology in the population examined. Biological tests for homology often fail for reasons other than lack of genetic similarity (2). Accordingly, genetic relatedness can best be measured by comparing the genetic macromolecules themselves. For definitive studies on biological relatedness, genetic homology per se must be determined. Genetic homology is defined in biological terms as the similarity between sequences of loci, but in physical terms, it is the similarity between nucleotide sequences within deoxyribonucleic acid (DNA) molecules.

The genetic potential of an organism is encoded in the linear sequence of the nucleotides in its DNA. These nucleotide sequences determine the sequences of amino acids in structural or enzymic proteins which directly or indirectly constitute the phenotype of the cell. Accordingly, evolutionary divergence from a common ancestor proceeds as the progeny accumulate base substitutions in their DNA. Because of our increased capability to manipulate the DNA molecule, approaches to microbial classification other than traditional determinative systematics can be developed. In fact, an evolutionary approach to bacterial taxonomy, long hindered by lack of an adequate fossil record, now seems feasible.

II. DNA Nucleotide Composition

The Watson and Crick model of double-stranded DNA predicts that the guanine (G) content of a DNA molecule equals its cytosine (C) content; likewise, the adenine (A) and thymine (T) contents are also equal. However, the ratio of $(A+T)/(G+C)$, or the mole percent of guanine + cytosine [% GC = $(G+C)100/(A+T+G+C)$] may vary from one species to another. As indicated earlier, closely related organisms have very similar nucleotide sequences, and therefore very similar GC ratios. Similarity in % GC does not necessarily indicate genomic similarity but, unlike % GC of DNA from two organisms, establishes that they are not identical.

The relationships among selected actinomycetes, with special emphases on streptomycetes, nocardiae, and mycobacteria, have been determined, based upon the nucleotide composition of their DNA (Table I). The streptomycetes constituted a homogeneous group whose DNA contained between 69 and 74 mole % guanine + cytosine (% GC). The nocardial and mycobacterial DNA preparations contained 61–69% GC. The nocardial strains fell into either a 62–64% GC group or a 67–69% GC group (22).

III. DNA:DNA Association

Because the phenotype of an organism is determined by the nucleotide sequence in its DNA, a comparison of the nucleotide sequences of DNA preparations from two organisms should give a definitive evaluation of their relatedness. Unfortunately, complete direct sequence analysis of a DNA molecule is not currently possible. The complementary nature of the DNA double helix, however, can be used to circumvent these

TABLE I
NUCLEOTIDE COMPOSITION OF SELECTED
ACTINOMYCETE DNA SAMPLES

Source	% GC[a]
Actinomadura	68
Actinomyces	58
Actinoplanes	71–73
Ampullariella	72
Dactylosporangium	71–73
Micromonospora	71–73
Mycobacterium	64–69
Mycococcus	68
Nocardia	61–69
Oerskovia	74–76
Planobispora	70–71
Planomonospora	72
Spirillospora	71–72
Streptomyces	69–74
Streptosporangium	69–71
Streptoverticillium	69–71
Thermomonospora	67–69

[a] Values compiled from our published works (4,5,9,14).

technical difficulties. It has been established that the two strands of the DNA helix can be separated and specifically reassociated or annealed (19).

The phenomenon of association of complementary strands of DNA provides a powerful tool for exploring the relationships among microorganisms (18). Investigators analyzing DNA preparations with 60–70% GC encountered many problems with the earlier hybridization methods (27). Accordingly, my colleagues and I (9, 13) have modified the method of Warnaar and Cohen (24) for quantitative assay of hybridization between denatured DNA fixed to nitrocellulose membrane filters and free, denatured DNA. We have used this modified technique to assess, on a molecular level, the relationships among representatives of many genera of actinomycetes, in particular *Mycobacterium, Nocardia,* and *Streptomyces* (Table II). The relationships among these organisms are of particular interest because of their special relevance for industry, medicine, and agriculture. Moreover, the taxonomy of the actinomycetes remains a subject for active study and debate.

A difficult problem in using nucleic acid association for taxonomic studies is the selection of a system that will give the needed information. There are two general methods to choose between (a) association involv-

TABLE II

ASSOCIATION OF DNA FROM *Nocardia asteroides* STRAINS 300 AND 330 WITH DNA FROM SELECTED NOCARDIAE[a]

Test DNA from	Percent homology with	
	300	330
N. asteroides 300	100 (42)	93
N. asteroides 323	0	0
N. asteroides 324	86	92
N. asteroides 330	94	100 (58)
N. asteroides 333	100	78
N. asteroides 334	60	17
N. asteroides 462	82	91
N. asteroides 571	0	80
N. brasiliensis 301	0	26
N. brasiliensis 473	13	29
N. caviae 421	0	18
N. erythropolis 520	19	0
N. gypsoides 515	0	0
N. phenotolerans 514	3	9
Nocardia sp. 304	0	44
Nocardia sp. 468	29	0
N. transvalensis 516	4	12

[a] By the method of Enquist and Bradley (13) using nitrocellulose membrane filters. The reaction mixtures, containing ca. 13,000 count min^{-1} in 0.5 µg of labeled DNA, were incubated at 75°C for 18 hours. The absolute percent binding of the reference DNA is given in parentheses. Background counts of 32 to 39 count min^{-1} were subtracted prior to calculation.

ing denatured DNA from one source immobilized in an agar matrix (23) or on a membrane filter surface (20) and (b) association in which both test denatured test DNA samples are free in solution (7). Either procedure will yield specific and reproducible data when properly applied. Nevertheless, both methods have limitations and certain applications are better suited to one or the other of these techniques. Association of DNA in free solution constitutes a particularly useful method for estimating genome size and the degree of shared nucleotide sequences in DNA samples from organisms whose DNA can be isotopically labeled only with great difficulty (Table III). However, competition experiments cannot be carried out in free solution systems.

The rate of association of complementary single-stranded DNA molecules is dependent upon a number of experimental conditions: (a) it is proportional to the square root of the molecular weight of the single-stranded DNA; (b) it is slightly faster for DNA with a high % GC than

TABLE III
Association of DNA from *Mycobacterium bovis* Strain TMC-410 with DNA from *Mycobacterium tuberculosis*[a]

Sample A	Sample B	Percent homology
M. bovis BCG	*M. bovis* TMC-410	95 ± 6
M. tuberculosis H37Rv	*M. bovis* TMC-410	92 ± 8
M. tuberculosis H37Ra	*M. bovis* TMC-410	90 ± 8
M. tuberculosis H37Ra	*M. tuberculosis* H37Rv	98 ± 3

[a] Optical reassociation of DNA in free solution using the methods described by Bradley (8). Percent homology = 100 + 100 ($Cot_{1/2}$ A + $Cot_{1/2}$ B − 2 $Cot_{1/2}$ mix)/($Cot_{1/2}$ A + $Cot_{1/2}$ B).

in DNA with a low % GC; (c) it is dependent upon the salt concentration of the diluent; (d) it is inversely proportional to the viscosity of the diluent; (e) it is dependent on the concentration of the DNA; and (f) it is dependent upon the purity of the DNA preparation. The rate of association of complementary single-straded DNA is independent of pH between 5 and 9 at high salt concentrations (10).

IV. The Cot Concept

Britten and Kohne (11) showed that specific hybrid formation is a function of the initial concentration of each DNA species and the time of incubation. They introduced the acronym Cot, derived from the product of initial concentration (Co) and time (t). The units are generally moles of nucleotides per liter and seconds. Cot values describe the association of DNA when the temperature of incubation, salt concentration, and DNA fragment size are defined. The Cot value is readily calculated using the numerically equivalent statement: Cot = $\frac{1}{2}$ (A_{260}) (incubation time in hours = moles of nucleotide liter^{-1} seconds) where A_{260} is the initial absorbance of the DNA solution.

It is generally assumed that DNA association follows second-order reaction kinetics because the process involves the collision of two complementary strands. The conventional Cot plot (Cot value vs percent reassociated DNA) does not constitute a reliable confirmation that the reaction is proceeding by uncomplicated second-order kinetics, as is generally alleged (10). It should be noted that one of the most convenient ways to measure percent reassociation is as absorbance at 260 nm. Because dissociated DNA absorbs more ultraviolet light than reassociated DNA, simply following the decrease in absorbance with time will provide the necessary information on the degree of reassociation.

From the Cot plot, the Cot value at which half of the initially denatured DNA has reassociated can be determined graphically. This point has been designated as $Cot_{1/2}$. When the DNA is half reassociated, $C/Co = \frac{1}{2} = 1/(1 + k\, Cot)$; by rearrangement of the equation, we arrive at the statement $Cot_{1/2} = 1/k$. In the absence of repeated nucleotide sequences, i.e., redundancy, k is inversely proportional to the bacterial genome size (11).

V. Optical Reassociation

The degree of association between denatured DNA samples of diverse origins can be determined quantitatively from renaturation rates of the individual DNA preparations and their mixture; these DNA reassociation rates are calculated from spectrophotometric data (12,25). To determine the kinetics of the reassociation process optically, actinomycete DNA is dissolved in a salt solution and sheared by sonic oscillation. Formamide is added to the sheared DNA sample to lower the thermal stability of the double-stranded DNA. The DNA is heat denatured in a recording spectrophotometer; after the DNA is completely denatured, the temperature is rapidly adjusted to 25°C below the Tm of the preparation. Because formamide absorbs at 260 nm, denaturation and reassociation are monitored at 270 nm where DNA still has a large absorption but the absorption by formamide is low (8). The absorbance at 270 nm is monitored until the initially denatured DNA is more than 75% reassociated, at which time the Tm of the annealed sample is determined (Fig. 1).

The second-order renaturation rate constant (k_2) may be calculated from experimentally measured absorption values (A_N = absorbance of the native DNA; A_D = absorbance of the denatured DNA; A_T = absorbance of the mixture at time t). According to Wetmur and Davidson (25):

$$1/(A_T - A_N) = 2.04 \times 10^{-4} k_2 t + 1/0.36\, A_N \quad (1)$$

therefore

$$0.36\, A_N/(A_T - A_N) = 7.34 \times 10^{-5} k_2 t\, A_N + 1 \quad (2)$$

Because $0.36\, A_N = A_D - A_N$

$$(A_D - A_D)/(A_T - A_N) = 7.34 \times 10^{-5} k_2 t\, A_N + 1 \quad (3)$$

$$k_2 = (1.36 \times 10^4/t A_N) \left(\frac{A_D - A_N}{A_T - A_N} - 1 \right) \quad (4)$$

The units of k_2 are liter mole^{-1} second^{-1} or M^{-1} second^{-1}.

To evaluate critically whether renaturation is proceeding by second-order kinetics, $(A_D - A_N)/(A_T - A_N)$ is plotted against time (Fig. 2). This plot should generate a straight line; moreover, the calculated second-

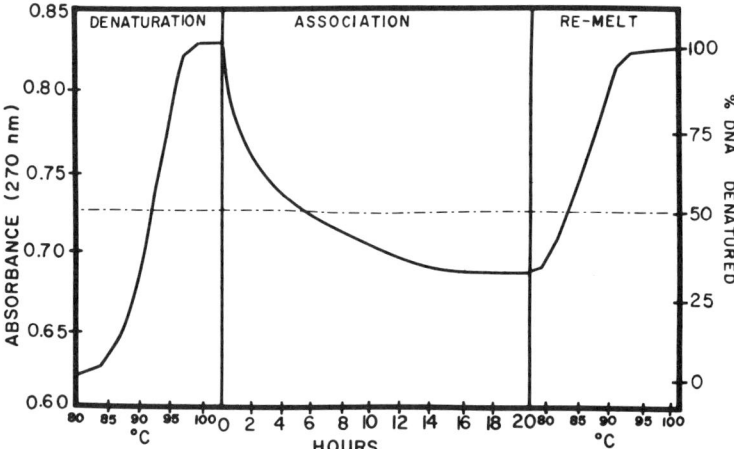

FIG. 1. Denaturation and association of *Streptomyces venezuelae* DNA. This DNA association experiment was carried out in 25% formamide dissolved in 6 × SSC. The incubation temperature during renaturation was 65°C. The changes in absorbance were monitored at 270 nm. Note the change in scale of the abscissa from temperature to time and back to temperature.

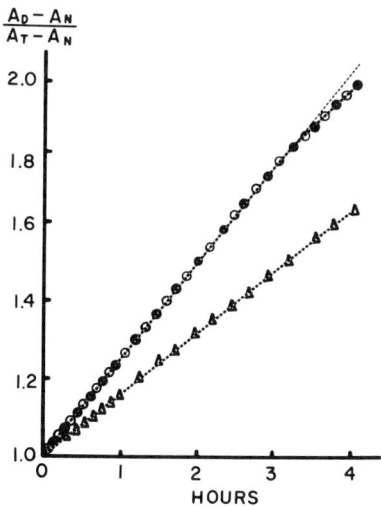

FIG. 2. Kinetics of DNA reassociation. Sheared denatured DNA of *Mycobacterium tuberculosis* H37Ra and *M. bovis* BCG renatured according to second-order kinetics. The second-order rate constant k_2 was 2.1 ± 0.1 liter mole^{-1}s^{-1} for H37Ra DNA samples having initial reactant concentrations of $A_{260} = 0.3$ (△———△) or 0.45 (○———○) and for BCG DNA having an initial reactant concentration of 0.45 (●———●). A_D = absorbance of denatured DNA; A_N = absorbance of native DNA; A_T = absorbance of reassociating DNA at time t.

order rate constant (k_2) should be the same regardless of the initial reactant concentration (25).

Optical reassociation appears to possess a number of desirable features for determining relatedness among actinomycetes: (a) no radiolabeled DNA is required; (b) high association temperatures can be used (without the complication of leaching of the immobilized DNA from the membrane); and (c) absorbancy accurately measures reassociated sequences per se whereas radioassay methods measure not only reassociated sequences, but also unassociated (single-stranded) loops and free ends. The percent homology is calculated from optical reassociation data using the expression

$$\% \text{ homology} = 100 + 100\,(\text{Cot}_{1/2}\,A + \text{Cot}_{1/2}\,B - 2\,\text{Cot}_{1/2}\,\text{mix})/(\text{Cot}_{1/2}\,A + \text{Cot}_{1/2}\,B)$$

This equation is derived from one developed by Seidler and Mandel (21). Based upon optical DNA association results, *M. bovis* is not a species distinct from *M. tuberculosis* (Table III). Accordingly, I propose that *M. bovis* be returned to synonymy with *M. tuberculosis* and reduced to a subspecies *M. tuberculosis* subspecies *bovis*.

VI. Phylogenetic Implications

In order to make taxonomic inferences based upon nucleic acid association data, the extent of DNA nucleotide sequences held in common between a test DNA and a reference DNA must be compared. The degree of shared sequences becomes a quantitative index of relatedness. However, DNA samples from related organisms usually contain identical nucleotide sequences, a spectrum of partially matched sequences and totally dissimilar sequences; therefore, it is not possible to arrive at a single value that defines the absolute relatedness of one organism to another. This apparent complication actually provides a basis for deducing phylogenetic relationships among organisms.

The exactness of base pairing, as well as the extent of association between single-stranded DNA preparations from different strains, can be measured experimentally. In one method used extensively, association is allowed to proceed at two different incubation temperatures. At the higher temperature only well matched sequences should form duplexes. At the lower temperature, partially matched and exactly matched sequences should form duplexes (6). In the second method, the thermal stability of the DNA duplexes formed at one incubation temperature is determined. The data may be graphically depicted as the amount of test DNA released at each elution temperature or as the cumulative amount of test DNA released at a given elution temperature and all lower elution temperatures. Accordingly, the thermal stability of an annealed DNA duplex is characterized by its Tm,e (elution temperature at which 50% of the DNA duplexes have been disassociated). It should be noted that the Tm,e obtained by the release of DNA fragments due to complete strand separation is not necessarily equivalent to the midpoint of the hyperchromic shift upon heating (optical Tm). In practice, however, the Tm,e of reassociated homologous DNA is usually the same as the optical Tm of the native DNA.

The difference between the Tm,e value of an interspecific duplex and that of the homologous reference duplex is designated as the ΔTm,e value. There appears to be a direct correlation between the ΔTm,e and the percentage of unpaired bases in an annealed duplex. As a generalization, 1% unpaired bases within a reassociated nucleotide sequence lower the Tm,e by 1°C (1) or 1.5°C (17).

When surveying the similarity of DNA samples from a large group of organisms, it is tedious to determine the thermal stability of each reaction at a number of incubation temperatures. In our laboratory, we find it adequate and convenient to use two incubation temperatures, one *exacting* and one *nonexacting*. Usually the *nonexacting* incubation temperature is set at 30°–35°C less than the Tm of the reference DNA and the exacting temperature at 15°C to 20°C less than the Tm of the reference DNA. The percent association relative to the homologous reaction is determined for each incubation temperature. The ratio of (the relative binding at the exacting temperature)/(the relative binding at the nonexacting temperature) has been designated the Divergence Index (DI). Divergence index values are useful in gauging the presence or the absence of closely related genetic material. A ratio close to 1.0 indicates that all the sequences that bind the reference DNA are almost identical to it whereas a DI value approaching 0.0 indicates that the test DNA possesses almost no regions of identity with the reference DNA but possesses many similar nucleotide sequences. There appears to be a correlation between ΔTm,e and DI.

Divergence index values can be interpreted at the molecular level in terms of the distribution of nucleotide divergence. We refer to the nucleotide divergence occurring more or less randomly distributed throughout the genome as *dispersed divergence*. This is in contrast to *localized divergence* or *localized conservation*, where changes occur in specific regions only. The duplexes formed during exacting incubation conditions are only those of closely matched sequences (b) whereas duplexes formed during nonexacting conditions are composed of both incompletely matched (a) and closely matched (b) sequences. Symbolically, DI equals $b/(a+b)$. The values of a and b are subject to the following limitations: $a + b < 100$ or $a + b = 100$; $a + b > 0$ or

$a + b = 0$; $b < a + b$ or $b = a + b$. The total number of sequences available to react is $a + b + c = 100$, where c is the percent of DNA sequence unable to anneal because the nucleotide sequences in the mobile DNA and the immobilized DNA are extensively mismatched.

From a graphical presentation of data of this type, certain proposals about the mechanism by which genetic diversity arose can be developed (Fig. 3). The grid formed by coordinates DI $[b/a + b)]$ versus % b $[100b/(a + b + c)]$ can be bisected by a line from the origin to the coordinates DI = 1 and % b = 100. This diagonal has been referred to

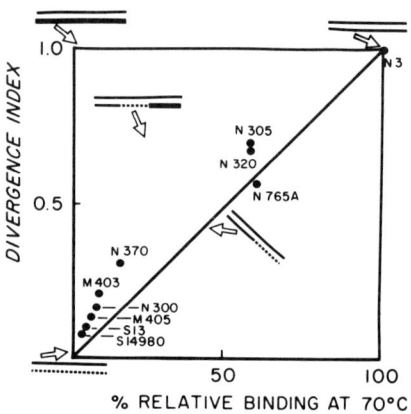

FIG. 3. Reassociation of various actinomycete DNA preparations with *Nocardia erythropolis* 3 reference DNA under exacting (70°C) and nonexacting (50°C) conditions. The line of maximum divergence is obtained by setting the binding at the nonexacting incubation temperature at 100%. N3, N305, N320, and N765A refer to strains of *N. erythropolis;* N370 to *N. corallina;* N300 to *N. asteroides;* M403 and M405 to *Mycobacterium smegmatis;* S13 to *Streptomyces venezuelae,* and S14980 to *S. violaceoruber.* Superimposed on the graph are diagrams of the proposed genome structures. Exactly matched nucleotide sequences ═══ ; partially matched sequences ----- ; unmatched sequences ▬▬▬ .

as the line of maximum divergence. The line of maximum divergence is generated by setting $a + b = 100$, or by setting $c = 0$. No experimental points should fall below the line of maximum divergence because b cannot be greater than $a + b$. Experimental points having coordinates near DI = 1 and % b = 100 are generated by DNA preparations that associate exactly, i.e., a and c are very small. Experimental points having coordinates between DI = 0 to 1 and % b = 0 are generated by DNA preparations which form very few duplexes, i.e., a and b are very small. Coordinates DI = 0.5 and % b = 50 reflect DNA preparations which lack c and consist of an equal number of DNA fragments of type a and type b, i.e., are capable of forming equal amounts of exactly matched and partially matched duplexes. Coordinates DI = 0.8 and % b = 20 would reflect a DNA preparation in which c is large and a is small.

When DNA samples from selected actinomycetes are analyzed and plotted in this way, the experimental values generate a line that approximately corresponds to the line of maximum divergence. These results indicate that genetic diversity in the actinomycetes is primarily due to dispersed nucleotide divergence. Accordingly, random mutation is the probable mechanism by which genetic diversity has arisen in this group of microorganisms. Conversely these results indicate that gene transfer involving unique nucleotide sequences is not a major factor in the evolution of the actinomycetes. Moreover, there is no evidence that particular regions of the actinomycete genome are unusually susceptible to mutation, thereby creating extensive localized nucleotide divergence.

VII. Neutral Mutations

A basic premise in DNA association analyses is that the ability to form specific DNA duplexes with samples from two organisms constitutes evidence for relatedness. Conversely the lack of specific hybrid duplex formation is interpreted to indicate some degree of unrelatedness. That base mispairing indicates phenotypic dissimilarity, however, is not absolutely established. It is conceivable that phenotypically similar organisms might possess substantial amounts of genomic diversity. In such instances, there would be a significant degree of base mispairing during DNA association.

King and Jukes (16) have proposed that there may be random neutral genetic mutations that have no effect upon the competitive fitness of the organism. Wright (26) has suggested that neutral mutations can become fixed as evolutionary changes through the action of genetic drift or by selection for neutrality.

A molecular basis for neutral mutations can be proposed. It is well established that there is more than one codon for each amino acid (3). Kimura (15) examined the codon dictionary and suggested that in the 61 amino acid-specifying codons, there are 549 possible single base substitutions with one-fourth of these (134) being substitutions to synonymous codons. Moreover, of the three positions in a codon, mutations occurring in the first two positions usually cause amino acid substitutions and are frequently eliminated by natural selection whereas mutations in the third position are often selectively neutral and would be maintained (16). It is possible, therefore, that substantial nucleotide divergence has accumulated within a species with little change in overt phenotype. It is conceivable that nucleotide divergence as a result of neutral mutations might proceed rapidly.

Acknowledgment

The new data presented in this report are based upon research supported by Public Health Service Grant AI-09098 from the National Institute of Allergy and Infectious Diseases.

References

1. Bautz, E. K. F. (1965). In "Evolving Genes and Proteins" (V. Bryson and H. J. Vogel, eds.), Academic Press, New York.
2. Bradley, S. G. (1965). Int. Bull. Bacteriol. Nomencl. Taxon. 15, 239–241.
3. Bradley, S. G. (1966). Advan. Appl. Microbiol. 8, 29–59.
4. Bradley, S. G. (1970). In "Host-virus Relationships in Mycobacterium, Nocardia and Actinomyces" (S. E. Juhasz and G. Plummer, eds.), pp. 179–188. Thomas, Springfield, Illinois.
5. Bradley, S. G. (1971). Advan. Front. Plant Sci. 28, 349–362.
6. Bradley, S. G. (1971). In "Recent Advances in Microbiology" (A. Pérez-Miravette and D. Peláez, eds.), pp. 3–7. Asociacion Mexicana de Microbiologia, Mexico.
7. Bradley, S. G. (1972). Amer. Rev. Resp. Dis. 106, 122–124.
8. Bradley, S. G. (1973). J. Bacteriol. 113, 645–651.
9. Bradley, S. G., G. H. Brownell, and J. E. Clark. (1973). Can. J. Microbiol. 19, 1007–1014.
10. Brenner, D. J. (1970). Develop. Ind. Microbiol. 11, 139–153.
11. Britten, R. J., and D. E. Kohne. (1968). Science 161, 529–540.
12. DeLey, J., H. Cattoir, and A. Reynaerts. (1970). Eur. J. Biochem. 12, 133–142.
13. Enquist, L. W., and S. G. Bradley. (1970). Advan. Front. Plant Sci. 25, 53–73.
14. Farina, G., and S. G. Bradley. (1970). J. Bacteriol. 102, 30–35.
15. Kimura, M. (1968). Genet. Res. 11, 247–269.
16. King, J. L., and T. H. Jukes. (1969). Science 164, 788–798.
17. Laird, C. D., B. L. McConaughy, and B. J. McCarthy. (1969). Nature (London) 224, 149–154.
18. McCarthy, B. J., and E. T. Bolton. (1963). Proc. Nat. Acad. Sci. U.S. 50, 156–164.
19. Marmur, J., R. Rownd, and C. L. Schildkraut. (1963). Progr. Nucl. Acid. Res. 1, 231–300.
20. Monson, A. M., S. G. Bradley, L. W. Enquist, and G. Cruces. (1969). J. Bacteriol. 99, 702–706.
21. Seidler, R. J., and M. Mandel. (1971). J. Bacteriol. 106, 608–614.
22. Tewfik, E. M., and S. G. Bradley. (1967). J. Bacteriol. 94, 1994–2000.
23. Tewfik, E., S. G. Bradley, S. Kuroda, and R. Y. Wu. (1968). Develop. Ind. Microbiol. 9, 242–249.
24. Warnaar, S. O., and J. A. Cohen. (1966). Biochem. Biophys. Res. Commun. 24, 554–558.
25. Wetmur, J. G., and N. Davidson. (1968). J. Mol. Biol. 31, 349–370.
26. Wright, S. (1966). Proc. Nat. Acad. Sci. U.S. 55, 1074–1081.
27. Yamaguchi, T. (1967). J. Gen. Appl. Microbiol. 13, 63–71.

Current Status of Nomenclature of Antibiotic-Producing Bacteria

ERWIN F. LESSEL[1]

American Type Culture Collection, Rockville, Maryland

I. Introduction 71
II. Antibiotic-Producing Bacteria as Problems to Nomenclature 71
III. Two Major Nomenclatural Problems 72
IV. Major Changes in International Code of Nomenclature.... 73
V. Requirement for Deposition of Type Strains 74
VI. Effect of Changes in International Code 75
VII. Speciation of *Streptomyces* 75
References 76

I. Introduction

For a number of years I was actively engaged in a close scrutiny and evaluation of the scientific names of the bacteria while preparing material for inclusion in the *Index Bergeyana* (1). Most impressive was the fact that very few names survived a rigid application of the rules of nomenclature unscathed. Indeed, all of the rules have been violated at one time or another, and some in rather bizarre fashion. However, rather than present a series of case histories on the subject, I will discuss current developments in bacteriological nomenclature, with special reference to the antibiotic-producing bacteria.

II. Antibiotic-Producing Bacteria as Problems to Nomenclature

As a class, the antibiotic-producing bacteria are no different from other bacteria with respect to problems associated with their nomenclature. However, in the recent past it became increasingly evident that new scientific names published in the patent literature, and many of these were names of antibiotic producers, were especially troublesome to bacterial systematists for two reasons: (1) the existence of many of the names was unknown to the general scientific community either because of unfamiliarity with the patent literature or because the names were buried in the text of the patents and did not appear in the title or in an index to the patent literature, and (2) in many cases, cultures of the newly named organism were unavailable, and an adequate description of the taxon was not given in the patent so that it was not possible

---

[1] Present address: Lederle Laboratories, a Division of American Cyanamid Co., Pearl River, New York.

to know precisely to what the name referred. To combat these problems, two steps were taken. First, the International Code of Nomenclature of Bacteria (2) was modified (3) so as to render the scientific names of bacteria published in a patent application or in an issued patent as not effectively published; however, such names could be regarded as effectively published if subsequently published elsewhere so as to meet the requirements for effective publication. For purposes of priority, the date of issuance of the patent is recognized as the date of publication of the name. Second, in order to assure the adequacy of the descriptions of newly named streptomycetes published in the patent literature, it was recommended that certain characters be included in these descriptions, and a list of these characters was published in the *International Journal of Systematic Bacteriology* (IJSB) (4).

For obvious reasons, some of the strains for which proposals of new species names are made are not available to other investigators. However, this particular problem is not unique to those working with antibiotic-producing microorganisms; as is well known, those studying other kinds of bacteria sometimes do not make their cultures available until after they have completed all their intended studies on the organism (*ASM News*, Nov. 1971, p. 17; Feb. 1972, pp. 110–112; Apr. 1972, pp. 221–223).

III. Two Major Nomenclatural Problems

The difficulties associated with the patent literature have prompted bacterial systematists to focus on two major problems in bacterial nomenclature: (1) How can the numerous new names of bacteria proposed throughout the world's scientific literature be kept track of? (2) What can be done to assure that there is no question as to the application of the new names?

In discussing these and related problems in a symposium held in 1970 at the Virginia Polytechnic Institute and State University (7), I made several proposals which, in my opinion, not only would place bacterial systematics on a solid foundation from which taxonomists could at least operate on common grounds even if they did not agree with each other but also would save considerable time, energy, and money that must otherwise be expended merely for the purpose of determining whether an organism had been previously named—an activity that is, frankly speaking, simply not worth the effort.

I am sure that many of you are familiar with the arduous task of going back to the older literature and attempting to determine whether an organism you have isolated and believe to belong to a new species has been named before. This can be a time-consuming, and frequently a frustrating, experience. In most cases it is found that there is insufficient information provided in older descriptions to permit comparisons with new isolates in terms of contemporary technology. If you yourself do not take the time to examine the older literature carefully to ensure that you are not unnecessarily introducing a new name into the bacteriological literature, the chances are that someone else eventually will, and the result could be that the name you proposed is relegated to that ignominious state referred to as synonymy. At present, failure to examine all the earlier literature in order to determine whether an isolate has been previously named simply means that a scientific piece of work has not been thoroughly done. However, in many cases where earlier synonyms are discovered, conservation of the later name is indicated anyway because it has become widely used and accepted and because the earlier name has only rarely been used and is almost completely unknown. In any event, the proposals I put forward three years ago were incorporated in a proposed revision of the Bacteriological Code (6); this revision was accepted by the International Committee on Systematic Bacteriology of the International Association of Microbiological Societies at the First International Congress for Bacteriology, held in Jerusalem. The newly revised Code will be published in 1975.

IV. Major Changes in International Code of Nomenclature

The primary innovations made in the new Code were cited in the Introduction to the proposed revision of the Code, published in the January 1973 issue of the *International Journal of Systematic Bacteriology* (IJSB). These changes are as follows, and I quote:

(1) A new starting date (1 January 1980 rather than 1 May 1973) for the nomenclature of bacteria will be instituted so as to put into practice more meaningful requirements for the valid publication of names. Among these are:

(a) New names and combinations must be published in the IJSB or, if published previously elsewhere, an announcement of such publication must be made in the IJSB.

(b) A description or a reference to a previously and effectively published description of the named taxon must also be given in the IJSB.

(c) The type of each named taxon must be designated.

Furthermore, it will be recommended that the description of the named taxon contain at least those characters cited for the taxon in the list of minimal standards which will be compiled by the members of the various ICSB taxonomic subcommittees and by other experts. In addition, it will be recommended that, in the case of cultivable organisms, cultures of the type strains of newly named species and subspecies be deposited in culture collections from which they would be available.

(2) For names published prior to 1 January 1980, Approved Lists of Names of Bacteria will be compiled by the members of the taxonomic subcommittees and by other experts for approval by the Judicial Commission and the ICSB. Only

the names of bacteria which are adequately described and for which there is a type or neotype strain, if the organism is cultivable, will be placed on the approved lists. In determinations of priority after 1 January 1980, then, only those names which appear on the approved lists of names or which are validated by publication in the IJSB after 1 January 1980 need be taken into consideration. Thus it will no longer be necessary to conduct extensive, frequently difficult literature searches merely for the purpose of determining the earliest name which was used for a bacterial taxon. Most important, however, will be the fact that after 1 January 1980 all of the validly published names for the bacteria will have clear and precise applications because the names will be associated with adequate descriptions and type or neotype strains.

Now I should like to stress two points:

1. The approved lists are merely lists of names whose applications are clear because the organisms are adequately described and because, where possible, the names are associated with living cultures of type or neotype strains with which the descriptions can be updated as the need arises; the fact that all cultures of the type strains may die or may change does not necessarily mean that the name associated with the type will be removed from the list. The lists are *not*, I repeat, not, intended to imply a fixed taxonomy. Many of the names may be found by subsequent work to be synonyms of other names on the lists, and there is nothing to prevent the transfer of any of the species on the lists from one genus to another if the data so indicate.

2. The lists are open-ended; that is, anyone may, so long as he observes the new requirements for valid publication of names, resurrect any name published prior to January 1, 1980 and request placement of the name on the approved lists. The ICSB will maintain the lists and will publish them in the IJSB.

V. Requirement for Deposition of Type Strains

In my estimation there is one final step that must be taken to ensure the success of this effort to place bacterial systematics on a sound and workable basis, and that is to *require* rather than merely to recommend the deposition of type strains in culture collections from which the type strains would be available. There are some questions which have been raised in connection with making this a requirement, such as: "What constitutes a culture collection?" "How can the deposition be verified?" and "What is the status of the name if the cultures of the type strain change or are lost?" However, I do not regard these problems as insurmountable, and I believe a workable rule can be formulated. Without cultures of the type strains, the species descriptions cannot be updated, and in a few short years our presently adequate descriptions may be just as inadequate and meaningless as many of the early descriptions of bacterial taxa presently are to us.

I would suggest that, for the valid publication of the name of a species or subspecies, it be required that the type strain be deposited in a culture collection from which cultures of the type strain would be available. In fact, I would even go so far as to require that the type strain be deposited in at least one of a number of specified major international culture collections, such as the American Type Culture Collection, the Northern Regional Research Laboratory, the Centraalbureau voor Schimmelcultures, and the collections of the United Kingdom and of the Japanese and German federations of culture collections. It also should be required that the appropriate culture collection accession numbers for the type strain be cited in the original publication of the name. In those cases in which the organism is not cultivable, this requirement would not pertain. Furthermore, the loss of the culture of a type strain by any or all culture collections in which the type had been deposited would not affect the validity of publication of the name; this would merely mean that a neotype strain should be designated to replace the type.

VI. Effect of Changes in International Code

The net effect, then, of the recently approved changes in the Bacteriological Code is that beginning on January 1, 1980, it will be necessary to consult only one source for the available names of bacteria, the *International Journal of Systematic Bacteriology*. The acceptable names published prior to January 1, 1980 will appear in the IJSB on the approved lists, and all new names published after January 1, 1980 must also be published in the IJSB either in an original paper or as a notice of publication elsewhere. Because modern descriptions will be associated with each of these names and because each species will have a type or neotype strain, there will be no doubt whatever as to the application of names. Thus, in the detection of new species and subspecies, the chores of having to go through all the older literature and of trying to comprehend the earlier descriptions will be obviated, and the path will then be clear for real progress in bacterial taxonomy.

VII. Speciation of *Streptomyces*

With respect to the bacteria that produce antibiotics, many are members of the genus *Streptomyces*. Even though hundreds of nomen species have been placed in this genus, there is every indication that the taxonomic and nomenclatural problems associated with these organisms may be more satisfactorily resolved in the near future. A large international group of streptomycetologists have cooperated in a venture known as the "International *Streptomyces* Project." These workers have, over the past several years, assembled and described a collection of the type strains of over 450 species (5,8–12), primarily streptomycetes, and are now planning to study these type strains in an effort to determine how many valid species of *Streptomyces* there are and what are their correct names. It seems, therefore, that the taxonomy and nomenclature of most of the antibiotic-producing bacteria will be in excellent shape in the near future, certainly in time for including the names of the acceptable species on the first approved list of names, to be published on January 1, 1980.

REFERENCES

1. Buchanan, R. E., J. G. Holt, and E. F. Lessel. (1966). "Index Bergeyana." Williams & Wilkins, Baltimore, Maryland.
2. Editorial Board of the International Committee on Bacteriological Nomenclature. (1958). "International Code of Nomenclature of Bacteria and Viruses." Iowa State Univ. Press, Ames.
3. Editorial Board of the Judicial Commission of the International Committee on Nomenclature of Bacteria. (1966). *Int. J. Syst. Bacteriol.* **16**, 459–490.
4. Gottlieb, D. (1963). *Int. Bull. Bacteriol. Nomencl. Taxon.* **13**, 169–170.
5. Gottlieb, D., and E. B. Shirling. (1967). *Int. J. Syst. Bacteriol.* **17**, 315–322.
6. Lapage, S. P., W. A. Clark, E. F Lessel, H. P. R. Seeliger, and P. H. A. Sneath. (1973). *Int. J. Syst. Bacteriol.* **23**, 83–108.
7. Lessel, E. F. (1970). *Int. J. Syst. Bacteriol.* **20**, 339–344.
8. Shirling, E. B., and D. Gottlieb. (1966). *Int. J. Syst. Bacteriol.* **16**, 313–340.
9. Shirling, E. B., and D. Gottlieb. (1968). *Int. J. Syst. Bacteriol.* **18**, 69–189.
10. Shirling, E. B., and D. Gottlieb. (1968). *Int. J. Syst. Bacteriol.* **18**, 279–392.
11. Shirling, E. B., and D. Gottlieb. (1969). *Int. J. Syst. Bacteriol.* **19**, 391–512.
12. Shirling, E. B., and D. Gottlieb. (1972). *Int. J. Syst. Bacteriol.* **22**, 265–394.

Microorganisms in Patent Disclosures

IRVING MARCUS[1]

U.S. Patent Office, Washington, D.C.

I. Introduction 77
II. Responsibilities and Requirements 78
    Appendix 83

I. Introduction

The substance of this paper was presented as an exercise in bringing curators of culture collections up to date on practice in the Patent Office regarding the requirements for an adequate disclosure of an organism in connection with its description in a patent application. The problem was contentious then but is now much clearer. We have published guidelines that have assisted those prosecuting patent applications and the examiners in handling the subject. And it is hoped that before too long there will be, in addition, legislation[2] that prescribes the requirements. In addition, the subject is being considered by the international community[3] to facilitate handling when the invention is made outside the country where the patent application is filed.

Article 1, Section 8 of the Constitution provides "to promote the progress of Science and Useful Arts, by securing for limited times to Authors and Inventors, the exclusive right to their respective writings and discoveries." Patent laws were enacted based on this provision, and most seriously concerned with the subject we discuss today in Section 112 of our Patent Statute (Title 35 USC) paragraph 1:

> The specification shall contain a written description of the invention, and of the manner and process of making and using it, in such full, clear, concise and exact terms as to enable any person skilled in the art to which it pertains, or with which it is most nearly connected, to make and use the same, and shall set forth the best mode contemplated by the inventor of carrying out his invention.

This brings me directly to the subject involved. How does one adequately describe a microorganism, particularly one that is hitherto unknown and unavailable? The importance of the subject in the field of

---

[1] Present address: 8411 Spencer Court, Chevy Chase, Maryland.
[2] S.2504 October 1, 1973, Section 112(f) (see Appendix).
[3] World Intellectual Property Organization (WIPO), Geneva Meeting, April 23–26, 1974. Since the paper was delivered, two meetings of the Committee of Experts have been held at WIPO in Geneva, Switzerland (April and July, 1974) and a draft treaty on "The Deposit of Microorganisms for the Purposes of Patent Procedure" has been prepared and forwarded to the member countries for comment. Another meeting is scheduled in Geneva, April 22–29, 1975. The author was a representative at both meetings in 1974.

scientific achievement is apparent from the fact that most of the patents in the field involve the use of microorganisms in the production of products and, while not limited to the production of antibiotics, this is the major field involved. The situations possible are several. First, an applicant may have discovered a hitherto unknown microorganism and produces from it an antibiotic, also hitherto unknown. Second, this unknown microorganism may produce a known antibiotic in a process invention, and finally, a known and available organism may be used to produce a new product. In the latter case, the specification may adequately describe the microorganism and no real problem of identification exists. In the first two situations, the prior practice, for the most part, involved a rejection of the patent application as insufficient under 35 USC 112 since the description did not enable a person skilled in the art to reproduce the invention. There were variations among the examining divisions in the early 1950s and there was at first an acceptance, to overcome this rejection, of a deposit of a specimen of the microorganism with certain depositories which maintain these for later dissemination to the public.

Several applications involving these rejections reached the Patent Office Board of Appeals for decision, this being the Patent Office tribunal that hears appeals from examiners' refusal to allow claims. Until the recent case *In re* Argoudelis *et al.*[4] the practice was to permit a deposit in a recognized depository of a culture of the microorganism which deposit would not be available to the public until a patent was granted. Such deposit must have been made no later than the filing of the patent application since the existing case law, *Ex parte* Kropp, 143 USPQ 148, held that subsequent deposit could not cure an insufficient disclosure.

The depositories at the time were located at the Northern Regional Laboratory in Peoria, Illinois (NRRL), the American Type Culture Collection, in Washington, D.C., (now Rockville, Maryland) (ATCC), and Rutgers University, New Brunswick, New Jersey. Each had its own contract with the applicant and the Patent Office was not privy thereto and had no jurisdiction over the parties. Our experience, as later discovered, indicated that these contracts had various strings attached and many were the complaints our Office received that patented products could not be reproduced since samples of the microorganisms were carefully controlled by the patentee even after the patent issued.

II. Responsibilities and Requirements

It is with this background that we at the Patent Office are asked: (1) What is the responsibility of a Curator of a Culture Collection?

[4] *In re* Argoudelis *et al.*, 168 USPQ 99 (December 17, 1970).

or the Depositor? (2) What does the Patent Office require of the Depository insofar as the durability of the deposit, the authenticity of the culture deposited, and the type of contract? (3) Why is all this required? It is with the answers to these questions that the balance of this discussion relates. I have referred above to a particular section of the Patent Statute 35 USC 112 since it is this section that is the nub of the controversy about cultures of microorganisms. As stated in *Ex parte* Kropp, mentioned above, "If appellant were dealing with a known organism which had a well defined source and which had been obtained and used by others before, or even with an organism which was merely known and available to persons skilled in the art, the present question of the sufficiency of the disclosure would not arise."

It is apparent that this prior knowledge and availability is a question of fact. An organism of known taxonomy and morphology can be adequately described; it is available and it can be described in the patent specification. The statute is satisfied. The subject of deposits of microorganisms became the style in the 1950s when there was no other adequate description. There is no statutory provision at present for deposits of microorganisms, although provision was made for same in the proposed Patent Reform Act S.643 and in a bill to revise our patent laws, S.2504, and now S.23, 94th Congress. Perhaps enabling statutory provision, long overdue, would have made the present discussion moot. But the problem existed—What is required so the person skilled in the art would have before him the information to carry out the invention? The innovation of depositing was conceived by the applicants for patents and their assigns. Culture collections known for their public and/or private aims both in this and foreign countries became the recipients of a multitude of deposits in connection with patent applications. In no case was the Patent Office privy to the contract the depositor made with the depository, and there was little uniformity in the nature of the contract. One thing was certain, however. The deposit was not available to the public during the pendency of the application, was ostensibly available after the patent grant, but even then under conditions set down by the depositor. And then the controversy began.

The Patent Office had accepted as adequate disclosures in patent applications descriptions using a microorganism which had been prior deposited in a culture collection which was presumed to be one that would make the culture available to the public upon grant of the patent, but the culture was restricted to nominees of the depositor, or the Patent Office if it requested it, during the pendency of the application. And all this without any authority in 35 USC 112 for anything but a written description as satisfying the enabling disclosure requirements thereof. And then, as mentioned above, the name Argoudelis became well known.

In an application for patent filed in 1961, the applicants claimed a new antibiotic "sparsogenin" which could be obtained from a species of actinomycetes which they designated Streptomyces sparsogenes var. sparsogenes. In addition to some description of the microorganism, they disclosed how to use it to prepare the antibiotic as well as how to separate the product from the culture medium. A subculture of the variety of microorganism was stated to be obtainable from the permanent culture collection of the Northern Regional Research Laboratory at Peoria, Illinois (with the accession number NRRL 2940), where the microorganism had been desposited, but of course it did not state by whom this subculture could be obtained.

The Patent examiner rejected the product claims over the disclosure in the *Journal of Antibiotics,* an article by Anzai *et al.,* which described the antibiotic tubercidin obtainable from the fermentation broth of a strain of actinomycetes isolated from the soil somewhere in Japan. The antibiotic was identified by its physical and chemical properties and was apparently acknowledged to be the same as that claimed. The culture procedure was described as was the separation, but it was argued that no one could reproduce the product from the description or chemically, absent the microorganism. Further since there was no enabling disclosure to reproduce the product, it was urged that the reference was not properly applied. The examiner had of course allowed the process claims. The ultimate decision of the Patent Office Board of Appeals held the reference was inapplicable but that applicants' disclosure was no better than that of the reference. In effect, it was inadequate absent a culture of the microorganism and, since the culture was not available to the public at the time of filing of the application, it did not comply with the aforesaid 35 USC 112. To put the issue plainly, the restricted type of disclosure was not enabling. The Patent Office then modified its stance regarding disclosures of cultures of microorganisms and required that they be freely available to the public prior to filing else the disclosure of their deposit in a culture collection did not complete the written description and satisfy Section 112. This was the issue that was appealed to the Court of Customs and Patent Appeals (CCPA) and ultimately reversed in the Argoudelis *et al.* case.

It was at this point that the depositories realized that their contracts with the depositors required modification. And applicants for patents had to follow the decision of the Board of Appeals and deposit cultures without restriction or risk the filing of an application defective ab initio. This situation existed until December 17, 1970, when the CCPA held that deposits restricted until the grant of the patent, as described in the specification, met the enabling disclosure requirements of 35 USC 112. And now we have two types of contracts by the depositories, one that continues to accept deposits without restriction and the other that accepts the restricted type of deposit found proper in the Argoudelis decision.

With the decision now the case law to be followed, the Patent Office was immediately swamped with inquiries as to how we would enforce the deposit requirements, and we published some simple guidelines[5] that we will require applicants to follow to enable the disclosure to comply with 35 USC 112 as interpreted by *In re* Argoudelis. These are:

(1) No later than the effective United States filing date of the application, the applicant must have made a deposit of a culture of the microorganism in a depository affording permance of the deposit and ready accessibility thereto by the public if a patent is granted, under conditions which assure

   (a) that access to the culture will be available to one determined by the Commissioner to be entitled thereto (Rule 14 and 35 USC 122), and

   (b) that all restrictions on the availability to the public of the culture so deposited will be irrevocably removed upon granting of the patents;

(2) such deposit is referred to in the body of the specification as filed and is identified by deposit number, name and address of the depository, and the taxonomic description to the extent available is included in the specification; and,

(3) the applicant or his assigns has provided assurance of permanent availability of the culture to the public through a depository meeting the requirements of (1). Such assurance may be in the form of an averment under oath or by declaration by the applicant to this effect. It is also possible that a copy of the applicant's contract with the depository may be required by the examiner to be made of record as evidence of making the culture available under the conditions stated above.

Thus the Patent Office for the first time is setting down conditions that impose an obligation on both the depositor and the depository. The quid for a patent to advance the useful arts is the giving of your discovery to the public, after your legal monopoly expires. In the inventions involving the use of hitherto unknown microorganisms, the only way to accomplish this is to make these available through a technically competent organization that will receive, store, and then disseminate the culture. Thus, even if the patent owner is no longer interested in his cultures after the patents expire, they are available to the public for research or otherwise through the public depository. This then imposes obligations on the depositories as follow:

Permanence. The culture should be in the public domain forever, in spite of the fact that no one has any assurance it will remain alive.

Authenticity. There must be reasonable assurance that the culture deposited is in fact the correct one identified and that it has not mutated during storage.

Responsibility. Is it the Curator or the depositor that ensures these prior requirements? Are the depositories servants of the public or are

[5] Guidelines for Deposit of Microorganisms, April 29, 1971, 886 O.G. 638.

they so tied up contractually to the depositor that the public may still be at a loss to practice the invention even after the patent issues? To say that these things can no longer occur overlooks the fact that many deposits were made prior to the Argoudelis case and thus tied up by contracts made prior to these now being used.

It would appear that the deposit of the culture, for at least the seventeen-year patent life, must be available publicly to protect the patentee and, as stated above, forever after for the public benefit and the aforementioned quid for the patent monopoly. This is the reason why the proposed prior legislation referred to an approved deposit in a public depository designated by the Commissioner of Patents by publication. Incidentally, since January 1, 1971, such a regulation was put into effect in Japan. Foreign deposits such as with the ATCC in the United States are no longer acceptable in Japan. The standards for the nature of the deposit are even set forth and include confirmation of the identity of the culture with that specified in the patent application. I cannot say at this time what the exact nature of our conditions will be, but those set forth in Japan are indeed rigorous. In Great Britain, the report of the Banks Committee recommends legislation for deposits, and I assume that the other nations will follow. There is one point in the Argoudelis decision that could well be considered controversial. In discussing the applicants' compliance with depositing, the court stated: "(1) a public depository was used and (2) the depository is operated by a department of the United States Government. . . ." We, at this time, do not know whether this latter was merely an observation of the Court of a state of facts or a limiting of the deposits, by the court, to a Government-operated depository. I can say, however, that there are, at this time, no preconceived notions at the examiner level to so limit deposits absent express direction from the Commissioner of Patents or specific case law.

Now WIPO (World Intellectual Property Organization) has before it a study on the requirements for deposit of microorganisms for the purposes of patent applications. The study was made during 1973 and was reported in November 1973. The objective is a multilateral convention according to which each contracting state would recognize the culture deposit collections of the other contracting states, would free from import restrictions cultures deposited in the other contracting states in connection with priority based patent applications, and would ensure that cultures deposited in their own collection were released in appropriate cases. Or there may be established an International Culture Collection. And there are suggestions between these two possibilities. Meetings were held in Geneva in April and July of 1974. A draft treaty was prepared and considered by the member countries in April, 1975.

Appendix

S. 2504
Section 112. Specification

"(f) When the subject matter sought to be patented relates to a process involving the action of a microorganism not already known and available to the public or to a product of such a process, the written description required by subsection (a) of this action shall be sufficient as to said microorganism if-

"(1) not later than the date that the United States application is filed, an approved deposit of a culture of the microorganism is made by or on behalf of the applicant or his predecessor in title, and

"(2) the written description includes the name of the depository and its designation of the approved deposit and, taken as a whole, is in such descriptive terms as to comply with subsections (a) and (b) of this section.

"(g) For the purpose of subsection (f)(1) of this section, an approved deposit shall be a deposit which-

"(1) is made in any public depository in the United States which shall have been designated for such deposits by the Commissioner of Patents by publication, and

"(2) is available, except as otherwise prohibited by law, in accordance with such regulations as the Commissioner may prescribe-

"(A) to the public upon issuance of a United States patent to the applicant or his predecessor or successor in title which refers to such deposit, or

"(B) prior to issuance of said patent, as specified in sections 122 and 132(e) of this title"

Section 119. Benefit of earlier filing date in foreign country; right of priority

"(d) When the application claiming priority under this section discloses an invention relating to a process involving the action of a microorganism not already known and available to the public or to a product of such a process and an approved deposit is made under section 112(f) of this chapter, the approved deposit shall be considered to have been made on the earliest date that an application in a foreign country, the priority of which is being claimed, contains a reference identifying a deposit of the same microorganism made in a public depository."

Microbiological Control of Plant Pathogens

Y. Henis and I. Chet

*Department of Plant Pathology and Microbiology,
The Hebrew University of Jerusalem, Faculty of Agriculture,
Rehovot, Israel*

I.	Introduction	85
II.	Naturally Occurring Microbiological Control	86
III.	Induced Microbiological Control of Plant Pathogens	87
	A. Direct Control	87
	B. Indirect Control	91
IV.	Mechanisms of Microbiological Control	98
	A. Competition	98
	B. Antibiosis	100
	C. Lysis	101
	D. Germination and Subsequent Decline of Resting Structures	103
	E. Parasitism and Predation	104
V.	Integrated Control	105
	A. Integration with Pesticides	105
	B. Integration with Fumigants and Heat Treatments	106
	C. Multiple Integration	107
	References	107

I. Introduction

In his brilliant essay on biological control of soil-borne plant pathogens, Garrett (1965) defines biological control as "any conditions under which, or practice whereby, survival or activity of a pathogen is reduced through the agency of any other living organisms (except man himself), with the result that there is reduction in incidence of the disease caused by the pathogen." We shall adopt this definition of biological control in its broader sense, rather than the narrow one that takes into account only the direct use of microorganisms in disease control (Norris, 1971).

This review covers some aspects of microbiological control of plant pathogens, including its occurrence in nature, the most common microbial groups and the mechanisms involved, with emphasis on its practical implications.

Microbiological control of plant diseases can be achieved either directly, through inoculation, or indirectly, by changing the conditions prevailing in the plant's environment, and thus the microbiological equilibrium of its ecosystem, or by a combination of both approaches.

Changing conditions in the plant's environment may affect the parasite, its host, or the saprophytic microorganisms potentially involved in disease control. No attempt will be made to cover all the vast literature dealing with cultural practices, agricultural techniques, and empirical studies on the response of crops to plant pathogens under various conditions. These aspects have been comprehensively discussed in books (Garrett, 1970; Baker and Cook, 1974), symposia (Baker and Snyder, 1965; Preece and Dickinson, 1971; Parmeter, 1970), and reviews (Wood and Tveit, 1955; Mitchell, 1973; Papavizas, 1973; Wilhelm, 1973; Henis and Katan, 1975). Biological control of plant parasitic nematodes deserves separate treatment and is not covered here either. The reader is referred to excellent reviews on the subject (Boosalis and Mankau, 1965; Duddington and Wyborn, 1972). Finally, changes in disease severity following herbicide application and resulting from indirect changes in microbiological equilibrium in the plant ecosystem have been reviewed by Katan and Eshel (1973) and will not be discussed here either.

II. Naturally Occurring Microbiological Control

Microorganisms parasitic to plants constitute a tiny fraction of the microflora and microfauna inhabiting the vicinity and surfaces of plant organs. It has been frequently observed that disease severity is greatly increased when the pathogen is reintroduced into its presterilized infection site, whether on the roots (Baker and Snyder, 1965; Garrett, 1970) or on leaves (Preece and Dickinson, 1971), indicating that the saprophytic microorganisms inhabiting the surfaces of plant organs may serve as a biological buffer zone, preventing the pathogen from infecting its host.

The phenomenon of soil fungistasis in spore-forming soil fungi, either pathogens or saprophytes, described by Dobbs and Hinson (1953), which has been shown (Ko and Lockwood, 1967; Hsu and Lockwood, 1973) to be related to microbial activity is an example of another widespread, naturally occurring, microbiological suppression. In this case, however, soil-borne plant pathogens may gain advantage from the observed inhibition in spore germination, as it prevents their germination in the absence of the host, thus prolonging their survival in the soil (Jackson, 1965).

Although microbiological control of animal pathogens is outside the scope of this review, its comparison with the control of plant pathogens is of some interest. Impairment of the biological equilibrium of the naturally occurring microflora inhabiting animal skin or the digestive system, as observed in some cases of a prolonged and extensive use of antibiotics, may result in augmentation of potential skin respiratory tract and intestinal pathogens (Alexander, 1971; Savage, 1972).

These examples lead to the conclusion that microbiological control of pathogens is a naturally occurring, widespread phenomenon, turning the academic hope of utilizing it in practice into a feasible one.

III. Induced Microbiological Control of Plant Pathogens

A. Direct Control

Potential antagonists to plant pathogens are abundant in the soil, in the air, and on plant surfaces. Their effect on the parasite can be easily demonstrated and studied on growth media (Wood and Tveit, 1955). However, introduction of potential antagonists into the undisturbed ecosystem does not usually bring about a significant and consistent control of the pathogen. In general, the already existing microflora resists any newcomer, preventing its establishment in the ecosystem (Alexander, 1971; Wilhelm, 1973). In some cases, however, attempts to control disease by inoculation have been partly successful. Some typical examples are given below.

1. Viruses

Until recently, there have been relatively few reports on viruses infecting fungi. During the last decade, viruses were found in over 60 species from some 50 genera of fungi of the major taxonomic groups (Lemke and Nash, 1974), including plant pathogens. In some cases, plant diseases caused by viruses have been controlled by cross protection with attenuated or related strains (Matthews, 1970) or by early inoculation procedures (Broadbent, 1964). Lemaire et al. (1970) reported a decline in take-all disease of wheat caused by *Ophiobolus graminis* as a result of a virus infection of the mycelium of the pathogen.

2. Eubacteria, Myxobacteria, and Actinomycetes

Cordon and Haenseler (1939) were among the first to use an antagonistic strain of *Bacillus simplex* to inhibit *Rhizoctonia solani*. Addition of a bacterial suspension to greenhouse soil gave appreciable control of seed decay and damping-off in cucumber and pea seedlings. Working with lettuce seedlings, Wood (1951) achieved a significant control of damping-off of lettuce caused by *R. solani* by addition of cultures of *Streptomyces* and *Bacillus* sp. which had been isolated from soil and inhibited *R. solani* on agar media. Similar results were obtained with sugar beet grown in soil preinoculated with *B. subtilis* (Dunleavy, 1952). Olsen and Baker (1968) obtained an enrichment of *B. subtilis* which produces antibiotics against *R. solani* in steamed soil. Direct inoculation of *B. subtilis* isolates to presteamed soil also depressed the pathogen. Aldrich and Baker (1970) controlled stem rot and wilt of carnation caused by *Fusarium roseum* 'Culmorum' by dipping carnations cuttings in suspension of antagonistic isolate of *B. subtilis* prior to planting. Cuttings treated with *B. subtilis* showed fewer symptoms, and had more extensive root development and stem elongation than controls inoculated with *Fusarium* without the bacteria.

Bacteria, isolated from soil, were antagonistic to *F. roseum* 'Culmorum' both *in vitro* and *in vivo*. Unrooted carnation cuttings were inoculated with antagonistic bacteria and rooted in the presence of the pathogen. The bacteria reduced the number and development of basal-stem lesions, and did not inhibit rooting. Carnation cuttings which were inoculated with the most effective bacterial isolate (a pseudomonad) were free of lesions 4 months after inoculation of the cuttings with *F. roseum* 'Culmorum', yet the pathogen was recovered from these cuttings (Michael and Nelson, 1972). The authors suggested that the antagonistic bacteria might have protected the cut surface at the base of the cutting until it formed a protective layer of callus tissue. The pseudomonad used in these experiments produced a fungistatic antibiotic which inhibited the growth of the pathogen *in vitro*.

Many soil and rhizosphere-dwelling myxobacteria of the genera *Cytophaga* (Webley *et al.*, 1967), *Myxococcus* (Haska and Norén, 1967), and *Sorangium* (Gillespie and Cook, 1965) are prolific producers of extracellular lytic enzymes. According to Peterson *et al.* (1965) production of chitinase but not cellulase enables some of the *Sorangium* isolates to lyse fungi without injuring host plants. Hocking and Cook (1972) inoculated fumigated peat with isolates of *Cytophaga johnsonii* and *Sorangium* sp. before sowing seeds of four species of conifer. The myxobacteria actively colonized the rhizosphere of the developing seedlings without affecting seedlings' development. In subsequent inoculation tests with pathogenic fungi, such as *Pythium intermedium, Rhizoctonia solani, Fusarium oxysporum,* and *F. solani*, there was significantly less mortality, chlorosis, and stunting than in controls without myxobacteria. It is not clear whether those myxobacteria actively destroyed the pathogen. However, the incidence of disease and the ease of isolation of pathogenic fungi from seedlings inoculated with the myxobacteria indicate that this destruction is incomplete. This may be due to low concentrations of the extracellular enzymes and to their poor distribution. The authors believe that further studies on myxobacterial colonization will possibly increase the protective value of these organisms.

An interesting possibility of biological control of crown gall by an isolate of the saprophyte *Agrobacterium radiobacter* var. *radiobacter* was reported by New and Kerr (1972), who achieved a full control of the pathogen *A. radiobacter* var. *tumefaciens* by dipping roots of young peach seedlings in a cell suspension of the antagonist before planting in soil infested with the pathogen. Kerr (1972) was also able to protect young plants by inoculating seeds with the saprophytic isolate. He found that total gall incidence on plants grown from inoculated seeds was 31% as compared with 79% in uninoculated seeds. Similarly, Hsieh and Buddenhagen (1974) recently demonstrated a suppressing effect of the saprophyte *Erwinia herbicola* on infection of *Xanthomonas oryzae* and on development of symptoms of bacterial blight in rice.

Mehrota and Caludius (1972) isolated antagonists from the rhizosphere of *Lens culinaris* and used them in the control of root rot and wilt disease caused by *S. rolfsii* and *F. oxysporum*, respectively. Using inoculated pots, they obtained a full control of *F. oxysporum* f. sp. *lentis* with *Trichoderma viride, Streptomyces gougeroti,* 2 other strains of *Streptomyces*, and 2 bacterial isolates. In the case of *S. rolfsii* not a single antagonist could fully eradicate the disease.

Plant inoculation with bacteria other than *Rhizobium* to improve seed germination, plant development, and crop yield have been practiced in the USSR for many years. Mishustin and Naumova (1962), however, in their review of *Azotobacter* and other bacterial fertilizers criticized many of the inoculation experiments which showed an increase in crop yield, because of lack of adequate statistical analysis. Still, significant response was often obtained, especially when treated seed was sown in soil where untreated seed developed poorly. The beneficial effects of *Azotobacter* inoculation on plant growth and crop yield may include nitrogen fixation, production of growth-regulating substances, and disease control through the direct or indirect antagonistic effect of *Azotobacter* toward plant pathogens (Brown et al., 1968). Azotobacter has been recently reported to produce an antifungal antibiotic, active *in vitro* against *Fusarium* (Lakshmi-Kumari et al., 1972). A detailed study is still required, however, to assess its antagonistic effect toward plant pathogens.

3. Fungi

Members of the genus *Trichoderma* are active as hyperparasites (Barnett and Binder, 1973) and antibiotics producers (Dennis and Webster, 1971). Thus, *T. viride* produces antifungal antibiotics, lytic enzymes (Baker and Cook, 1974), as well as volatile metabolites which inhibit sporulation (Hutchinson and Cowan, 1972). Jaarsveld (1942) inoculated seedlings of Chinese cabbage with the pathogen *R. solani* and with some antagonistic fungi. Most efficient control was obtained with *T. viride* followed by *Pyronema confluens, Penicillium expansum,* and *Cladosporium herbarum*. Fedorinchik (1971) demonstrated the ability of *Trichoderma lignorum* to control *Verticillium* and *Hypocrea* attacking cucumber and *Helminthosporium* and *Fusarium* attacking wheat. He developed a method to induce a massive sporulation of *T. lignorum* within 4 days for mass inoculation.

When trees of *Juglans regia* were grown in soil infested with *Phytophthora cinnamomi*, a most efficient suppression of the pathogen was obtained after sterilization by methoxyethylmercuric chloride and inoculation with *T. viride* to avoid reinfestation (Morquer and Touvet, 1972).

Wood-inhabiting fungi quickly invade wood exposed by natural or artificial wounds inflicted on trees. Grosclaude *et al.* (1973) reported that nursery tests on 2-year-old plum trees (*Prunus mariana*) revealed the effectiveness of *Trichoderma viride* in the protection of pruning wounds against the basidiomycete *Stereum purpureum*. *T. viride* was applied as a spore suspension by ordinary spraying and also through pruning shears designed especially for that purpose, to achieve further labor saving. In both application methods, complete protection was obtained. Chemical protection was equally effective in controlling the disease. As for the mode of action of *T. viride*, the authors suggested the involvement of the following factors: (a) trophic competition in favor of the organism first introduced; (b) a higher growth rate of *T. viride*; (c) release of growth-inhibiting substances by *T. viride* spores or mycelium; (d) surface action of ungerminated *T. viride* spores.

Partial control of *Fusarium* wilt in tomato caused by *Fusarium oxysporum* f. sp. *lycopersici* was achieved not only by other fungi, but also by other formae of the same species (Davis, 1968). Singh (1972) reported that *Tridentaria tylota* attack *Pythium* oospores and decrease *Pythium* leaf spot disease in maize. More recently Dhingra and Khare (1973) tested 38 fungi isolated from the rhizoplane of urid bean (*Phaseolus mungo* L.) for their ability to antagonize *Rhizoctonia bataticola* in vitro and in vivo. *Arachinotus* sp., *Aspergillus aculeatus*, *Cephalosporium humicola*, and *Trichoderma lignorum* were most active *in vitro*. *T. lignorum* did not show activity in nonautoclaved field soil in soil columns. *Arachinotus* sp. or *Aspergillus aculeatus* applied to the seed controlled pre- and postemergence damping off of urid bean, giving a total stand of 73% and 69.7%, respectively, in nonautoclaved field soil, as compared to 1.2% in controls. One of the few cases in which biological control by direct inoculation was successfully employed at a commercial scale is the inoculation of pine stumps with oidia of *Peniophora gigantea* which compete with the pathogenic fungus *Fomes annosus* (Rishbeth, 1963). Another example of a direct use of a fungal competitor is the control of chestnuts blight. The hypovirulent strains of *Endothia parasitica* are effective in suppressing the French virulent strain of the parasite which causes a blight in American chestnut trees (Anagnostakis and Jaynes, 1973).

A different aspect of microbiological control is the use of mycorrhiza. It has been demonstrated by Marx (1969a,b) that ectomycorrhizae formed by different fungal symbionts on *Pinus echinata* and *P. taeda* seedlings were resistant to infections caused by either zoospores or mycelium of *Phytophthora cinnamomi*. Nonmycorrhizal roots were heavily infected by both forms of the pathogen. Wingfield (1968) observed that ectomycorrhizae formed by *Pisolithus tinctorius* improved the survival of loblolly pine seedlings grown in the presence of *Rhizoctonia solani*. Marx (1973) found that shortleaf pine seedlings with ectomycorrhizae formed by *P. tinctorius* or *Cenococcum graniformae* grew as well in nonsterile soil with *P. cinnamomi* as did nonmycorrhizae pine seedlings in nonsterile soil without the pathogen. Nonmycorrhizal seedlings exhibited significant symptoms of feeder roots. Ectomycorrhizal development was inversely related to the number of susceptible, nonmycorrhizal feeder roots available for attack by *P. cinnamomi*. The mechanism of protection is not known yet. Marx (1973) suggested that the propagules of the pathogen may have been killed on the root surface by antimicrobial volatile substances produced by the ectomycorrhizae (Krupa and Fries, 1971), or by antagonistic rhizosphere microorganisms. The ectomycorrhizae, therefore, not only contribute to plant nutrition, but also play a role in the control of root infection.

4. Unidentified Factors

The presence in soil of biological factors, probably of microbial origin, antagonistic to soil-borne plant pathogens has been reported in several cases. Menzies (1959) demonstrated the occurrence in older irrigated soils of a biological, transferable factor that suppresses potato scab, which could be fortified by amendments of alfalfa meal. Shipton *et al.* (1973) found factors antagonistic to *Ophiobolus graminis* in fields subjected to long periods of drought and in irrigated plots, but not in grassland and virgin soils. The antagonistic properties of the active soils were eliminated with fumigation by methyl bromide. Addition of as little as 1% of the antagonistic soil to the fumigated soil provided excellent restoration of antagonistic properties. In contrast, inoculation of fumigated soil with virgin soil provided no restoration of antagonism. The authors suggested that the antagonism was due to the activity of non-spore-forming bacteria and soil fungi. In another study (Vojinović, 1973), a decline of the pathogen *Ophiobolus graminis,* observed in a prolonged monoculture of wheat, was related to an increase in the antagonistic microflora near the rhizosphere, 90% of which being actinomycetes.

B. Indirect Control

1. Physical Factors

The effect on biological control of altered physical conditions of soil, such as temperature, moisture, composition of soil atmosphere, tillage, deep ploughing, and immersion of soil, have been discussed in detail by Sewell (1965). Among these factors, soil moisture seems to be of a special importance, since it also affects soil temperature and its atmosphere. According to Sewell, moisture may either affect the disease by its direct effect on the pathogen, or through its indirect effect on the host symptoms, or on its predisposition to the infection. Stover (cited by Sewell, 1965) successfully controlled banana wilt caused by *Fusarum oxysporum* f. sp. *cubense*, by flooding. The factors involved in this case were oxygen depletion and CO_2 excess (Newcombe, 1960) rather than the production of toxic metabolites by the anaerobic soil microflora (Smith and Cook, 1974).

2. Inorganic Amendments

Inorganic fertilizers commonly used in agriculture, as well as microelements, may affect disease incidence and severity either directly by suppressing the pathogen, or indirectly through their effect on the microbiological equilibrium and on host tolerance to disease (Henis and Katan, 1975). Using amendments for manipulating soil pH as a means of control of plant pathogens is possible on the following conditions: (1) that the pathogen is capable of growing and inciting disease only at a relatively narrow pH range; (2) that the host plant is capable of growing at a pH range which is not suitable for the pathogen; (3) that the buffering capacity of the environment is not too great. Soil pH may also affect disease severity by changing the disease susceptibility of the host or by favoring the activity of the antagonistic microflora. Addition of sulfur to soil has been successfully used for reducing soil pH to values below 5.2–5.5 and, consequently, suppressing plant pathogenic actinomycetes such as *Streptomyces scabies*, the causative agent of potato scab and *S. ipomea*, which causes soil rot of sweet potato (Baker and Cook, 1974). Henis and Chet (1967) found that high dosages (300 ppm) of ammonia were lethal to sclerotia of *S. rolfsii* and concluded that the direct toxic effect of ammonia on *S. rolfsii* was a function of high pH and time. However, at low concentrations, ammonia may affect plant pathogens indirectly by changing the microflora in the soil and on the sclerotial surface (Henis and Chet, 1968). Similar observations were made also with other nitrogenous amendments such as urea, ammonium nitrate, calcium nitrate, ammonium acetate as well as chitin and peptone (Henis and Chet, 1968). In these treatments, decrease in germinability was correlated with significant increased numbers of antibiotic producing organisms especially bacteria and actinomycetes associated with sclerotia. Sclerotia buried in nontreated soil contained a negligible number of associated microorganisms as compared with those buried in amended soils. Ethanol extracts of sclerotia from amended soils inhibited fungal growth and decreased germinability of nontreated sclerotia. The utilization of the sclerotium as a nutrient source by the soil microflora may take place after the death of its cells, as a result of antagonistic action. In the case of *S. rolfsii,* effective amendments may serve as nitrogenous sources for the soil microorganisms, enabling them to attack the carbonaceous sclerotial wall (Chet et al., 1967). Ammonium nitrate amendments enhanced the competitive saprophytic activity of *Rhizoctonia* but suppressed it when combined with cellulose, oat straw, or soybean amendments (Davey and Papavizas, 1963). Similarly, inorganic nitrogen, mainly nitrate, decreased the populations of microorganisms antagonistic to the pathogens *Fusarium solani* f. *phaseoli, Rhizoctonia solani, Thielaviopsis basicola,* and *Verticillium albo-atrum* in bean rhizosphere as well as the lytic activity of the amended soil (Papavizas, 1963).

Competition between pathogenic fungi and antagonistic microflora toward nitrogen may occur in amended soil. An increase in the antagonistic microflora and a suppression of the pathogen is often observed when the pathogen is a weak competitor or cannot utilize the supplied nitrogen source (Henis and Katan, 1975). Huber (cited by Papavizas, 1973) controlled take-all disease of wheat caused by *Ophiobolus graminis* by fertilization. He found that application of nitrogen fertilizer in fall and planting of wheat at spring increased disease severity, whereas adding the same fertilizer just before seeding in the spring reduced disease incidence and increased the yield. It is interesting to note that addition of $(NH_4)_2SO_4$ decreased incidence whereas NH_4NO_3 increased it (see also Huber, 1972).

Inorganic amendments may affect survival, growth, and virulence of fungal pathogens, as demonstrated for nitrogen with *F. solani* f. sp. *phaseoli* (Maier, 1968) and for micronutrients with *F. oxysporum* f. sp. *lycopersici* (Woltz and Jones, 1968).

Nitrite may accumulate in urea- or ammonium-fertilized soil as a result of differential inhibition of nitrogen oxidizers by high content of ammonia (Alexander, 1971). The accumulating nitrite may affect plant pathogens, as shown by Sequeira (1963), who found that urea added to soil suppressed *F. oxysporum* f. sp. *cubense* because of accumulation of nitrite in soil resulting from inhibition of nitrite-oxidizing bacteria. Nitrite was also found to inhibit phycomycetes, such as *Phytophthora* (Zentmyer and Bingham, 1956) and *Pythium* (Grover and Sidhu, 1966).

Growth inhibition of plant pathogens at relatively low concentration of soluble manganese have also been demonstrated in the case of *S. scabies* (McGregor and Wilson, 1966; Mordvedt et al., 1961).

Calcium ions affect virulence of many pathogens such as *R. solani* (Bateman, 1964) by rendering pectic acid in host tissue less available to the pectolytic enzymes produced by the pathogen, and *S. rolfsii*, by rendering insoluble the oxalic acid produced by the fungus (Bateman and Beer, 1965).

Trace elements suppress germination and production of enzymes and toxins involved in development of fusarial wilt. This aspect has been extensively reviewed by Sadasivan (1965).

Stotzky and Martin (1963) observed a relationship between the distribution of some clay minerals (particularly montmorillonite) and the spread of banana wilt caused by *Fusarium oxysporum* f. sp. *cubense*. Stotzky and Rem (1966a,b,c) related the restriction of the pathogen spread to the buffering capacity of montmorillonite which allowed for the development of an abundant antagonistic soil microflora. A relationship between distribution of clay minerals and prevalence of the human pathogen *Histoplasma capsulatum* has been also noted by Stotzky and Post (1967). These interesting observations may have a wide potential implications in future use of clay minerals for disease control (Kerr, 1965).

3. Organic Amendments

Microbial decomposition of crop residue and other organic soil amendments is accompanied by a temporary increase in counts of specific zymogenous groups. The effect of organic amendments on plant pathogens and on disease incidence and severity varies greatly and depends on the type of the amendment, its concentration, C/N ratio, and its degradability. Pathogen virulence, survival, and saprophytic activity may be affected either directly or indirectly through the activity of the soil microflora.

a. Crop Residues. In agricultural practice, the use of crop rotation to control plant pathogens is partially based on the effect of the recent crop residues on disease incidence and severity in the next crop (Cook and Watson, 1970; Papavizas, 1973; Baker and Cook, 1974). Snyder *et al.* (1959) found that barley amendments could control bean root rot whereas additions of nitrogen nullified this control. Further studies revealed that wide C/N ratio of amendment decreased disease incidence by depleting available nitrogen, thus inhibiting the germination of *Fusarium* chlamydospores (Maurer and Baker, 1965). Papavizas (1963) tested the lysis of 4 assay plant pathogenic fungi in culture (*Fusarium solani* f. *phaseoli*, *R. solani*, *Thielaviopsis basicola*, and *V. albo-atrum*), following addition of treated soil. Most lysis was caused by soil amended with oat straw, as compared with autoclaved or ammonium nitrate-amended soil. High N applications significantly reduced lysis in both unamended and oat straw-amended soil.

Huber *et al.* (1965) examined the effect of organic residues on nitrification in enrichment media, using samples from the amended soils as inocula. The residues were selected on the basis of their reported effects on root rot caused by *Fusarium* and *Rhizoctonia*. They found that residues reported to decrease disease severity (alfalfa, soybean, pea, corn) also increased the rate of nitrification, whereas cellulose, barley and oats which increase disease severity, also reduced nitrification. They suggested that organic amendments may modify disease severity by affecting nitrification, and that this effect was independent of the C/N ratio.

Papavizas and Davey (1960) and Davey and Papavizas (1960) found a significant increase in counts of soil microorganisms as well as counts of actinomycetes antagonistic to *R. solani* during suppression of *Rhizoctonia* disease of bean by organic amendments. Similar results were reported by Manning and Crossan (1969). Henis et al. (1967) demonstrated a relationship between the efficiency of various organic amendments in decreasing the infection index in bean seedlings and in suppressing the sarpophytic activity of *Rhizoctonia* in the soil, and the increase in total counts and *Rhizoctonia*-antagonistic soil microorganisms.

Root rot of peas caused by *Aphanomyces euteiches* was significantly reduced by incorporating into the soil of stem and leaf tissues of cruciferous crops such as cabbage, kale, radish, mustard, and Brussels sprouts (Papavizas, 1966). Cabbage amendments also reduced root rot of bean and sesame caused by *Thielaviopsis basicola* (Papavizas et al., 1970). During microbial degradation, cruciferous amendments release antifungal volatile sulfur-containing substances, such as mercaptans, sulfides, and isothyiocyanates (Papavizas, 1966, 1967). Lewis and Papavizas (1971) studied the effect of several sulfur-containing volatile compounds and vapors from degrading cabbage on growth, zoospore formation, germination and motility of *Aphanomyces euteiches*. They found that vapors from the cabbage tissue adversely affected growth and morphology of the fungus, whereas vapors arising from corn amendment, which did not affect pea root rot caused by this pathogen, had no effect on the fungus either.

Excellent control of avocado seedlings root rot caused by *Phytophthora cinnamomi* was obtained with alfalfa meal amendments by Zentmyer (1963), who observed a large increase in saprophytic fungi and bacteria in treated soil in comparison with a nontreated soil. Gilpatrick (1969) also obtained elimination of *P. cinnamomi* from infected avocado roots by alfalfa amendments and noted that the toxicity of the amended soil to the pathogen correlated with the level of ammonia produced during alfalfa decomposition. A concentration of 17 ppm ammonia in phosphate buffer prevented zoospore germination and killed mycelium of *P. cinnamomi in vitro*. He concluded that both fungitoxicity of ammonia to

P. cinnamomi and its phytotoxicity to the host roots, which made them unattractive to the pathogen, were involved in the disease control by alfalfa amendments.

Alfalfa tissues contain saponins that are surface-active agents toxic to fungi (Goodwin *et al.*, 1929; Assa *et al.*, 1972). Zentmyer and Thompson (1967) showed that both zoospore germination and sporangia formation by *Phytophthora cinnamomi* were strongly inhibited by saponin fractions obtained from alfalfa meal at concentrations comparable to those found in soil amended with alfalfa meal. According to Zentmyer and Thompson (1967) other factors, undoubtedly involved in this control, include competition for nutrients and the possible production of antibiotics resulting from significant increase in the activity of the soil microflora.

Menzies (1959) reported a suppression synergistic effect of alfalfa amendment and the antiscab biological factor found in soils of old irrigated potato fields. Menzies and Gilbert (1967) obtained an increase in respiration rate, bacterial counts, and vegetative growth of fungus mycelia following exposure of soil to volatile substances from plant residues, mainly alfalfa. Gilbert and Griebel (1969) found that exposure of soil containing microsclerotia of *Verticillium dahliae* to volatile substances obtained from alfalfa hay caused an initial increase followed by elimination of the pathogen from soil. The volatile substances also caused quantitative and qualitative changes in the soil microflora; this finding suggested that competition and/or antagonism contributed to the eventual decline of *V. dahliae*.

Owens *et al.* (1969) isolated from alfalfa and other plant residues more than 20 volatile compounds, mostly aliphatic aldehydes and alcohols. The major active component, which accounted for almost half of the soil respiration enhancement and caused preferential increase in numbers of soil microorganisms, was acetaldehyde. Soil fungi were also stimulated by alcohols having 3 carbons or more. Recently, Linderman and Gilbert (1973) showed that mycelial growth from sclerotia of *Sclerotium rolfsii* in natural soil was stimulated when exposed to vapors of alfalfa distillate, but was adversely affected if the soil was previously exposed to the volatiles, apparently as a result of the increased activity of the soil microflora.

Weinhold *et al.* (1964) found that potato scab was prevented in the field by growing soybeans as a green manure crop in the fall after potato harvest and turning the crop under before planting the potatoes. The effect was rather specific since barley cover crop used as a green manure increased disease incidence.

b. Polysaccharides. Mitchell and Alexander (1961, 1962, 1963) selectively stimulated in the soil strains of *Bacillus* and *Pseudomonas* capable of lysing fungal cell walls, by the addition of chitin. Chitin amendments at a rate of 200 lb per acre markedly reduced the severity of the root rot diseases caused by *Fusarium solani* f. *phaseoli* or *F. oxysporum* f. *conglutinans*. Neither cellulose nor inorganic nitrogen produced similar beneficial effects. Mitchell and Alexander suggested that the influence of chitin may result from an increase in the number of mycolytic microorganisms. Chitin amendments increased actinomycete population in the soil (Baker and Cook, 1974), this group being particularly active in antibiotic production. The content of antifungal substances in aqueous extracts of chitin-treated soils was far greater than in unamended soils. Addition of chitin neither affected *Pythium debaryanum*, a phycomycete that does not have chitin in its cell wall, nor the crown gall bacterium *Agrobacterium tumefaciens* (Mitchell and Alexander, 1963). Moreover, when chitinous material was amended to suppress *Fusarium* wilt in tomato, an increase in disease incidence caused by *Pythium aphanidermatum* was observed (Singh and Pande, 1965). According to Kerr and Bumbieris (1969) chitin amendments stimulated *Pythium*, which used chitin as a nutrient. Control of various pathogenic fungi was also obtained with chitin amendments by Lloyd et al. (1963), Khalifa (1965), and Lloyd and Lockwood (1966). Sneh et al. (1971) followed the saprophytic activity and pathogenicity of *Rhizoctonia solani* in natural, chitin-amended soil. Addition of lignin had no significant effect on pathogenicity of *R. solani*, nor did it increase the inhibitory effect of chitin. No parasitism or lysis of *R. solani* hyphae by soil organisms could be observed in the chitin-amended soil. The fungus was not inhibited in sucrose-amended soil in which CO_2 was produced in quantities greater than in the chitin-amended soil. Sneh and Henis (1972) found a correlation between the inhibition of saprophytic activity of *R. solani* in chitin-amended soil and the increase in antifungal activity of *n*-butanol extracts as compared with nonamended soil. Extracts of chitin in amended soil inhibited the growth of 17 other species of fungi. The authors postulated that the active factors may be nonspecific cell constituents of microbial origin.

Laminarin, a β-D-(1,3)-glucan, is a common component of fungal cell walls (Chet et al., 1967). The addition of this polymer to soil resulted in decline in the severity of diseases caused by several soil-borne fungal pathogens such as *Fusarium oxysporum* and *Fusarium solani* but not *Pythium debaryanum*. Laminarin treatment resulted in actinomycetes suppression, indicating that antibiosis resulting from the activity of soil actinomycetes was not a significant factor in the suppression of fungi by this glucan. It seems possible that cell wall-degrading enzymes (glucanases), rather than antibiotics, were involved in the suppression of *Fusarium* spp., which contain β-D(1,3)-glucans in the cell wall (Mitchell, 1963).

c. *Nitrogenous Organic Compounds.* According to Bumbieris and Lloyd (1967), live fungal hyphae lyse faster in soil previously supplemented with peptone or blood and bone meal than in nonsupplemented soil, probably because of their favorable effect on the mycolytic microflora. Peptone amendments also decrease germinability of sclerotia of *S. rolfsii* and increase the antagonistic microflora in the treated soil (Henis and Chet, 1968).

IV. Mechanisms of Microbiological Control

The dynamic biological equilibrium prevailing in natural ecosystems depends, among other factors, on the relationships between the microbial inhabitants. In arable soil, counts of living organisms may reach a level of 10^8 to 10^9 per gram, whereas on root surfaces they can soar to values of much higher orders of magnitude (Alexander, 1961). Individual organisms can support each other's growth by turning unavailable nutrients into available ones, by synthesizing and excreting necessary nutrients, such as amino acids or vitamins, by changing physical conditions from nonoptimal into optimal ones, or by decomposing or neutralizing toxic substances. On the other hand, they may repress each other's growth by competing for nutrients, oxygen, or space, by excreting inhibitory metabolic products or by using each other as a prey. These complex interactions fall into neutral, beneficial and detrimental categories (Baker, 1968; Mitchell, 1973; Alexander, 1971).

In their struggle for survival, soil microorganisms compete for colonization of plant residues. This aspect has been thoroughly studied and brilliantly analyzed by Garrett (1956, 1970), who proposed the term "competitive saprophytic ability" and defined it as "the summation of physiological characteristics that make for success in competitive colonization of dead organic substrates." This ability is of great importance for the survival, growth, and spread of soil-borne pathogens. In this review we shall deal mainly with the use of detrimental types of interactions, i.e., competition, antibiosis, parasitism, and predation which may be exploited for biological control of plant pathogens.

A. Competition

This interaction may be defined as suppression of growth, which takes place whenever two or more organisms struggle for nutrients, space, or any other environmental factors that become limiting for growth (Park, 1960).

Any essential requirement of microorganisms can serve as potentially possible basis for a competitive interaction. However, although competition for oxygen and minerals is well known, its role in the biological control of plant pathogens is mostly limited to nitrogen (Baker, 1968; Alexander, 1971), which is one of the chief limiting factors in soils.

Competition between microorganisms for available nitrogen, carbon, or minerals can easily be increased by soil amendments. Blair (1943) attributed the depressive effect or organic amendments on *Rhizoctonia solani* to the increased activity of the soil microflora which resulted in nitrogen starvation of the pathogen, and suggested that further inhibition in the growth of the pathogen probably resulted from the increased production of carbon dioxide in the soil.

Snyder *et al.* (1959) controlled bean root rot by barley residues. High C/N ratio inhibited both germination of *F. solani* chlamydospores and the penetration of the pathogen into its host. Supply of nitrogen nullified the suppression.

Maurer and Baker (1965) working on the effect of cellulose and glucose amendments on *Fusarium* bean root rot found that nitrate was better than ammonium in reducing the disease, and that the addition of nitrification inhibitors, such as N-serve, increased disease severity. They concluded that competition for available nitrogen between the pathogen and soil fungi plays an important role in the control of bean root rot by organic amendments.

1. Competition between Bacteria and Fungi

Marshall and Alexander (1954) demonstrated a suppression of the growth of *F. oxysporum* by two species of bacteria in autoclaved soil. The suppressive effect became apparent by the addition of glucose and could be eliminated by adding inorganic nitrogen. They related the superior competitive capacity of the bacteria to their better ability to utilize simple inorganic nutrient sources. Blakeman and Frazer (1971) found that germination of *Botrytis cinerea* conidia on chrysanthemum leaves was inhibited by saprophytic bacteria. The results obtained by Sztejnberg and Blakeman (1973) support the hypothesis that competition for nutrients between the bacteria and the germinating conidia, rather than production of antibiotics, was the mechanism involved.

2. Competition between Different Fungi

Demonstration of competition *sensu stricto* between fungi in a well defined experimental system has been reported only in a few cases. The saprophytic behavior of *Rhizoctonia solani* in the soil was extensively studied by Papavizas (1970), who concluded that this pathogen was a relatively weak competitor. He also observed a positive relationship between intrinsic growth rate and saprophytic ability in different isolates of *R. solani*. Waksman and Hutchings (1937) showed that *T. viride*, a powerful cellulose decomposer, preferred proteins and hemicellulose when grown in pure culture, but readily attacked cellulose when grown in the same medium with noncellulose decomposers such as *Rhizopus* and *Cunninghamella,* which successfully competed with *T. viride* for the other components of the growth medium. Lindsey (1965) compared different pairs of fungi for their ability to compete with each other. He found that *Fusarium solani* f. sp. *phaseoli* previously infested in glucose-amended autoclaved soil, colonized soil microbiological sampling tubes at a very low frequency in the presence of *Fusarium roseum* f. sp. *cerealis,* but this growth-suppressing effect was nullified by addition of nitrate nitrogen to the soil.

B. Antibiosis

In its broader sense antibiosis may be defined as the inhibition of one organism by a metabolite of another (Baker and Cook, 1974). Usually, this definition excludes common metabolic products, such as carboxylic acids, ethanol, and CO_2. Production of antibiotics on growth media is common among soil fungi, actinomycetes, and bacilli. However, the role of antibiosis in biological control, their production and ecological significance in average soil have prompted a considerable debate (Jackson, 1965; Alexander, 1971; Baker and Cook, 1974).

Antibiotics produced in or added to soil undergo inactivation as a result of degradation (Jackson, 1965) and adsorption on clay (Pinck, 1962) and on organic matter (Martin and Gottlieb, 1952). Still, the importance of antibiosis in the control of plant diseases has been demonstrated, especially in acidified or in partially sterilized soil. Weindling and Fawcett (1936) studied the role of *Trichoderma viride* in the biological control of *Rhizoctonia solani* in citrus seedbed. They found that *T. viride* actively parasitized the pathogen and suppressed its growth in acidified soil. Weindling and Emerson (1936) demonstrated the production by *T. viride* in growth media of the acid-stable antifungal antibiotic gliotoxin and a second antifungal antibiotic viridin, which was detected later (Brian and McGowan, 1945). This suggests the mediation of parasitism through antibiotic action (Garrett, 1965). Olsen and Baker (1968) found that among saprophytes which survived in soil after heat treatment were strains of *Bacillus subtilis* producing antibiotics effective against *R. solani.* In another study Weinhold and Bowman (1968) found that a strain of *B. subtilis,* which produced a subtilinlike antibiotic, was involved in the control of potato scab in soybean plots infested with *Streptomyces scabies.* Henis and Chet (1968) studied the effect of nitrogenous amendment on the germinability of sclerotia of *Sclerotium rolfsii* and on their accompanying microflora. A significant decrease of sclerotial germination in amended soil was observed. This decrease was correlated with increased numbers of antibiotic-producing organisms associated with sclerotia. Ethanol extracts of these sclerotia inhibited fungal growth and decreased germinability of nontreated sclerotia. It was suggested that effective amendments suppressed sclerotial germinability indirectly by increasing the antagonistic activity of the microorganisms at the soil-sclerotium interface. Hutchinson and Cowan (1972) identified volatile metabolites produced by *Trichoderma* which inhibited sporulation of *Aspergillus*. In laboratory experiments, *Azotobacter* was found to produce a thermolabile, ether-soluble fungistatic substance which significantly inhibited the linear growth and conidial germination of *Fusarium moniliforme* (Lakshmi-Kumari et al., 1972).

The involvement of antibiosis in microbial interactions between plant pathogens and the saprophytic microflora of the leaf has been demonstrated in several cases (Barnes, 1971).

Most convincing evidence concerning the role of antibiotics in biological control is presented in the excellent series of studies of Marx (1969a,b, 1973) and Marx and Davey (1969a,b). These authors studied the effect of ectotropic mycorrhizal fungi of shortleaf pine (*Pinus echinata* Mill.) on the resistance of pine roots to pathogenic infections. In agar plate tests, 5 ectotropic mycorrhizal fungi inhibited the growth of nearly half of 48 different fungal pathogens. *Leucopaxillus cerealis* var. *piceina* inhibited 92% of the test pathogens, and its culture filtrates were inhibitory to growth of *Phytophthora cinnamomi* and soil bacteria and completely inhibited zoospore germination of this pathogen. An antifungal and antibacterial antibiotic was extracted, isolated, and identified as diatryne nitrile. The ectotrophic mycorrhizal fungus produced diatryne antibiotics in mycorrhiza and rhizospheres of shortleaf pine seedlings in aseptic culture and prevented seedling infection by *Phytophthora cinnamomi*. However, in a later work Marx (1973) suggested that antagonistic microorganisms and inhibiting volatiles produced by the ectomycorrhiza rather than antibiotics were mainly involved in the control of *Phytophthora cinnamomi* with ectomycorrhiza formed by *Pisolithus tinctorius* or *Cenococcum graniforme*.

It seems that although the control of plant pathogens by antibiotic-producing microorganisms has not always been successful, there is still much evidence indicating that antibiotics may play an important role in microbiological equilibrium, especially in small locules immediately surrounding the active organisms (Alexander, 1961).

C. Lysis

Microorganisms capable of lysing other organisms are widespread in natural ecosystems (Webley and Jones, 1971). They seem to play an important role in microbial equilibrium in soil and can serve as a powerful tool for microbiological control. Counts of mycolytic bacteria have been reported within a range of 10^4 to 10^8 per gram of soil (Alexander, 1971).

Many strains of *Bacillus, Pseudomonas*, and *Streptomyces* isolated from soil are effective in lysing cell walls of pathogenic fungi under laboratory conditions (Skujinš and Potgieter, 1964; Chet and Henis, 1969; Horikoshi, 1959; Baker, 1968). In some cases, no correlation between susceptibility to lysis in culture and in soil could be observed (Lloyd *et al.*, 1965). Fungal mycelium can be digested in soil either by autolysis, i.e., through the self-digestion by intracellular enzymes, or by extracellular enzymes of mycolytic microorganisms, i.e., heterolysis (Lloyd and Lockwood, 1966). In this review we shall refer only to *heterolysis* as a possible mechanism in microbiological control.

Horikoshi and Iida (1958) and Horikoshi (1959) found that lytic enzymes produced by *Bacillus circulans* and *Streptomyces* sp. digested hyphal cell walls of *Aspergillus oryzae*. Mitchell (1963) and Mitchell and Alexander (1961, 1962, 1963) demonstrated that microorganisms can destroy fungal mycelium in soil through their lytic activity. They amended the soil with the fungal cell wall components chitin and laminarin and achieved a decrease in diseases caused by those soil-borne pathogens.

Lloyd *et al.* (1963) isolated 92 actinomycetes from soil, 88 of which dissolved the mycelium of *Glomerella cingulata*. In chitin-amended soil the lytic activity of the antagonistic microorganisms significantly increased, whereas in sterile soil no lytic activity was detected. Lloyd *et al.* (1965) suggested that lysis of *Glomerella* by chitinase-producing streptomycetes is a secondary process induced by toxic metabolites. Chitinase-containing cell-free extracts had no lytic activity on fungal cell walls. Khalifa (1965) obtained a 12% reduction in *Fusarium* wilt of peas by adding chitin to soil at the time of planting as compared to 45–25% reduction when chitin was applied 3–8 weeks before planting. He also observed an increase in *Fusarium*-antagonistic and lytic microorganisms following chitin amendment. Sneh and Henis (1972) extracted antifungal substance active against *Rhizoctonia solani* from chitin-amended soil. Thus, it is possible that both antibiosis and lysis are involved in the control of fungal pathogens by chitin amendments.

Chet *et al.* (1967) and Chet and Henis (1968) found β-(1,3)-glucan and chitin in both hyphal and sclerotial walls of *Sclerotium rolfsii*. Sclerotial walls but not hyphal walls contain melanin also. A *Pseudomonas* sp. capable of lysing *S. rolfsii* walls was isolated from soil by enrichment techniques and its lytic enzyme was identified as β-(1,3)-*endo*-glucanase (Chet and Henis, 1969). In this work, cell walls were prepared from hyphae and from the outer layers of sclerotia of *S. rolfsii* grown in the presence or absence of catechol (which induces melanin deposition in the hyphal walls) or in the presence of the sodium salt of ethylenediaminetetraacetic acid (Na_2EDTA), which prevents melanin synthesis by the sclerotia. The bacterial glucanase solubilized 25.2% and 6.3% of ordinary and melanin-containing hyphal walls and 5.9% and 10.3% of ordinary and melanin-free sclerotial walls, respectively. This work supported former indications that melanin acquires to fungal cell walls a resistance to microbial lysis (Potgieter and Alexander, 1966; Bloomfield and Alexander, 1967; Chu and Alexander, 1972).

It should be noted, however, that melanin is not the only factor that affects resistance of fungi to lysis. Ballesta and Alexander (1968, 1971) found that the cell walls of the non-melanin-containing fungus *Zygorhynchus vuillaminii* are resistant to lysis by mycolytic actinomycetes, apparently because of a high content of an unusual fucose-uronic acid polysaccharide. In another study (1972) these authors compared the susceptibility of several basidiomycetes, including the plant pathogens *Armillaria mellea* and *Fomes annosus* to enzymic preparations containing glucanase and chitinase. The most resistant portions of the walls of these fungi contained mannose and xylose as well as glucose, suggesting that their resistance to lysis may be associated with the presence of wall polysaccharides containing monomers in addition to glucose and N-acetylglucosamine.

D. GERMINATION AND SUBSEQUENT DECLINE OF RESTING STRUCTURES

The walls of fungal resting structures are more resistant to lytic enzymes than the vegetative hyphae (Alexander, 1971). Stimulating germination of these structures results in the exposure of the emerging hyphae to the lytic activity of soil microflora. Lysis following germination of dormant structures in the soil was first reported by Mitchell *et al.* (1941), who observed an elimination of *Phymatotrichum omnivorum* sclerotia following organic amendments. Menzies and Griebel (1967) reported an increased sensitivity to drought following sporulation of *Verticillium dahliae* microsclerotia. Germination and subsequent decline of *V. dahliae* microsclerotia in soil exposed to volatile substances from alfalfa were reported by Gilbert and Griebel (1969). Sequeira (1962) obtained an elimination of *F. oxysporum* f. *cubense* propagules in soil following addition of sucrose or glucose, and related this to enhanced germination of chlamydospores following exposure of the hyphal wall to lytic organisms, and to the formation of microspores, which are less resistant to lysis.

Adams and Papavizas (1969) studied the survival of *Thielaviopsis basicola* in natural and alfalfa-amended soil. Incorporation of ground alfalfa hay into soil stimulated germination of both endoconidia and chlamydospores, followed by lysis of germ tubes within 48 hours. They concluded that at least two mechanisms were involved in the biological control of *T. basicola* by alfalfa hay, i.e., an initial rapid germination followed by lysis of propagules and subsequent increase in soil fungistasis.

E. Parasitism and Predation

In spite of the large volume of data accumulated on this subject, not much progress has been made toward the practical application of parasitism and predation in biological control of plant pathogens (Baker and Cook, 1974). Much attention has been focused on pathogenic fungi and nematodes. The latter group, however, is not dealt with here.

1. Mycoparasitism

This term refers to the phenomenon of one fungus parasitic on another (Boosalis and Mankau, 1965). Host–parasite relationships among fungi have been recently reviewed by Barnett and Binder (1973). Attempts to use mycoparasitism to control soil-borne pathogens have so far been very disappointing, mainly because they did not give consistent results and were impractical on a commercial basis (Boosalis and Mankau, 1965). Baker and Cook (1974) emphasized the limited value of hyperparasitism (i.e., mycoparasitism) in controlling pathogens that infect rapidly by multiple primary infections from high densities of soil-borne propagules as well as primary infections by aerial pathogens. They have postulated that hyperparasites should be most effective against survival or secondary spread of pathogens, because of the longer contact with the hyperparasite. Numerous examples are available regarding invasion into and destruction by hyperparasites of fungal structures. *Rhizoctonia solani* and *Penicillium vermiculatum* penetrate and parasitize hyphae of many fungi (Boosalis and Mankau, 1965). Among sclerotial parasites identified by Karhuvaara (1960) were *Trichoderma viride* and *Gliocladium roseum* on *Sclerotinia trifoliorum*, *Penicillium frequentans* on *S. borealis*, and *Mucor hyemalis* on *Claviceps purpurea*. *Coniothyrium minitans* attacks *S. trifoliorum* (Tribe, 1957) and *S. sclerotiorum* (Jones and Watson, 1969) sclerotia, whereas *T. viride* can also parasitize sclerotia of *S. sclerotiorum* (Jones and Watson, 1969). *Trichoderma harzianum* rotted sclerotial of *Sclerotium rolfsii* (Wells et al., 1972). Recently, Coley-Smith et al. (1974) reported that dried sclerotia of *Sclerotium delphinii* were rotted in moist soil by *Trichoderma hamatum*.

Siegle (1961) found that severity of take-all was reduced by simultaneous inoculation of wheat seedlings with the pathogen *Gaeumannomy-*

ces graminis (=*Ophiobolus graminis*) and the hyperparasite *Didymella exitialis*, which also produced amino acids that decreased the pathogenicity of *G. graminis*. Examples of parasitic fungi attacking leaf pathogens include *Eudarluca caricis* on rusts (Eriksson, 1966) and *Cephalosporium* on *Drechslera teres* (*Helminthosporium teres*) (Kenneth and Isaac, 1964). Their potential use in biological control has not been evaluated yet.

2. Soil Fauna

Although predation seems to contribute to the disappearance of microorganisms from soil, its role in biological control is doubtful (Boosalis and Mankau, 1965). Fungi have been shown to be preyed upon by mites (Smith, 1960; Saichuae *et al.*, 1972), and protozoa feeding on bacteria are abundant in soil (Alexander, 1961). Despite extensive studies by early microbiologists, the role of protozoa in the biological equilibrium in the soil is not fully understood. It is not likely, however, that their feeding habit will be so specific as to result in the reduction in soil of any specific population of a plant pathogenic bacterium.

3. Parasites of Plant Pathogenic Bacteria

a. Bacteriophage. Phages that attack phytopathogenic bacteria are widespread in soil (Darpoux, 1960). However, attempts to control with phages bacterial diseases of plants have so far been unsuccessful (Boosalis and Mankau, 1965).

b. Bdellovibrio. The small comma-shaped bacterium *Bdellovibrio* is a parasite on other gram-negative bacteria (Stolp and Starr, 1963), and is probably widespread in soil. However, its role in microbial ecology, especially as a possible antagonist of plant pathogens, has not been evaluated yet. Recently, Scherff (1973) achieved a control of bacterial blight of soybean with a strain of *B. bacteriovorus* which had been isolated from the rhizosphere of soybean roots. The bacterial parasite inhibited development of local and systemic symptoms of bacterial blight when inoculated onto soybean with *Pseudomonas glycinea* at the ratios of 9:1 and 99:1, respectively. These experiments add to our hope to achieve economically feasible control of plant pathogen by a combined inoculation and manipulation of the conditions prevailing in the plant ecosystem.

V. Integrated Control

A. INTEGRATION WITH PESTICIDES

Selective pathogen suppression can be achieved by the use of specific chemicals, such as fungicides (Erwin, 1973), which are applied at relatively low dosages. However, because of their specificity, these compounds are liable to favor the development of tolerant mutants of the pathogen. Their integration with biological control might reduce this hazard. Moreover, some fungicides have been reported to control disease indirectly, through their effect on the microbiological equilibrium in the plant ecosystem. For example, according to McKeen (1949) the control of *Aphanomyces cochlioides* and *Pythium aphanidermatum* of sugar beets with thiram was due only in part to its direct effect on these pathogens. Pinckard (1970) observed a profound change in the soil microflora as a result of application of subinhibitory concentrations of the monosodium salt of hexachlorophene, which resulted in the appearance of antifungal bacterial antagonists in the treated soil. Stánková-Opočenska and Dekker (1970) obtained a protection of cucumber seedlings against damping off, caused by *Pythium debaryanum*, by soaking the seeds in a solution of 1 ppm of 6-azauracil (AzU), a systemic fungicide that does not affect this pathogen directly. They noted a decrease in the fungal population and an increase in bacterial count in the rhizosphere of the developing treated seeds, and concluded that the control of damping off was an indirect one, possibly via the microflora of the root. Chinn (1971) controlled root rot of wheat caused by *Helminthosporium sativum* with subinhibitory doses of methylmercury dicyandiamide, and attributed the control to the indirect effect of the fungicide on the saprophytic soil microflora.

B. Integration with Fumigants and Heat Treatments

In contrast to most other pesticides, soil treatment with fumigants, such as chloropicrin, methyl bromide or formalin, results in eradication of both the pathogen and most of the saprophytic microflora. The decision to use such an extreme treatment is affected by economic considerations. Therefore, these treatments are mostly used in catch crops and greenhouses. Their integration with biological control may increase their efficiency and reduce the amount needed for an efficient disease control, thus lowering the cost and allowing their use in less profitable crops. Another advantage that lies in the integration of fumigation with biological control is the prevention of a rapid spread of the pathogen in the reinfested soil, in the absence of the antagonistic soil microflora. In the case of the control of *Armillaria mellea* by CS_2 treatment in orchards, the already existing natural fungal antagonists, especially *Trichoderma viride* which survives this treatment, increase in the treated soil (Bliss, 1951; Garrett, 1970; Wilhelm, 1973) and control the pathogen. Similarly, in soil treated with aerated steam the saprophytic spore-forming soil bacilli are not eradicated, multiply rapidly after treatment, and lessen the hazard of reinfestation (Baker and Cook, 1974). A part of this control has been attributed by Smith and Cook (1974) to the formation of fungistatic concentrations of ethylene by surviving soil clostridia, which may be enhanced by the addition of organic amendments.

C. Multiple Integration

Analysis of the cases reported so far on the successful use of biological means in controlling plant diseases leads to the conclusion that, in most cases, only a partial control has been achieved. This is not surprising in view of the complex nature and the dependence of biological control on so many biotic and abiotic factors. So far, biological control as a single treatment has not replaced other means of control, but may be rather considered as an important complement of these means, which include proper cultivation practices, use of disease-resistant varieties, reduction of inoculum by using pathogen-free plants and seeds, soil sanitation, fumigation, and specific suppression of the pathogen by highly selective pesticides. Papavizas (1973) relates to integrated control as "the approach where all methods of biological control may be brought into operation to reduce pathogenic activities to a tolerable permissible level, with chemicals applied only when absolutely necessary." It is hoped that further research along this line will help farmers to increase the production of food for the benefit of mankind.

Acknowledgment

The authors wish to express grateful appreciation to their colleague Dr. J. Katan and Professor R. Mitchell for their valuable suggestions and criticism during preparation of this review.

References

Adams, D. B., and Papavizas, G. C. (1969). *Phytopathology* **59**, 135–138.
Aldrich, J., and Baker, R. (1970). *Plant Dis. Rep.* **54**, 446–448.
Alexander, M. (1961). "Introduction to Soil Microbiology." Wiley, New York.
Alexander, M. (1971). "Microbial Ecology." Wiley, New York.
Anagnostakis, S. L., and Jaynes, R. A. (1973). *Plant Dis. Rep.* **54**, 225.
Assa, Y., Gestetner, B., Chet, I., and Henis, Y. (1972). *Life Sci.* **11**, 637–647.
Baker, K. F., and Cook, R. J. (1974). "Biological Control of Plant Pathogens." Freeman, San Francisco, California.
Baker, K. F., and Synder, W. C., eds. (1965). "Ecology of Soil-borne Plant Pathogens." Univ. of California Press, Berkeley,
Baker, R. (1968). *Annu. Rev. Phytopathol.* **6**, 263–294.
Ballesta, J. P. G., and Alexander, M. (1968). *Bacteriol. Proc. p.* G119.
Ballesta, J. P. G., and Alexander, M. (1971). *J. Bacteriol.* **106**, 938–945.
Ballesta, J. P. G., and Alexander, M. (1972). *Trans. Brit. Mycol. Soc.* **58**, 481–487.
Barnes, G. (1971). *In* "Ecology of Leaf Surface Micro-Organisms" (T. F. Preece and C. H. Dickinson, eds.), pp. 557–565. Academic Press, New York.
Barnett, H. I., and Binder, F. I. (1973). *Annu. Rev. Phytopathol.* **11**, 273–292.

Bateman, D. F. (1964). *Phytopathology* **54**, 438–445.
Bateman, D. F., and Beer, S. V. (1965). *Phytopathology* **55**, 204–211.
Blair, I. D. (1943). *Ann. Appl. Biol.* **30**, 118–127.
Blakeman, J. P., and Frazer, A. K. (1971). *Physiol. Plant Pathol.* **1**, 45–54.
Bliss, D. E. (1951). *Phytopathology* **41**, 665–683.
Bloomfield, B. J., and Alexander, M. (1967). *J. Bacteriol.* **93**, 1276–1280.
Boosalis, M. G., and Mankau, R. (1965). *In* "Ecology of Soil-borne Plant Pathogens" (K. F. Baker and W. C. Snyder, eds.), pp. 374–391. Univ. of California Press, Berkeley.
Brian, P. W., Hemming, H. G., and McGowan, J. C. (1945). *Nature (London)* **155**, 637–638.
Broadbent, L. (1964). *Ann. Appl. Biol.* **54**, 209–224.
Brown, M. E., Jackson, R. M., and Burlingham, S. K. (1968). *In* "The Ecology of Soil Bacteria" (T. R. Gray and D. Parkinson, eds.), pp. 531–551. Liverpool Univ. Press, Liverpool.
Bumbieris, M., and Lloyd, A. B. (1967). *Aust. J. Biol. Sci.* **20**, 1169–1172.
Chet, I., and Henis, Y. (1968). *Can. J. Microbiol.* **14**, 815–816.
Chet, I., and Henis, Y. (1969). *Soil Biol. & Biochem.* **1**, 131–138.
Chet, I., Henis, Y., and Mitchell, R. (1967). *Can. J. Microbiol.* **13**, 137–141.
Chinn, S. H. F. (1971). *Phytopathology* **61**, 98–101.
Chu, S. B., and Alexander, M. (1972). *Trans. Brit. Mycol. Soc.* **58**, 489–497.
Coley-Smith, J. R., Ghaffar, A., and Javed, Z. U. R. (1974). *Soil Biol. & Biochem.* **6**, 307–312.
Cook, R. J., and Watson, R. E., eds. (1970). *Wash., Agr. Exp. Sta., Bull.* **716**, 1–32.
Cordon, T. C., and Haenseler, C. M. (1939). *Soil Sci.* **47**, 207–215.
Darpoux, H. (1960). *In* "Plant Pathology" (J. G. Horsfall and A. E. Dimond eds.), Vol. 3, pp. 521–565. Academic Press, New York.
Davey, C. B., and Papavizas, G. C. (1960). *Phytopathology* **50**, 522–525.
Davey, C. B., and Papavizas, G. C. (1963). *Soil Sci. Soc. Amer., Proc.* **24**, 164–168.
Davis, D. (1968). *Phytopathology* **58**, 121–122.
Dennis, C., and Webster, J. (1971). *Trans. Brit. Mycol. Soc.* **57**, 25–30 and 41–48.
Dhingra, O. D., and Khare, M. C. (1973). *Phytopathology* **76**, 23–29.
Dobbs, C. G., and Hinson, W. H. (1953). *Nature (London)* **172**, 197.
Duddington, C. L., and Wyborn, C. H. E. (1972). *Bot. Rev.* **38**, 545–565.
Dunleavy, J. M. (1952). *Phytopathology* **42**, 465 (abstr.).
Eriksson, O. (1966). *Bot. Notis.* **119**, 33–69.
Erwin, D. C. (1973). *Annu. Rev. Phytopathol.* **11**, 389–422.
Fedorinchik, N. S. (1971). *Mikro. Fitopatol.* **5**, 699–705.
Garrett, S. D. (1956). "Biology of Root Infecting Fungi." Cambridge Univ. Press, London and New York.
Garrett, S. D. (1965). *In* "Ecology of Soil-borne Plant Pathogens" (K. F. Baker and W. C. Snyder, eds.), pp. 4–47. Univ. of California Press, Berkeley.
Garrett, S. D. (1970). "Pathogenic Root Infecting Fungi." Cambridge University Press, London and New York.
Gilbert, R. G., and Griebel, G. E. (1969). *Phytopathology* **59**, 1400–1403.
Gillespie, D. C., and Cook, F. D. (1965). *Can. J. Microbiol.* **11**, 109–113.
Gilpatrick, J. D. (1969). *Phytopathology* **59**, 973–978.
Goodwin, W. E., Salmon, J., and Ware, W. M. (1929). *J. Agr. Sci.* **4**, 151–156.
Grosclaude, C., Richards, J., and Dubos, B. (1973). *Plant Dis. Rep.* **57**, 25–28.
Grover, R. K., and Sidhu, J. S. (1966). *Sydowia* **19**, 231–239.
Haska, G., and Norén, B. (1967). *Physiol. Plant.* **20**, 851–861.

Henis, Y., and Chet, I. (1967). *Phytopathology* **57**, 425–427.
Henis, Y., and Chet, I. (1968). *Phytopathology* **58**, 209–211.
Henis, Y., and Katan, J. (1975). *In* "Biology and Control of Soil-Borne Plant Pathogens" (G. W. Bruehl, ed.), pp. 100–106. The American Phytopathological Society, St. Paul, Minn.
Henis, Y., Sneh, B., and Katan, J. (1967). *Can. J. Microbiol.* **13**, 643–656.
Hocking, D., and Cook, F. D. (1972). *Can. J. Microbiol.* **18**, 1557–1560.
Horikoshi, K. (1959). *Nature (London)* **183**, 186–187.
Horikoshi, K., and Iida, S. (1958). *Nature (London)* **181**, 917–918.
Hsieh, S. P. Y., and Buddenhagen, I. W. (1974). *Phytopathology* **64**, 1182–1185.
Hsu, S. C., and Lockwood, J. L. (1973). *Phytopathology* **63**, 334–337.
Huber, D. M. (1972). *Phytopathology* **62**, 434–436.
Huber, D. M., Watson, R. D., and Steiner, G. W. (1965). *Soil Sci.* **100**, 302–308.
Hutchinson, S. A., and Cowan, M. E. (1972). *Trans. Brit. Mycol. Soc.* **59**, 71–77.
Jaarsveld, A. (1942). *Phytopathol. Z.* **14**, 1–45.
Jackson, R. M. (1965). *In* "Ecology of Soil-borne Plant Pathogens" (K. F. Baker and W. C. Snyder, eds.), pp. 363–373. Univ. of California Press, Berkeley.
Jones, D., and Watson, D. (1969). *Nature (London)* **224**, 287–288.
Karhuvaara, L. (1960). *Acta Agr. Scand.* **10**, 127–134.
Katan, J., and Eshel, Y. (1973). *Residue Rev.* **45**, 145–177.
Kenneth, R., and Isaac, P. K. (1964). *Can. J. Plant Sci.* **44**, 182–187.
Kerr, A. (1965). *In* "Ecology of Soil-borne Plant Pathogens" (K. F. Baker and W. C. Snyder, eds.), pp. 248–251. Univ. of California Press, Berkeley.
Kerr, A. (1972). *J. Appl. Bacteriol.* **35**, 493–497.
Kerr, A., and Bumbieris, M. (1969). *Annu. Rep., Waite Agr. Res. Inst. (S. Aust.)* pp. 77–78.
Khalifa, O. (1965). *Ann. Appl. Biol.* **56**, 129–137.
Ko, W. H., and Lockwood, J. L. (1967). *Phytopathology* **57**, 894–901.
Krupa, S., and Fries, N. (1971). *Can. J. Bot.* **49**, 1425–1431.
Lakshmi-Kumari, M., Vijayalakshmi, K., and Subbarao, N. S. (1972). *Phytopathol. Z.* **75**, 27–30.
Lemaire, J. M., Lapierre, H., and Jouan, B. (1970). *C.R. Acad. Agr. Fr.* **56**, 1134–1137.
Lemke, P. A., and Nash, C. H. (1974). *Bacteriol. Rev.* **38**, 29–56.
Lewis, J. A., and Papavizas, G. C. (1971). *Phytopathology* **61**, 208–214.
Linderman, R. G., and Gilbert, R. G. (1973). *Phytopathology* **63**, 359–362.
Lindsey, L. O. (1965). *Phytopathology* **55**, 104–110.
Lloyd, A. B., and Lockwood, J. L. (1966). *Phytopathology* **56**, 595–602.
Lloyd, A. B., Noveroske, R. I., and Lockwood, J. L. (1963). *Phytopathology* **53**, 881 (abstr.).
Lloyd, A. B., Noveroske, R. I., and Lockwood, J. L. (1965). *Phytopathology* **55**, 871–875.
McGregor, A. J., and Wilson, G. C. S. (1966). *Plant Soil* **25**, 3–16.
McKeen, W. E. (1949). *Can. J. Res.* **27**, 284–311.
Maier, C. R. (1968). *Phytopathology* **58**, 620–625.
Manning, W. J., and Crossan, D. F. (1969). *Plant Dis. Rep.* **53**, 227–231.
Marshall, K. C., and Alexander, M. (1954). *Plant Soil* **12**, 143–153.
Martin, N., and Gottlieb, D. (1952). *Phytopathology* **42**, 294–296.
Marx, D. H. (1969a). *Phytopathology* **59**, 153–163.
Marx, D. H. (1969b). *Phytopathology* **59**, 411–417.
Marx, D. H. (1973). *Phytopathology* **63**, 18–23.
Marx, D. H., and Davey, C. B. (1969a). *Phytopathology* **59**, 549–558.

Marx, D. H., and Davey, C. B. (1969b). *Phytopathology* **59**, 559–565.
Matthews, R. E. F. (1970). "Principles of Plant Virology." Academic Press, New York.
Maurer, C. L., and Baker, A. (1965). *Phytopathology* **55**, 69–72.
Mehrotra, R. S., and Caludius, G. R. (1972). *Plant Soil* **37**, 657–664.
Menzies, J. D. (1959). *Phytopathology* **49**, 648–652.
Menzies, J. D., and Gilbert, R. G. (1967). *Soil Sci. Soc. Amer., Proc.* **31**, 495–496.
Menzies, J. D., and Griebel, G. E. (1967). *Phytopathology* **57**, 703–709.
Michael, A. A., and Nelson, P. E. (1972). *Phytopathology* **62**, 1052–1056.
Mishustin, E. N., and Naumova, A. N. (1962). *Microbiology (USSR)* **31**, 543–555.
Mitchell, J. E. (1973). *Soil Biol. & Biochem.* **5**, 721–728.
Mitchell, R. (1963). *Phytopathology* **53**, 1068–1071.
Mitchell, R., and Alexander, M. (1961). *Nature (London)* **190**, 109–110.
Mitchell, R., and Alexander, M. (1962). *Soil Sci. Soc. Amer., Proc.* **26**, 556–558.
Mitchell, R., and Alexander, M. (1963). *Can. J. Microbiol.* **9**, 169–177.
Mitchell, R. B., Hooton, D. R., and Clark, F. E. (1941). *J. Agr. Res.* **63**, 535–547.
Mordvedt, J. J., Fleischfresser, M. H., Berger, K. C., and Darling, H. M. (1961). *Amer. Potato J.* **38**, 95–100.
Morquer, R., and Touvet, A. (1972). *C.R. Acad. Sci., Ser. D.* **274**, 234–239.
New, P. B., and Kerr, A. (1972). *J. Appl. Bacteriol.* **35**, 279–287.
Newcombe, M. (1960). *Trans. Brit. Mycol. Soc.* **43**, 51–59.
Norris, J. R. (1971). *In* "Microbes and Biological Productivity" (D. H. Hughes and A. H. Rose, eds.), pp. 197–229. Cambridge Univ. Press, London and New York.
Olsen, C. M., and Baker, K. F. (1968). *Phytopathology* **58**, 79–87.
Owens, L. D., Gilbert, R. G., Griebel, G. E., and Menzies, J. D. (1969). *Phytopathology* **59**, 1468–1472.
Papavizas, G. C. (1963). *Phytopathology* **53**, 1430–1436.
Papavizas, G. C. (1966). *Phytopathology* **56**, 1071–1075.
Papavizas, G. C. (1967). *Plant Dis. Rep.* **51**, 125–129.
Papavizas, G. C. (1970). *In* "*Rhizoctonia solani*: Biology and Pathology" (J. A. Parmeter, Jr., ed.), pp. 108–122. Univ. of California Press, Berkeley.
Papavizas, G. C. (1973). *Soil Biol. & Biochem.* **5**, 709–720.
Papavizas, G. C., and Davey, C. B. (1960). *Phytopathology* **50**, 516–522.
Papavizas, G. C., Lewis, Z. A., and Adams, P. B. (1970). *Plant Dis. Rep.* **54**, 114–118.
Park, D. (1960). *In* "The Ecology of Soil Fungi" (D. Parkinson and J. S. Waid, eds.), pp. 148–159. Liverpool Univ. Press., Liverpool.
Parmeter, J. A., Jr. (1970). "*Rhizoctonia solani*: Biology and Pathology." Univ. of California Press, Berkeley.
Peterson, E. A., Katznelson, H., and Cook, F. D. (1965). *Can. J. Microbiol.* **11**, 595–596.
Pinck, L. A. (1962). *Clays Clay Miner.* **9**, 520–529.
Pinckard, J. A. (1970). *Phytopathology* **60**, 1308 (abstr.).
Potgieter, H. J., and Alexander, M. (1966). *J. Bacteriol.* **91**, 1526–1537.
Preece, T. F., and Dickinson, C. H., eds. (1971). "Ecology of Leaf Surface Micro-Organisms." Academic Press, New York.
Rishbeth, J. (1963). *Ann. Appl. Biol.* **52**, 63–71.
Sadasivan, T. S. (1965). *In* "Ecology of Soil-borne Plant Pathogens" K. F. Baker and W. C. Snyder, eds.), pp. 460–469. Univ. of California Press, Berkeley.
Saichuae, P., Gerson, U., and Henis, Y. (1972). *Soil Biol. & Biochem.* **4**, 155–164.
Savage, D. C. (1972). *In* "Microbial Pathogenicity in Man and Animals" (H. Smith and J. H. Pearce, eds.), pp. 25–57. Cambridge Univ. Press, London and New York.

Scherff, R. N. (1973). *Phytopathology* **63**, 400–402.
Sequeira, L. (1962). *Phytopathology* **52**, 976–982.
Sequeira, L. (1963). *Phytopathology* **53**, 332–336.
Sewell, G. W. F. (1965). In "Ecology of Soil-borne Plant Pathogens" (K. F. Baker and W. C. Snyder, eds.), pp. 479–495. Univ. of California Press, Berkeley.
Shipton, P. J., Cook, R. J., and Sitton, J. W. (1973). *Phytopathology* **63**, 511–517.
Siegle, H. (1961). *Phytopathol. Z.* **42**, 305–348.
Singh, R. N. (1972). *Mycopathol. Mycol. Appl.* **46**, 147–248.
Singh, R. S., and Pande, K. R. (1965). *Indian Exp. Biol.* **3**, 146–147.
Skujinš, J. J., and Potgieter, H. (1964). *Bacteriol. Proc.* p. G98.
Smith, A. M., and Cook, R. J. (1974). Cited in Baker and Cook (1974, p. 342).
Smith, G. (1960). "Industrial Mycology." Arnold, London.
Sneh, B., and Henis, Y. (1972). *Phytopathology* **62**, 595–600.
Sneh, B., Katan, J., and Henis, Y. (1971). *Phytopathology* **61**, 1113–1117.
Snyder, W. C., Schroth, M. N., and Christou, T. (1959). *Phytopathology* **49**, 755–756.
Stánková-Opočenska, E., and Dekker, J. (1970). *Neth. J. Plant. Pathol.* **76**, 152–158.
Stolp, H., and Starr, M. P. (1963). *Anthonie van Leewenhoek; J. Microbiol. Serol.* **29**, 217.
Stotzky, G., and Martin, R. T. (1963). *Plant Soil* **18**, 317–338.
Stotzky, G., and Post, A. H. (1967). *Can. J. Microbiol.* **13**, 1–7.
Stotzky, G., and Rem, L. T. (1966a). *Can. J. Microbiol.* **12**, 547–563.
Stotzky, G., and Rem, L. T. (1966b). *Can. J. Microbiol.* **12**, 831–848.
Stotzky, G., and Rem, L. T. (1966c). *Can. J. Microbiol.* **12**, 1235–1246.
Sztejnberg, A., and Blakeman, J. P. (1973). *J. Gen. Microbiol.* **75**, 15–22.
Tribe, H. T. (1957). *Trans. Brit. Mycol. Soc.* **40**, 489–499.
Vojinović, Ž. D. (1973). *Org. Eur. Med. Protein Plantes Bull.* **9**, 91–101.
Waksman, S. A., and Hutchings, I. J. (1937). *Soil Sci.* **43**, 77–92.
Webley, D. M., and Jones, D. (1971). *Soil Biochem.* **12**, 446–485.
Webley, D. M., Follet, E. A. C., and Taylor, I. F. (1967). *Antonie van Leeuwenhoek; J. Microbiol. Serol.* **33**, 159.
Weindling, R. (1932). *Phytopathology* **22**, 837–848.
Weindling, R., and Emerson, O. H. (1936). *Phytopathology* **26**, 1168–1070.
Weindling, R., and Fawcett, H. S. (1936). *Hilgardia* **10**, 1–16.
Weinhold, A. R., and Bowman, T. (1968). *Plant Soil* **28**, 12–24.
Weinhold, A. R., Oswald, J. W., Bowman, T., Bishop, J., and Wright, T. (1964). *Amer. Potato J.* **41**, 265–273.
Wells, H. C., Bell, D. K., and Jaworsky, C. A. (1972). *Phytopathology* **62**, 442–447.
Wilhelm, S. (1973). *Soil Biol. & Biochem.* **5**, 729–737.
Wingfield, E. B. (1968). Ph.D. Thesis, Virginia Polytechnic Institute, Blacksbury.
Woltz, S. S., and Jones, J. P. (1968). *Phytopathology* **58**, 336–338.
Wood, R. K. S. (1951). *Ann. Appl. Biol.* **38**, 217–230.
Wood, R. K. S., and Tveit, M. (1955). *Bot. Rev.* **21**, 441–491.
Zentmyer, G. A. (1963). *Phytopathology* **53**, 1383–1387.
Zentmyer, G. A., and Bingham, F. T. (1956). *Phytopathology* **46**, 121–124.
Zentmyer, G. A., and Thompson, C. R. (1967). *Phytopathology* **57**, 1278–1279.

Microbiology of Municipal Solid Waste Composting[1]

MELVIN S. FINSTEIN AND MERRY L. MORRIS

*Department of Environmental Science, Cook College,
Rutgers University, New Brunswick, New Jersey*

I.	Introduction	113
II.	Self-Heating Variations	116
	A. Occurrence and Processes	116
	B. Moisture and the Origin of the Heat	117
III.	The Temperature Ascent	121
	A. Microbe–Temperature Interactions	121
	B. Fungi	127
	C. Actinomycetes	130
	D. Bacteria	132
IV.	The Temperature Descent	134
	A. Temperature	134
	B. Organisms	135
	C. Inoculation	136
V.	Batch and Continuous Composting	138
VI.	Operational Factors	141
	A. Carbon/Nitrogen Ratio	141
	B. pH	141
	C. Interactions among Moisture, Aeration, and Temperature	143
VII.	Conclusion	148
	References	148

I. Introduction

Composting is the microbial degradation of organic solid (nonaqueous) material that involves aerobic respiration and passes through a thermophilic stage. It yields the stabilized end-product compost. Some introductory remarks of a quasi-technical nature are intended to put this process into perspective as a means of treating municipal solid wastes. Discussions are available on the components of municipal waste which are compostable (Francis, 1967; Anonymous, 1970). For a general discussion of the solid waste literature, see Golueke (1973a).

The following solid waste management objectives can be achieved through composting. *Sanitation:* Pathogenic microorganisms, larvae, and weed seeds are killed rapidly during composting. Nuisances, such as odors, insect breeding, and vermin harborage, can be prevented through the rapid decomposition of the putrescible organic materials. *Mass and bulk reduction:* Part of the organic carbon is mineralized to carbon dioxide and lost from the material. The bulk density of compost is higher than that of the starting material (Schulze, 1965); therefore compost

---

[1] Paper of the Journal Series, New Jersey Agricultural Experiment Station.

makes a more physically stable landfill than the original material. *Resource recovery:* Most uses for compost are agronomic. The process affords a means of returning to the soil mineral plant nutrients in organic form that might otherwise be lost. Since composting narrows the C/N ratio, adequate processing reduces the likelihood of damaging crop or ornamental plants through the microbial immobilization of soil nitrogen. Compost produced explicitly for reasons of soil husbandry, usually from crop residues and animal excreta, is termed artificial manure. Municipal solid waste compost was used successfully as a livestock feed in preliminary trials (McClure *et al.*, 1970). Nonagricultural uses that have been proposed for compost include the manufacture of low-grade wallboard (Malin, 1971) and use as an absorbant to clean up oil spills (Vaux *et al.*, 1972).

In poor, densely populated areas lacking water carriage sewerage systems, the disposal of human excrement and the preparation of manures through composting are perfectly complementary. Two critical needs, sanitation and nutrient conservation, can thereby be satisfied with hand labor. Such conditions in India led to the development in the 1920s of the Indore Process (Howard, 1940), which is still widely used (Bond and Straub, 1973). Although this can be taken as the first modern process, there are earlier records of systematized composting (Boussingault, 1845).

In Western societies, however, most large-scale composting has until recently been associated with agricultural rather than domestic wastes. The process finds more use in the intensive agriculture of Europe, where nutrient conservation and the maintenance of soil structure tend to be emphasized, than in the extensive agriculture characteristic of America. With regard to the purchase of the major plant nutrients, compost is generally noncompetitive with inorganic fertilizer as a source of nitrogen, phosphorous, and potassium (Toth, 1968; Mays *et al.*, 1973). The cost per unit element in compost is high, and application of this bulky material is troublesome. Few commercial farmers consider the benefits derived from the organic matter sufficient to overcome the economic disadvantages.

Compost plays a unique role in certain specialized practices, including hotbed gardening, which requires self-heating organic matter, and as a substrate for edible-mushroom cultivation. For these purposes compost derived from municipal solid waste can substitute for the traditional horse manure preparation (Block, 1965; Anonymous, 1968). Compost can be an economic soil conditioner for high-value crops, such as vegetables and flowers grown out of season.

Unlike wastes from agriculture and industry, which are potentially uniform, the composition of municipal solid waste cannot be controlled in detail. Consequently there is a degree of unpredictability regarding possible toxic effects. Sewage sludge, which is commonly combined with waste for composting, can be a source of heavy metals (Anonymous, 1973a; Peterson *et al.*, 1973). Mushroom spawn has been employed as an indicator of compost toxicity (Francis, 1967). On the positive side of this ledger, compost produced from municipal waste moderated metal toxicity in soil (Terman *et al.*, 1973), and some pesticides decomposed rapidly during the high-temperature stage of the process (Rose and Mercer, 1968).

Conservation-minded people have long advocated the large-scale composting of solid waste even though the economics have been distinctly unfavorable. Most such enterprises in the United States have been privately financed, in contrast to the public financing of municipal sewage treatment facilities. This crucial difference accounts in large measure for the outcome of many composting ventures—that of economic failure (Bond and Straub, 1973). Whereas income from treated effluent and sludge, in the rare cases where it exists, is incidental to the continued operation of wastewater treatment plants, the treatment of solid waste by composting necessarily emphasizes the production of a salable product. The market for compost has proved to be limited.

In the island of Jersey (United Kingdom) the use of compost in agriculture is subsidized (Francis, 1967). Utilization of compost for surface mine reclamation (Hortenstine and Rothwell, 1972; Scanlon *et al.*, 1973) seems unlikely to be carried out on a large scale in the United States without public financing.

Both the process and the product are popular in the home garden, where economic considerations are secondary. Unfortunately, enthusiasm for compost and other things organic sometimes reaches cultic proportions. An associated mystical element comes from the commonly held mistaken belief that special inoculums are essential for successful composting. Testimonials for commercial inoculums abound, but the bulk of the evidence is negative, as discussed in Section IV,C. In that section, nevertheless, the possibility of speeding the process through rationally timed manipulations related to inoculation will be explored. Wylie (1960) and Golueke (1972) rightly suggested that noncritical approaches to composting may have rendered a disservice to the promotion of its use as a process for the treatment of solid waste.

Composting has had a poor reputation among engineers and administrators in the solid waste field, who have tended to dismiss it as idealistic. This attitude stems from the history of financial difficulties and also, less justly, from associating the process with organic extremists. Reliable data are scarce (Golueke, 1973a), but there is no doubt that in the United States most municipal solid waste goes to the landfill, some in the form of incinerator residue. Numerous environmental constraints, however, are forcing rapid changes in the economics of solid waste management. Perhaps the potential usefulness of composting lies somewhere between the expectations of its uncritical advocates and those of its detractors.

There are signs of a reevaluation of the usefulness of composting under contemporary conditions. Among these may be noted a new receptivity on the part of decision makers to means of extending the life of landfills (Mason, 1969), the inclusion of composting as one of the components of a comprehensive solid waste recovery project (Franz, 1972b), increasing popularity of leaf composting on a municipal scale (Finstein and Arent, 1974), and the entrance of an industrial giant into the composting machinery market (Franz, 1972a). A process related to composting is being considered for the treatment of aqueous sewage sludge and animal excreta (Pöpel and Ohnmacht, 1972). It is appreciated that composting is a presently available technology for treating organic solid wastes that, properly managed, produces little pollutional side effect.

Existing accounts of composting emphasize operational aspects (Gotaas, 1956; Anonymous, 1971; Goleuke, 1972; Poincelot, 1972). The present chapter concentrates on the microbiology of the process and related topics exclusive of the problem of sanitation, which has been reviewed (Goleuke and Gotaas, 1954; Wiley, 1962a; Krige, 1964; Wiley and Westerberg, 1969). Although the composting of municipal solid waste is our point of reference, it would be restrictive (even if it were possible) to discuss only this type of starting material. Self-heating in contexts other than composting will be consulted where helpful.

II. Self-Heating Variations

A. Occurrence and Processes

In most ecosystems the heat liberated biologically (and nonbiologically) is too slight and is dissipated too quickly to cause significant temperature increases. Masses of decomposing organic materials are notable exceptions, within which intensive heat production and effective retention may coincide to bring about a temperature rise. Self-heating, as it is called, occurs spontaneously when organic materials are assembled provided there is sufficient mass for insulation, and that moisture, aeration, and nutrition are adequate.

Self-heating sometimes accompanies the spoilage of insufficiently dried hay, grain, wool, and other agricultural products. That it can lead to spontaneous combustion is widely appreciated. Mild self-heating is usefully employed in the processing of tobacco and cacao and for the upgrading of fodder (Cooney and Emerson, 1964). It was utilized experimentally in the retting of guayule for rubber production. Composting is the higher-temperature self-heating process variant used for the treatment of solid wastes.

B. MOISTURE AND THE ORIGIN OF THE HEAT

The water status of compost is usually evaluated in terms of the moisture content and is expressed on a wet weight basis. This basis of expression will be used herein. Although moisture content is a useful operational parameter, it does not describe the microbial growth environment in any fundamental way. As in other nonaqueous environments, in compost microbial growth actually takes place in aqueous solution. In food research, water activity (a_w), which is a fundamental property of aqueous solution, has become accepted as providing the best description of the water status (Scott, 1957; Christian, 1963). When the aqueous and gaseous phases are in equilibrium, a_w and relative humidity (RH) are equivalent, and it is the RH which has been measured in the interstitial atmosphere of compostlike systems (Section II,B,2) although apparently not in compost per se.

1. Heating through Biological Action

There is general agreement that almost all the heat liberated biologically in self-heating masses comes from the respiration of microorganisms attacking the organic substrates (Browne, 1929, 1933). With adequate moisture, plant cells may contribute slightly during the very earliest stages of the temperature rise while they remain functional, but this is quickly overshadowed by microbial heat production. In municipal solid wastes, where active plant cells are sparse compared to agricultural products, nonmicrobial heat output must be very slight.

When there is sufficient water for unrestricted growth, and other factors are favorable, self-heating masses commonly exceed 70°C within 2 or more days. Such a temperature rise is observed frequently in field-scale composts when the water content is within the range of 50% to 60% (Poincelot, 1972). Evidence that biological action can cause the temperature to rise to a maximum of 76°C comes in the form of a negative coefficient of heating reaction in wet wool over the 60°C to 78°C range (Rothbaum, 1961). Heat output declined at 74°C relative to lower temperatures, and there was none at 78°C. Nine strains of aerobic spore-forming thermophiles examined by Edwards and Rettger (1937) displayed growth maxima that fall between these values (75.6°C ± 1.1°C). Sterile wool maintained at 60°C gave a negligible heat output, which became appreciable after seeding (Rothbaum, 1961). Switching from air to oxygen when the wool was at 60°C reduced the heat output, a result that is reconcilable with biological activity but not chemical oxidations. Additional evidence linking microorganisms to self-heating will be found in other parts of this chapter.

Wool samples from one source were unusual in self-heating adiabatically (see Section III,A,2) to a maximum of 79.4°C ± 0.3°C under conditions suggestive of biological action (Walker and Harrison, 1960). This material demonstrated a negative coefficient of heating reaction between 78°C and 81°C. (A positive coefficient, indicative of nonbiologically mediated chemical reactions, was obtained over the 85°C to 93°C range.) Limited attempts to culture organisms potentially responsible for the heating to 79°C were unsuccessful (but see Brock and Freeze, 1969). There are scattered reports of composts heating to 80°C or 81°C (Straub, 1950; Glathe, 1959; von Klopotek, 1962; Schulze, 1965). Kane and Mullins (1973) reported that the average of three or more peak temperature observations was 78°C. The possibility that chemical reactions contributed to the attainment of such temperatures cannot be excluded, and some of the composts were exposed to the heat of sunlight. Nevertheless, coincidence seems to be an unlikely explanation for these scattered observations that have in common a peak of approximately 80°C, and this temperature may represent the limit to biological self-heating in organic masses.

2. Heating through Chemical Reactions

Temperatures above those attributable to biological action can be reached through exothermic chemical reactions, possibly leading to spontaneous combustion. The investigations of Walker and Harrison (1960), Rothbaum (1963), Dye and Rothbaum (1964), and Rothbaum and Dye (1964) have significantly clarified the conditions that permit self-heating to pass from a biological to a chemical causation. Two requirements involving water must be met. There must be sufficient water for a prior biologically medicated temperature rise to a point where the heat liberated through chemical reactions becomes appreciable. Yet the material cannot be so wet as to impart such a high thermal conductivity that the rate of heat loss from the mass equals the rate of its production. The rate of heat liberation through chemical reactions is slow until very high temperatures are reached. Both conditions were met in hay heating adiabatically when it was in equilibrium with interstitial atmosphere relative humidities of from 95% to 97% (Rothbaum, 1963). In wool the critical range was 94% RH to 98% RH (Dye and Rothbaum, 1964).

Measurements of RH in compost are lacking. However, judging from the usual moisture contents and the characteristic rapid approach to peak biological temperatures, it can be inferred that RH values are generally very close to 100% during the temperature ascent. Measurably unsaturated atmospheres are more likely to occur later in the process, after biological activity has subsided and the compost has entered the curing stage. Although no report was found of spontaneous combusion in compost, proposals to stockpile such material (Golueke, 1973b) should take this possibility into account. The small size of most compost stacks may preclude temperature increases much beyond the biological maximum. Browne (1929) and Rothbaum (1963) have discussed the relation between haystack size and spontaneous combustion, and Bowes (1956) has reviewed this subject with regard to sawdust heaps. There is at least one documented case of spontaneous combustion in stable manure (James et al., 1928). Dump and landfill fires are often ascribed to spontaneous combustion, but no supportive evidence was found in the literature.

3. Bacteria and Fungi in Damp Materials

Scott (1957) has organized the literature which puts into quantitative terms the commonplace observation that fungi grow better than bacteria in dry materials. Useful briefer discussions may be found elsewhere (Rothbaum and Stone, 1961; Christian, 1963). Optimum relative humidities are generally lower for fungi than for bacteria. Many bacteria grow best at RH values very close to 100%, whereas lower optima are common among the fungi. Many fungi can develop at less than 90% RH, but few bacteria can do so. Because of their less stringent water requirements, fungi are of particular importance in the spoilage of products whose preservation depends on dryness.

The interrelationships among RH, fungal and bacterial growth, and self-heating have been examined most completely in wool. At less than 94% RH fungi probably contributed to a slow temperature increase of this material to about 40°C (Dye and Rothbaum, 1964). Similarly, in hay at 86% RH a slow increase to 53°C may have been the result exclusively of fungal activity (Rothbaum, 1963). [The upper limit for the growth of thermophilic and thermotolerant fungi is approximately 60°C (Cooney and Emerson, 1964), and this probably marks the theoretical limit of the capacity of fungi to cause a temperature increase.] Observed higher temperatures in both materials attributable to biological action were associated with relative humidities more favorable to bacterial development. Fungal plate counts representing heating wool at 93% and 97% RH were over two orders of magnitude greater than those representing wool approaching 100% RH. In the wet wool (i.e., the RH measurement was unequivocally 100%), no fungi developed on the plates. The sensitivity limit was 10^4 colony-forming units per gram. Bacterial development was extensive throughout this range of RH values. The authors concluded that in the wet wool (the moisture content was 47%) the heating was caused entirely by bacteria.

In wool initially too dry for bacterial growth, this was initiated as metabolic water produced by fungi accumulated (Rothbaum and Dye, 1964). Accelerated heat output and the attainment of higher temperatures resulted.

In sterilized soil inoculated with a representative of both groups and incubated at a fixed temperature, development of the fungus was more strongly represssed by the bacterium at the higher moisture level than at the lower one (Finstein and Alexander, 1962). The more severe repression of the fungus in the wetter soil was apparently caused by enhanced bacterial competition for the limited supplies of carbon and, to a lesser extent, nitrogen.

On these grounds and others (see Section III), it can be argued that early studies (Miehe, 1907, 1930; Hildebrandt, 1927; Norman, 1930; Gaskill and Gilman, 1939; Christensen and Gordon, 1948) unduly emphasized the role of fungi in raising the temperature of wet materials. These reports include impressive demonstrations, carried out in insulated containers, of fungi increasing the temperature of sterilized materials liberally supplied with water. In nonsterilized organic materials, however, a sufficiency of water generally leads to bacterial dominance during the temperature ascent, as documented most completely by Dye and Rothbaum (1964), but convincingly by others also (Carlyle and Norman, 1941; Webley, 1947; Forsyth and Webley, 1948; Niese, 1959; Chang and Hudson, 1967). An apparent exception, where fungi persisted throughout a temperature rise that peaked at 62°C or 63°C for a few hours (Festenstein et al., 1965), involved hay that was dry by composting standards. The moisture content was reported as 40%, basis of expression not specified. Since the water status for forage is usually given as the dry weight, on a wet weight basis the moisture content presumably was 29%. Other hays containing rich fungal floras, some of them initially wetter (Gregory et al., 1963; Festenstein et al., 1965), were examined long after the peak temperatures had passed. Therefore the findings tell little about biological developments during the temperature ascents.

Some contemporary students of composting mycology tend to be equivocal regarding the possible contribution of fungi to the temperature ascent (Cooney and Emerson, 1964; Chang and Hudson, 1967; Kane and Mullins, 1973). Others assert that fungi are the main agents of heat production (Oesterle et al., 1963). In contrast, von Klopotek (1962) considered that, insofar as heat liberation is concerned, their role may have been exaggerated. We are in agreement with the latter assessment.

4. Bacteria and Fungi in Wet Materials

Bacteria flourish during self-heating when moisture is not restrictive of biological development (see Section III), and the temperature rises more rapidly than in drier materials. The example of adiabatically heating wool (Dye and Rothbaum, 1964) is a case in point. At 47% moisture content (100% RH), 70°C was exceeded in about 2.5 days, whereas twice this time was required at 31% moisture content (approaching 100% RH). Similar results were obtained with hay (Rothbaum, 1963). (It should be noted that when the RH is close to 100% it is altered perceptibly only by large changes in the moisture content.) It is evident that, regardless of the investigational viewpoint, any wet self-heating material permeated with air can be likened to compost.

III. The Temperature Ascent

Self-heating masses are dynamic with respect to moisture, oxygen, substrate, pH, and other abiotic factors. Temperature is clearly an overriding ecological determinant, however, and discussion of the associated organisms can be logically organized around this factor. On a practical level, rapid heating is essential to the success of municipal solid waste composting ventures. Potential nuisance-causing putrescible substances are thereby quickly destroyed.

A. Microbe–Temperature Interactions

1. Early Research

Miehe (1907) regularly found the mesophile *Escherichia coli* (*Bacillus coli*) in hay during the earliest stages of self-heating. The obligate thermophile *Bacillus stearothermophilus*, which he was first to describe and named *B. calfactor*, occurred later in the hotter material. These and related observations led Miehe (1907) to suggest that a sequence of mesophilic and thermophilic organisms developed during the temperature ascent. This attractive proposition was subsequently reiterated by a number of authors, but, until 1941, without benefit of any major experimental advance. Waksman pointed out (Waksman et al., 1939a,b) that compost research was slowed by the exclusion in some investigations of obligate thermophiles from population estimates. In reference to this and other aspects of compost research he admonished investigators to relate biology to function. Since then, Neilson et al. (1957) and Dye (1964) have examined the problem of enumerating aerobic thermophiles.

A more intractable experimental difficulty, the solution to which Norman pioneered (Norman et al., 1941), involved heat loss. This distorts the microbe–temperature interactions, for in self-heating masses the ascending temperature not only results from the past microbial release of heat, but it also largely determines the course of future population developments and heat release. The escape of small amounts of heat, for example, might delay the onset of the growth of obligate thermophiles, while complete loss would bar their development.

2. Adiabatic Experimentation

Ideal conditions for studying the sequence of events during self-heating would provide for unrestricted aerobic respiration and retention of all the heat produced within the mass. Although large field-scale organic masses are exceedingly well insulated within, these present a variety of experimental difficulties. There are many advantages to working with a small amount of material in a reaction chamber. Insulation alone, however, cannot prevent significant conductive losses, especially when heat production is slow (Norman, 1930). Also, the need for aeration must be reconciled with the prevention of convective losses and evaporative cooling.

Norman and associates (1941) were the first to circumvent these difficulties. They constructed an automatically controlled adiabatic apparatus for the investigation of self-heating with straw. A modified version of this device was employed later (Bartholomew and Norman, 1953). A similar apparatus was built at about the same time for the study of grain spoilage (Ramstad and Geddes, 1942). Manually controlled devices have been employed in pure culture investigations with corn (Wedberg, 1940) and in the study of self-heating hay and municipal solid waste (Niese, 1959). Walker and Harrison (1960) analyzed the shortcomings of five previously described automatic apparatuses (temperature differential too small, excessive random drift, lack of uniform performance at all temperatures, unreliability), and designed a model free of these defects. They provided detailed plans for its construction.

In the adiabatic apparatus conductive losses are minimized by placing the reaction chamber in a water bath (air incubators were used in the manual devices) and preventing the development of substantial temperature gradients between the contained self-heating material and the surrounding water. The water is heated electrically so that its temperature continuously increases in concert with that of the test substance. To prevent false heating, the bath is always at a slightly lower temperature. Convective losses associated with aeration are minimized by preheating the air to the bath temperature. To avoid drying of the experimental material and evaporative cooling, the air is prehumidified. The apparatus of Walker and Harrison (1960), in addition to making possible valid self-heating temperature ascents with manageable amounts of material, can provide reliable estimates of heat output.

The record of one self-heating temperature rise is reproduced in Fig. 1. The temperature increased in two stages, with heating rate maxima at approximately 40°C and 60°C, and the intervening minimum at 50°C. Remarkably similar values came from oat straw, meadow hay, mixtures of municipal solid waste and sewage sludge, and raw wools prepared

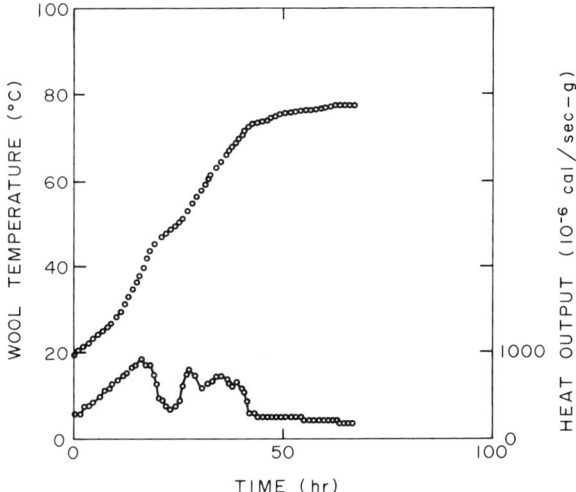

FIG. 1. Wet wool self-heating adiabatically. Reproduced from Walker and Harrison (1960), by approval of the authors and permission of N.Z.J. Agr. Res.

for market by four methods (Table I). Overall, the self-heating minima range from 44°C to 56°C. Some of the reports do not permit reliable estimation of the maxima.

Detailed examination of the mesophilic range with straw (Bartholomew and Norman, 1953) showed that the exact temperature at which the first maximum occurred was a function of the starting temperature. Apparently, at the lower starting temperatures the supply of readily available straw constituents became depleted before temperatures optimum for the mesophilic population were reached. This, however, did not alter the overall two-phase nature of the temperature ascent. Readily available substrates in raw wool are not quickly exhausted (Rothbaum, 1961), and this is probably also true of hay and of waste-sludge mixtures.

The two-staged temperature ascent revealed clearly by adiabatic experimentation was presaged in the report of Haldane and Makgill (1923). Imposition of a stepwise temperature increase on hay, without

TABLE I

RATE MAXIMA AND MINIMA IN WET MATERIALS SELF-HEATING ADIABATICALLY[a]

Material heating	Approximate moisture content (% wet wt)	Temperature[b] (°C)				Reference
		Maximum		Minimum	Maximum	
Straw	75	41 (40–41)		54 (52–56)	61 (60–65)	Carlyle and Norman (1941)[d]
Hay	71–72	—		46 (44–48)	—	Niese (1959)[d]
Waste[c]	41–47	—		48 (45–51)	—	Niese (1959)[d]
Wools	47–57	40 (39–41)		52 (50–52)	60 (59–61)	Walker and Harrison (1960)[e]
Hay	76	—		55	—	Rothbaum (1963)[d]
Wool	47	—		54	—	Dye and Rothbaum (1964)[d]

[a] Where relative humidities were given, data were taken only from trials involving materials in equilibrium with interstitial atmospheres that were unequivocally saturated.

[b] Average values (ranges in parentheses). Some values were estimated from graphs. Niese (1959) utilized a manually controlled apparatus and relied on aeration by unaided diffusion. The other investigators used automatically controlled devices with forced aeration (see text).

[c] Municipal solid waste 90%, sewage sludge 10%.

[d] Based on rate of temperature increase.

[e] Based on rate of temperature increase, and heat outputs. The averages noted were given in the original paper as representative of many trials. The ranges were estimated from five graphs.

reference to self-heating, resulted in two-staged oxygen uptake and carbon dioxide release. The timing of the imposed temperature increase was similar to that which occurs through self-heating. With the aid of insulation alone, two-staged temperature increases are sometimes demonstrated (Carlyle and Norman, 1941; Golueke et al., 1954; Wiley, 1955; Galler and Davey, 1971; Kane and Mullins, 1973), but the resolution is usually poor.

Carlyle and Norman (1941), Niese (1959), Walker and Harrison (1960), and Dye and Rothbaum (1964) enumerated viable bacteria in wet materials self-heating adiabatically. In the earlier investigations samples for the plate counts were removed from satellite reaction chambers, while in the two most recent studies it was possible to sample the chamber in which the heating record originated. In all the materials a mesophilic population expansion coincided with the burst of activity as indicated by temperature. The time-zero data are missing from one of the studies (Walker and Harrison, 1960), where the earliest count is close to the maximum observed. In all the materials, the mesophilic decline was established in advance of the temperature rate minimum, perhaps 3°C to 15°C prior to this point of reference. The decline was less sharp in the wools than in the other materials. Thermophilic proliferation commenced prior to the temperature rate minimum, and this population generally peaked in density before 60°C was reached.

The available temperature records and corresponding plate counts are consistent with Bausum and Matney's (1965) suggestion that the boundary between bacterial mesophilism and thermophilism lies between 44°C and 52°C.

Observations of bacterial proliferation during ascents from ambient to peak temperatures are summarized in Table II. Much of the disparity among the values no doubt results from differences in enumeration efficiencies. The mesophilic increment in municipal solid waste was exceptionally slight. This is probably related to the finding that the waste initially contained very high numbers of mesophilic bacteria. In trials that were comparable, the waste commenced self-heating with less delay than the hay, which initially yielded low counts of both mesophiles and thermophiles.

The adiabatic research indicates that the temperature ascent in wet materials is governed by microbe (bacterial)–temperature interactions as follows. The accumulation of microbial waste heat is at first stimulatory to the developing mesophilic organisms but turns inhibitory at temperatures within the 40°C to 50°C range (i.e., the temperature feedback switches from positive to negative). No doubt there is species variation with respect to the exact critical temperature, and facultative thermophiles and thermotolerant organisms (in the sense of Farrell and Camp-

TABLE II
Approximate Maximal Bacterial Plate Count Increments Recorded during Ascents of Wet Materials from Ambient to Peak Temperatures[a]

Mesophiles		Thermophiles		Material heating	Reference
Agar incubation temp. (°C)	Max increment[b]	Agar incubation temp. (°C)	Max increment[b]		
30	24	60	715	Straw	Carlyle and Norman (1941)
30	630	—	—	Straw	Carlyle and Norman (1941)
25	745	55	1845	Hay	Niese (1959)
25	1120	55	389	Hay	Niese (1959)
25	1.5	55	4260	Waste	Niese (1959)
25	1.5	55	1400	Waste	Niese (1959)
—	—	55	194	Waste	Niese (1959)
—	—	55	4×10^4	Lucerne	Niese (1959)
—	—	55	1800	Lucerne	Niese (1959)
—	—	60	8×10^{4c}	Wool	Walker and Harrison (1960)
30	10^4	50	4×10^{3c}	Wool	Dye and Rothbaum (1964)
22.5?	100	45?	10^5	Straw	Chang and Hudson (1967)
—	—	45?	10^{4d}	Straw	Chang and Hudson (1967)

[a] Adiabatic experimentation, except Chang and Hudson (1967), from outdoor compost.
[b] Highest count/starting count.
[c] Minimum value.
[d] Actinomycetes, no increase prior to the second day.

bell, 1969) may span the transitional range. Counts from plates incubated at 40°C and 50°C, representing wool (Dye and Rothbaum, 1964), are suggestive of the involvement of thermotolerant bacteria. The growth of obligate thermophiles, which depends on prior mesophilic heating, subsequently follows a similar pattern. This outline is helpful when one considers the differences, on a microbial level, between batch and continuous composting processes (see Section V).

The adiabatic experiments implicate bacteria in the temperature ascent largely through a correspondence between bacterial and temperature responses. Walker and Harrison (1960), in addition, provide the corresponding caloric outputs. Evidence of a more direct nature is available from Rothbaum's (1961) study at fixed temperatures. His data can account for the observed increase of a bale of wool from 17°C to 71°C in 48 hours in terms of bacterial heat output. Such an increase required an estimated average output of at least 3.5×10^{-4} cal/sec-gm. Outputs exceeding or equaling this value were attained during experimental incubations of wet wool at a number of mesophilic and thermophilic temperatures up to 70°C. Enumeration of the viable bacteria in the experimental samples permitted estimations of the outputs per cell, which ranged from 0.6 to 1.5×10^{-12} cal/sec-cell. The highest value per call, which was at 60°C, agreed well with the output of 1.25×10^{-12} cal/sec-cell measured at 60°C by broth-grown isolates from the wool (*B. coagulans* and *B. stearothermophilus*). Thus it was demonstrated that these bacteria, in conjunction with strains active at lower temperatures, are capable of heating the wool as observed.

B. FUNGI

1. *As a Group*

Fungi have often been observed in composts and related materials, but as Chang and Hudson (1967) noted, frequently without reference to the phase of heating. As discussed previously, moisture content is critical to the possible involvement of fungi in the temperature ascent, and values typical of composts favor bacteria over fungi during this stage of the process. Furthermore, the temperature rise itself is unfavorable to fungi. The uninterrupted presence of fungi on compost exteriors (Stoller *et al.*, 1937; Eastwood, 1952; Anonymous, 1953; von Klopotek, 1962; Fergus, 1964; Chang and Hudson, 1967), which tend to be relatively cool and dry, can be misleading with respect to their behavior within.

Eastwood (1952) apparently was the first to note that fungi disappeared from compost interiors at high temperatures. Her study, however, probably accounted only for mesophilic forms since, as Chang and Hudson (1967) pointed out, the lower incubation temperature (25°C) was too cool for thermotolerant fungi while the plates incubated at 60°C were too hot for these and thermophilic forms. [See Cooney and Emerson (1964) for definitions of thermophilic and thermotolerant as applied to fungi.] Von Klopotek (1962) and Chang and Hudson (1967) selected the more favorable temperatures of 25°C and 48°C, and 22.5°C and 45°C, respectively. In the latter investigation, in addition to dilution-plate counts, qualitative outgrowth procedures were systematically employed. Compost particles were incubated in moist chambers, and on agar. The results from all three procedures were in good agreement. They also enumerated bacteria in one trial, and these increased as fungi decreased. Reports documenting the behavior of fungi as a group in composts are summarized in Table III.

TABLE III
Approximate Temperatures at Which Cultivable Fungi Disappeared from Compost Interiors during Temperature Ascents

Material composted	Approximate moisture content (% wet wt.)	Temperature last positive sample (°C)	Temperature first negative sample (°C)	References
Grass	45	16	65	Eastwood (1952)[a]
Straw	45	20	64	Eastwood (1952)[a]
Waste	?	65	67[b]	von Klopotek (1962)[c]
Straw	75	48	68	Chang and Hudson (1967)[c]

[a] The apparent early disappearance was probably exaggerated by the long sampling intervals. Only mesophilic fungi were accounted for (see text).

[b] Sporadic appearance of fungi was ascribed to sample contamination.

[c] Early in the heating of the municipal solid waste, the mesophilic counts increased <10-fold, and in some trials with straw thermophilic counts increased similarly.

Similar results came from incubation of compostable materials at fixed temperatures representative of composting. Although this approach seriously departs from the continuous ascent characteristic of self-heating, it may afford an indication of the biological possibilities at the temperatures tested. Fungal development was followed through dilution-plate counts and the Rossi-Cholodny contact slide technique. Waksman et al. (1939b) and Kaila (1952) noted, in manure and straw, respectively, that fungi were abundant at 50°C but essentially absent at 65°C. In mixtures of dung and sterilized straw similar results were obtained at 50°C and 60°C (Henssen, 1957).

The studies of real and stimulated composts suggest that the disappearance of viable fungi is well advanced before the temperature reaches 60°C and is essentially complete by 65°C. Duration of exposure to unfavorable temperatures is an important determinant that has not been investigated in the present context. This combination of factors may have a bearing on the observation that fungi remained viable throughout a brief peak temperature of 62°–63°C in relatively dry hay (Festenstein et al., 1965).

2. Specific Fungi

Extreme environments characteristically have relatively homogeneous communities dominated by few species. These are at least resistant to the harsh condition and frequently are obligate for it (Alexander, 1971). That species diversity narrows as compost temperatures rise is evidenced in gross terms by the disappearance of the fungi. A few details on the restructuring of the fungal, actinomycete, and bacterial populations are available. Fungal identity is presented as in the original publication.

Solid waste from German cities contained an average of 1.5×10^6 mesophilic colony-forming units per gram (von Klopotek, 1962). Representatives of seven genera were encountered, among which *Geotrichum candidum* predominated. Thermophilic and thermotolerant fungi, especially *Aspergillus fumigatus*, averaged a few thousand per gram. *Geotrichum candidum* accounted for the bulk of the transient increase noted early in self-heating. This organism disappeared, however, between 45°C and 55°C, as did the less abundant *Mucor* sp. At this time, *Cladosporium* (which had not been detected in the raw waste) dominated the plates and continued to do so until 67°C, when fungi were no longer recoverable. *Penicillium duponti* appeared only on the plates representing the raw waste and the 55°C sample. Strangely, thermophilic and thermotolerant fungi gave no indication of proliferating as the temperature rose through their favorable range, and the relative position among the predominant representatives (*Aspergillus fumigatus*, *Mucor pusillus*) remained essentially unchanged until these disappeared from the compost between 55°C and 64°C.

Chang and Hudson (1967), in contrast, found that the mesophilic fungi on straw disappeared early in the temperature ascent. Only thermophilic and thermotolerant species (*Absidia ramosa*, and especially *A. fumigatus* and *M. pusillus*) were recoverable after the compost exceeded 40°C. The temperature of the subsequent sample was approximately 68°C, and it yielded no fungi.

Kane and Mullins (1973) cultured thermophilic and thermotolerant fungi from unprocessed municipal waste. Some of the samples became warm before examination, while stored in household refuse containers.

The study was carried out in Florida during the summer. Isolates included *A. fumigatus, Chaetomium thermophile, Humicola lanuginosa, M. pusillus, Thermoascus aurantiacus,* and *Torula thermophila.* No population sequence was observed during composting of the waste. Fungi disappeared from the compost interior.

In manure incubated at 28°C *Zygorhynchus, Humicola, Trichothecium, Cephalosporium,* and *Alternaria* developed (Waksman et al., 1939b). At 50°C *Thermomyces* and the *Monilia* or *Oidium* groups were found. As the 50°C incubation proceeded fungal mycelia tended to disappear, possibly through biological antagonisms. Fungi were rarely seen on the slides at 65°C and were absent from the plates. In a similar investigation (Henssen, 1957) *Thermomyces lanuginosus, A. fumigatus,* and *Humicola* sp. developed from some dung samples incubated at 50°C, but fungi were absent at 60°C.

C. ACTINOMYCETES

1. As a Group

The general tendency of actinomycetes to colonize fresh substrates more slowly than bacteria and fungi (Lacey, 1973) pertains to composts. In straw the count of mesophilic actinomycetes was erratic throughout the composting period (Chang and Hudson, 1967). Thermophiles proliferated, but late relative to the bacteria. These data are entered into Table II. Similar results came from hay self-heating in Dewar flasks (Festenstein et al., 1965). The late appearance of actinomycetes was a feature common to municipal waste composts regardless of the many manipulative and substrate variations tested (Anonymous, 1953). Their earlier than usual appearance could not be forced by inoculation with heavily colonized particles taken from an older compost. The actinomycetes disappeared from the introduced material within 24–36 hours of mixing with the fresh waste. Time of reappearance was independent of inoculation. Actinomycetes were more conspicuous on dry particles than on wet (e.g., paper compared to vegetable trimmings). Visible growth developed in 5–7 days where forced air entered the mass (Schulze, 1962a). In a windrow consisting of municipal waste, cellulase activity, possibly originating in a thermophilic actinomycete, increased markedly with time (Stutzenberger et al., 1970). At a fixed temperature of 50°C actinomycetes developed later than bacteria and fungi (Waksman et al., 1939a). An actinomycete raised the temperature of sterilized straw feebly compared to fungi (Norman, 1930).

Actinomycetes are notably adherent to their substrate, including compost particles (Erikson, 1952; Anonymous, 1953). Erikson (1952), being dubious about the recovery of these organisms from grass compost by dilution procedures (see also Waksman *et al.,* 1939a), utilized the time elapsed to visible outgrowth from incubated particles as a rough measure of relative abundance. The time elapsed decreased from 7 days early in the temperature rise to 1 or 2 days, after thermophilic temperatures (46°–65°C) had become established. Actinomycete growth was restricted to within 6 inches of the compost surface. Actinomycetes developed poorly in masses that were not well aerated (Anonymous, 1953). Development on the surface was suppressed by frequent turning. Visible surface growth of actinomycetes, possibly including fungi, has been referred to as a white limelike coating (Forsyth and Webley, 1948), and similarly as a chalky white encrustation (Erikson, 1952). Another descriptive term, fire-fang (Fergus, 1964), takes cognizance of its typical appearance during or after the peak-temperature stage.

There are somewhat conflicting data concerning the maximum temperature permitting actinomycete growth in composts. At fixed temperatures these flourished at 65°C but failed to develop at the next higher temperature of 75°C (Waksman *et al.,* 1939a,b). In Henssen's (1957) experience the comparable temperatures were 60°C and 70°C. Kaila (1952) observed actinomycetes at the highest incubation temperature tested (65°C). Tendler and Burkholder (1961) reported that 67°C was the upper growth limit for 59 of 503 thermophilic isolates tested in broth. An additional 119 strains grew at 65°C. Erikson's (1952) isolate from compost did not grow on any medium at temperatures between 62°C and 80°C.

These findings are not entirely compatible with Niese's (1959) extensive contact slide observations on self-heating hay and municipal waste. Actinomycetes were first seen at 55°C and again 65°C. They seemed more abundant at 73°C (including spores) and at the peak temperature of 75.2°C. There was little change as the temperature started to descend (observations were at 73.4°C and 70.8°C). Activity or viability was not determined. Although actinomycetes were conspicuous, the author judged that bacteria predominated. Chang and Hudson's (1967) plate counts of thermophilic actinomycetes appeared to increase as the temperature increased to the peak at 72°C. The possible growth of actinomycetes at such compost temperatures must be considered unresolved at the present time.

2. Specific Actinomycetes

Cross and Goodfellow (1973) and Lacey (1973) have provided partial guides through the nomenclatural maze that confuses many of the reports of actinomycetes in composts and related materials. In composts, suffice it to say, organisms forming single or a few spores have been frequently observed and sometimes obtained in culture (Waksman *et al.,* 1939a,b;

Forsyth and Webley, 1948; Erikson, 1952; Anonymous, 1953; Henssen, 1957; Stutzenberger et al., 1970; Stutzenberger, 1972). This characteristic is suggestive of the genera *Thermoactinomyces* and *Thermomonospora*, but secure assignment after the fact is not usually possible. Erikson's (1952) isolate from straw compost formed heat-resistant endospores and therefore belongs in the genus *Thermoactinomyces* according to the scheme of Cross and Goodfellow (1973). A cellulose decomposer from municipal waste compost (Stutzenberger, 1972) was identified as *Thermomonospora curvata*.

Lacey (1973) discussed the communities that developed in hay samples in Dewar flasks differing with respect to water content and degree of heating. With mild heating (45°–50°C), *Streptomyces* was dominant. The more thermophilic genera *Thermoactinomyces, Micropolyspora*, and *Thermomonospora* prevailed in the wetter, hotter hays, most of which, however, were relatively dry and cool by composting standards. The highest temperature recorded was approximately 63°C. Apparently these observations were made after the heating had subsided. Explicit studies of actinomycete succession during the temperature ascent seem to be lacking.

D. Bacteria

Eubacteria unlike fungi and actinomycetes, thrive during all stages of composting. The two latter groups are classifiable through morphology alone, but the bacteria in composts and other environments have few visually distinctive attributes. It is therefore more difficult to trace the succession of bacteria as the temperature rises.

In raw municipal solid wastes from four locations, nonspecific bacterial plate counts and counts of heat-resistant bacterial spores varied within surprisingly narrow ranges (Peterson and Stutzenberger, 1969). Average values per gram fresh weight were, respectively, 2.3×10^8 and 3.9×10^4. Coliform and fecal coliform densities were more variable but generally amounted to less than 1% and 0.1% of the plate counts, respectively. Isolates from a household solid waste (Cook et al., 1967) were tentatively identified as *Bacillus, Pseudomonas, Serratia, Achromobacter, Flavobacterium, Streptococcus*, and *Escherichia*, genera that are common to soils and water environments. Thermophilic bacilli and clostridia developed from material retrieved from an old landfill, but surprisingly not from the fresh waste.

Miehe's (1907) observation of an *E. coli–B. stearothermophilus* sequence in self-heating hay was noted previously. Carlyle and Norman (1941) observed that colony variety decreased as the temperature increased. It was reported (Anonymous, 1955) that as the composting of waste progressed there was a shift from a varied population (*Pseudomonas, Achromobacter, Flavobacterium, Micrococcus, Bacillus*) to one dominated by *Bacillus*, but few supporting details were given. In the experiments of Niese (1959) plates representing hay incubated at mesophilic temperatures yielded predominantly yellow, red, and white colonies of nonspore-forming short rods. From self-heating waste came, in addition, a citron-pigmented colony of similar cell type. These disappeared as the temperature reached 55°C. Three kinds of colonies arose on the thermophilic plates, all from spore-forming bacteria. Two were identifiable (*B. subtilis* and *B. stearothermophilus*), but the third was not.

Niese (1959) found that the strains of *B. subtilis* and the unidentified isolate grew on meat-extract–yeast agar at temperatures ranging between 25°C and 65°C, with the best development at approximately 50°C and 55°C, respectively. The *B. stearothermophilus* strains grew between 40°C and 73°C and 65°C was most favorable. A *Bacillus* isolated from compost (Hirano et al., 1958) grew at temperatures above 35°C and below 72°C and the optimum was approximately 65°C. All 87 strains of *B. stearothermophilus* entered into Gordon's (1973) compilation grew at 65°C, and 45 strains did so at 70°C.

A progressive simplification of the bacterial population structure in favor of sporeformers has been viewed on contact slides. Niese (1959), and especially Waksman et al. (1939b), commented on the variety of cell shapes, sizes, and types of agglomerations at low temperature, and the relative uniformity at higher temperatures. Niese's (1959) culture and extensive slide observations in this regard are in good agreement, to the extent that they can be compared. For example, following a brief increase, short rods disappeared from both plates and slides at about 55°C. Similarly, apparent sporeformers (longer rods) became increasingly conspicious on slides as the thermophilic count increased. (Not all the rods seen were necessarily sporeformers, however.) Spores became evident at 55°C, and increasingly so as the heating progressed. Waksman et al. (1939b) noted many sporeformers at 65°C, and at 75°C these essentially monopolized the slides, including clostridial types. Henssen (1957) saw only bacteria on slides removed from 70°C material. She isolated *Clostridium thermocellum* and a second thermophilic cellulose-deomposing *Clostridium*. Significantly, the reports of *Clostridium* came from investigations with manure and dung, which are prone to anaerobiasis. Forsyth and Webley (1948) isolated aerobic sporeformers from peak-temperature grass and straw composts and observed similar forms on contact slides. Future studies detailing the waxing and waning of bacterial species during self-heating could well consider the use of fluorescent antibody techniques to provide a positive link between culture and *in situ* observations.

A completely mixed laboratory-scale compost maintained between 55°C and 58°C, and recharged every 7 days with fresh waste, yielded only spore-forming bacteria on trypticase soy agar incubated at 55°C (Morris, 1975). At the end of a 7-day feeding cycle, the colonies that developed on the plates could be attributed about equally to vegetative cells and spores. This changed markedly upon introduction of fresh waste. Seven hours thereafter an average of 93% of the colonies arose from vegetative cells and the remainder from spores. With the passage of time the proportion of colony-forming units in the compost attributable to vegetative cells decreased while spores increased.

Although there can be no doubting the importance of thermophilic sporeformers in the high temperature ranges of self-heating, it would be premature to rule out the involvement of other bacteria. As noted previously (Section II,B,1) there is good evidence of biological heating to 79°C, which is beyond the range of known sporeformers. Inoculation of material that presumably self-heated through biological means to almost 80°C into broth hotter than 68°–70°C did not result in growth (Walker and Harrison, 1960). Under similar circumstances, there was no growth in 80°C broth, but growth occurred upon reincubation at 60°C (Dye and Rothbaum, 1964). Interestingly, Brock and Freeze (1969) isolated a nonspore-forming bacterium that grew at a maximum of 79°C. The minimum was approximately 40°C, and the optimum 70°C. This bacterium (*Thermus aquaticus*) was cultured from nutrient-poor hot aquatic environments. Enrichment by conventional procedures for thermophiles failed (nutrient-rich broth at 55°C). The use of dilute medium with respect to organic compounds and incubation within the 70°C to 75°C range was required for successful enrichment. The relevance of these findings to Walker and Harrison's (1960) observation that wool heated to 79°C is unknown.

IV. The Temperature Descent

A. Temperature

Withholding air from self-heating masses quickly reduces heat output and causes a drop in temperature (Walker and Harrison, 1960; Festenstein *et al.*, 1965). With an uninterrupted air supply the temperature declines when the readily available substrates become exhausted. In the absence of turning the decline may be temporarily reversed by a secondary ascending trend (Miehe, 1907; von Klopotek, 1962; Festenstein *et al.*, 1965; Chang and Hudson, 1967). Materials giving a porous structure, e.g., straw, seem more prone to secondary ascents than grass (Chang and Hudson, 1967). Old hay gave secondary peaks whereas the more recent crop did not (Festenstein et al., 1965).

Turning cools compost temporarily, but subsequently may result in a higher temperature than immediately before the disturbance (Wiley and Spillane, 1961). Although partial recovery was rapid, complete reattainment of the preturned temperature required from one-half to 2 days (Anonymous, 1953). The longer recovery periods occurred with turning during the overall declining temperature stage. In unsheltered windrows of unground municipal solid waste, abrupt cooling of more than 25°C was associated with rain and high wind (Anonymous, 1973b). Discussions are available on methods for judging when processing is finished and the compost ready for use (Jann et al., 1959).

B. Organisms

There is some information on fungal recolonization as the mass cools, but none on the restructuring of the actinomycete and bacterial populations during this stage of composting. Since fungi persist at the cool outer edges, these can serve as the base of reinvasion by species previously eliminated from the interior and also by propagules newly deposited from the air.

Recolonization of the interior by thermophilic fungi started soon after the retreat from peak temperatures commenced (Chang and Hudson, 1967). As the temperature fluctuated about the 50°C level, the thermophilic plate count increased to about 10^6 per gram. In contrast, mesophilic recolonization was delayed. In one compost, mesophiles reappeared 20 days after a peak temperature of 67°C was attained. The comparable values from another trial were 34 days and 72°C. Cooling to about 30°C seemed necessary for mesophilic reestablishment. The authors concluded that the course of recolonization is a function of many factors in addition to the maximum temperature, of which duration of exposure to high temperature is one.

Von Klopotek (1962) noted strong fungal reestablishment after the temperature fell below 60°C. As during the temperature ascent, there was the curious observation that mesophiles developed more strongly than thermophiles at temperatures seemingly appropriate to the latter (e.g., 59°C).

The data of Chang and Hudson (1967) have been arranged in Table IV to emphasize the recolonization sequence. *Mucor pusillus* (thermophilic) and *Absidia ramosa* (thermotolerant), which had been noted as the temperature ascended, did not reappear. This may reflect their propensity for simple organic substrates much less plentiful in the com-

TABLE IV
Fungal Recolonization of Straw Compost Interiors
after Peak Temperatures Passed[a]

Detected following the fungus-free peak temperature	Temperature relations[b]	Detected prior to fungus-free peak temperature
Aspergillus fumigatus	TT	Yes
Humicola lanuginosa	T	Occasional
H. insolens	T	Occasional
Chaetomium thermophile	T	Occasional
Malbranchea pulchella var. sulfurea	T	Occasional
Talaromyces duponti	T	Occasional
Sporotrichum thermophile	T	No
"Mycelia Sterilia C.t. 6"	T	No
Fusarium culmorum	M	Occasional
Stysanus stemonitis	M	No
Coprinus cinereus	M	No
C. megacephalus	M	No
Clitopilus pinsitus	M	No

[a] Adapted from Chang and Hudson (1967). Based on results from plating and outgrowth. The order of listing approximates the order of appearance. The composts were not turned.

[b] TT, thermotolerant; T, thermophilic; M, mesophilic (see Cooney and Emerson, 1964).

post than in straw (Chang, 1967). Of the mesophiles, only *Fusarium culmorum* had been encountered previously. Basidiomycetes (the last three entries) appeared late, in agreement with observations in municipal solid waste composts (von Klopotek, 1962). Another latecomer in the waste compost was the nematode-trapping fungus *Arthrobotrys* sp. along with its prey. As with straw, most of the recolonizers were new to this material. The original article should be consulted for an extensive list of isolates. As leaf compost cooled, slime mold, earthworms, mites, ant colonies, and other insects appeared (Poincelot and Day, 1973).

C. Inoculation

Alexander (1971) analyzed the factors contributing to the resistance of existing microbial communities to the establishment of alien strains. The few successful attempts to advance human objectives by altering community structure through inoculation probably reflected a predisposition of the environments toward the new organisms. Their establishment requires that abiotic factors be at least within the tolerance range, and that the appropriate niche be either unfilled, or filled with less efficient organisms.

1. Sawdust

The only convincing demonstration found in the literature that compost inoculum may be useful concerns the special case of sawdust. Incorporation of basidiomycete-rich decayed wood (Davey, 1953) or 1% of a similar precomposted sawdust (Wilde, 1958), in conjunction with nutrient supplementation, speeded production of a usable soil conditioner from sawdust. This operation clearly resulted in the early appearance of the fungus. No rigorous distinction was made, however, between the physical and nutritional effects of the carrier and that of the fungus propagules. In view of the low peak temperatures reached (approximately 40°C during the preparation of a 50-gallon batch of inoculum) and the dissimilarity of the starting materials, the possible relevance of these observations to municipal solid waste composting is limited to the cooling stage of the latter.

2. Commercial Products

Inoculums for household and garden wastes can be purchased in tablet, granular, and powder form. Content descriptions often allude to beneficial decay microorganisms and nuisance prevention. As noted by Wylie (1960) one composting process is purportedly based on a special inoculum. Another process is described as utilizing special inoculating procedures. Serious investigation, however, tends to contradict claims that inoculation of compost is useful (Golueke et al., 1954; Farkasdi, 1965; Obrist, 1966).

3. Discussion of the Controversy

It is useful to consider the question of compost inoculation in terms of specific operational objectives. These might include: speeding the attainment of peak temperatures, elevating the peak temperature, and promoting recolonization as cooling sets in. The plausibility of inoculation advancing each of these objectives will be examined.

Inoculation of mesophiles during early stages of heating seems superfluous in view of the diversity of starting microfloras (Section III,D), and futile in view of their impending demise from exposure to high temperatures. The question therefore focuses on the introduction of thermophilic bacteria and actinomycetes that might be absent from the starting material. With the possible exception noted below, such absence seems unlikely since both groups are widely distributed in soil and in materials having contact with soil (Gaughran, 1947; Cross, 1968). Many are endosporeformers and therefore likely to persist despite environmental stresses. Given favorable abiotic conditions, rapid self-heating is inevitable. There can be no doubting the capacity of native floras to bring this about.

Based on many observations it is also predictable that, given adequate nutrition and physical conditions, composts will attain 70°C or thereabouts. Organisms capable of bringing this about are evidently widespread. One likely candidate for this function, *B. stearothermophilus*, has been identified in some heating materials (Miehe, 1907, 1930; Hildebrandt, 1927; Niese, 1959), but not in others (Gregory et al., 1963; Festenstein et al., 1965). The latter authors commented on its absence as they reported finding *B. licheniformis*. [According to Farrel and Campbell (1969), however, *B. licheniformis* (thermophilic strains) has commonly been confused with *B. stearothermophilus*.] There is no evidence to suggest that the failure to attain usual peak temperatures is attributable to the lack of critical organisms.

In contrast, a temperature near 80°C or slightly higher is unusual (Sections II,B,1 and III,D) and the biological agents, presuming they exist, are of unknown distribution. This raises the possibility that seeding with dependable sources may be necessary to ensure their presence. The desirability of composting at such temperatures, however, is by no means certain.

Mycological observations have shown the cooling mass to be open to recolonization by selected previous inhabitants and to new forms. Bacterial and actinomycete recolonization has not been documented, but no doubt occurs. Conceivably, intervention to speed this process might promote the decomposition of resistant materials such as cellulose and lignin, against which some fungi are especially active. Well-timed turning may promote reinvasion of interior compost regions by competent organisms. It is reasonable to consider concurrent seeding with small amounts of rich soil or mature compost, as was done in sawdust. This should not be taken out of context, however. The purchase of inoculum to promote recolonization would be wasteful.

V. Batch and Continuous Composting

There are many proprietary processes for the treatment of municipal solid waste by composting (Anonymous, 1971; Bond and Straub, 1973). Some are named after their developers, who are often closely associated with their operation and commercial promotion. The processes offer a variety of mechanical and structural approaches to the related problems of aerating and mixing the composting mass. A requirement peculiar to the continuous type of process, that of translocating the mass during composting, is met by arranging the mixing to impart a movement in one direction. Certain nonbiological unit operations such as conveying, separation, size reduction, watering, dewatering, pelletizing, etc., usually support the biological treatment process. The combination of operations employed, and their sequence, varies widely. Schemes for composting municipal solid waste are highly individualistic with respect to mechanical, structural, and operational detail.

From a microbiological viewpoint, however, only two major process variants are evident. These are batch and continuous composting (Table V). Field-scale batches of organic solids presumably undergo a progression of microbe-temperature interactions akin to those discussed in Section III,A,2. The time required to ascend through the mesophilic into the thermophilic range is measured in days. In contrast, well-designed continuous systems eliminate the mesophilic stage and operate continuously at thermophilic temperatures. This is possible because as the fresh waste enters the digester it is heated by the resident compost (Schulze, 1965), forcing direct initiation of thermophilic growth. Massive seeding with active thermophiles, which is intrinsic to continuous processes, may help speed heat output at the expense of the incoming waste. Continuous operation affords a means of decomposing putrescible materials quickly under close process control. It is doubtful that the end product differs greatly as a consequence of the type of process employed.

Mechanically continuous operation does not ensure continuous thermophilic composting. A progression of temperatures was characteristic of an eight-deck silo, from approximately 38°C on the first deck to 67°C on the last (Anonymous, 1955). The material typically attained 50°C when it reached the third or fourth deck. In the Varro Conversion Process the desired temperatures were maintained with the aid of electric air heaters (Malin, 1971). The main design shortcoming of these processes apparently involved aeration. Air was circulated over the mass exclusively, and penetration appears to have been poor.

In contrast, in the Fairfield Process air is forced through the composting mass (Schulze, 1965). The concentration of oxygen in the interstitial gas and the temperature are continuously monitored automatically. Undesirably low oxygen levels are prevented, and the temperature is maintained within specified limits in the thermophilic range through controlled aeration. Aeration serves not only to supply oxygen, but also as the means of forcing out the exhaust gas and carrying off excess heat. The Dano Process, of different design, is also capable of continuous thermophilic operation (N. Mandelblit, Israel Dano Corp., Ltd., Haifa, personal communication).

The Metro Waste Conversion System was a batch process of elaborate design (Olds, 1968; Anonymous, 1968; Kane and Mullins, 1973). Digester structures and forced aeration were employed, and the oxygen content

TABLE V
COMPOSTING PROCESS VARIANTS FROM A MICROBIOLOGICAL VIEWPOINT

Process variant	Digester structure	Distinctive features				
		Temperature stages	Aeration	Decomposition rate	Capital investment[c]	Process example
Batch	Usually absent (not essential)	Mesophilic → Thermophilic → Cooling in place	Diffusion, usually supplemented by turning	Varies, tends to be low	Low	Windrow (Anonymous, 1971)
Continuous[a]	Essential	Thermophilic → Cooling outside digester[b]	Forced aeration and mixing	High	High	Fairfield (Schulze, 1965)

[a] In two senses: (1) There is a semicontinuous plug flow of material undergoing composting through the structure (entering fresh waste displaces the oldest compost). (2) As a consequence of this arrangement, continuous thermophilic operation is possible.
[b] The cooling stage (curing or maturation) is accomplished in batches, with little need of turning.
[c] See Bond and Straub (1973).

and temperature were monitored. The benefits of linking such sophistication and expense to a batch process are not obvious.

VI. Operational Factors

A. Carbon/Nitrogen Ratio

A nitrogen deficiency can slow the rate of decomposition. Poultry manure–sawdust mixtures having a C/N ratio of 40/1 (based on weight) consumed more oxygen than mixtures at 25/1 during the first day of composting, but thereafter the positions were reversed (Galler and Davey, 1971). After 4 days the cumulative oxygen uptake by the materials richer in nitrogen was about 1.7 times that of the poorer materials. Other reports concerning the cardinal C/N ratio values lack precise criteria. However, operational experience indicates that, for municipal solid waste, starting C/N ratios narrower than approximately 25/1 may lead to ammonia volatilization. This can result in nuisance odors and a lowering of the agricultural value of the compost. Ratios wider than perhaps 40/1 are suggestive of a potential nitrogen deficiency. Loss of carbon through carbon dioxide formation progressively narrows the C/N ratio during composting.

Increasing use of paper products and increasing popularity of garbage grinder-disposers appears to have widened the C/N ratio of the solid waste discarded in American cities (Anonymous, 1965), emphasizing the importance of considering nitrogen addition. A test designed to evaluate the need for nutrient supplementation is based on the response of carbon dioxide evolution to incremental additions of the nitrogen source (Fuller and Bosma, 1965). Raw and digested sewage sludge are favored nitrogen sources.

In some situations it may be advantageous to operate under conditions of nitrogen deficiency. Leaves can thereby be composted with few turnings, yet with reduced risk of serious odor production. The disadvantage of submaximal microbial activity is accepted to avoid gross anaerobiasis and its odors. This may also be accomplished through low moisture contents. Decomposition is slow, but turning expenses are minimized.

B. pH

Carnes and Lossin (1970) investigated the problem of determining the pH of compost. The results were affected by the ratio of compost to water and sample preparation (grinding and drying). There was considerable variation between otherwise comparable samples. The authors proposed standardized sampling and dilution procedures.

Different compost strata may have different reactions. A waste windrow after 31 days of composting (last turned on day 14) yielded the following pH values: top, 7.9; intermediate, 5.9 and 5.4; bottom, 7.5 (Wiley and Spillane, 1961). The pH of the mixture was 6.6.

Municipal solid waste (Golueke et al., 1954; Anonymous, 1955; Wiley and Pearce, 1957; Wiley, 1962b; Block, 1965; Stutzenberger et al., 1970; Kane and Mullins, 1973), poultry manure–sawdust mixtures (Galler and Davey, 1971), and grass (Forsyth and Webley, 1948; Erikson, 1952) composted in batches underwent characteristic pH changes. Most of the materials were initially acidic (approximately 4.5–6.0) and became somewhat more so after stacking. When the pH values were near the minimum, the odor of acetic acid was noticed (Golueke et al., 1954; Block, 1965). The odor of butyric acid has also been noted (Golueke et al., 1954). [During continuous composting, however, the maximum volatile acid content did not coincide with the minimum pH (Anonymous, 1955).] Within a few days the pH of the batches started to increase, possibly reflecting the loss of organic acids through volatilization and microbial decomposition (Anonymous, 1955), and the release of ammonia through mineralization of organic nitrogen (Forsyth and Webley, 1948; Hoyle and Mattingly, 1954; Anonymous, 1955; Block, 1965). The pH finally stabilized within the range of 7 to 9.

Straw did not demonstrate the pH profile described above, but varied between 6.5 and 7.4 (Chang and Hudson, 1967). Sewage sludge cake that was initially pH 11 because of chemical treatment became slightly acidic (pH 6.5) with composting (Shell and Boyd, 1970).

As curing progresses, nitrification may commence with the production of appreciable nitric acid. In straw-sewage sludge mixtures in which total nitrogen comprised 1.97% to 2.44% of the starting material, nitrate was detected in 6–9 weeks (Hoyle and Mattingly, 1954). Nitrate appeared later in the materials poorer in nitrogen, and little accumulated at the lowest nitrogen level tested (1.10%). After extended periods (49–109 weeks) the pH values ranged from 5.7 to 7.0. With minor exceptions, the more acidic composts were those with the higher initial nitrogen content and the higher final nitrate content.

Thomsen (1910) found that the development of ammonium-oxidizing and nitrite-oxidizing enrichment cultures, using compost as the inoculum, was retarded or suppressed by sea salts.

With continuous thermophilic, completely mixed laboratory-scale composting, the reaction stabilized at a level dependent on the feed rate (Schulze, 1962b). When the daily feed rate, based on the volatile matter (VM) of the fresh waste as a percentage of the resident compost VM, was 10.8%, a pH of 8 was maintained even though the incoming waste varied between 4.7 and 6.9. Increasing the rate to 25% caused a slow decline to pH 5.6 followed by a return to 6.1, which persisted until the experiment was terminated. The compost produced while feeding at the low rate was the superior product. This was attributed to the residence time (average 12.7 and 7.0 days, respectively) rather than to pH directly. Thermophilic temperatures were maintained throughout. The author considered pH to be a useful indicator of process loading.

In a pilot-scale continuous plug-flow unit (Anonymous, 1955) the pH of the waste prior to composting was approximately 4.6, while 6.1 was typical of the first deck. There was a steady increase throughout the process, until on the eighth and last deck a value of 8.6 was reached. In an operational-scale, plug-flow, continuously thermophilic unit (Schulze, 1965), most incoming pH values ranged from 4.5 to 5.6 while outgoing values were from 7.0 to 7.8. The acidity of the raw waste in both operations probably reflected some fermentation prior to composting.

In one respect there is an apparent contradiction between the results of Schulze (1962b) and those of Wiley (1956). Wiley found that additions of acetic acid to a batch of waste, in amounts that prevented an increase above pH 6.0, also prevented the attainment of thermophilic temperatures. Similarly, a slower initial temperature rise was associated with the more acidic waste samples, and the temperature remained below 45°C until pH 7.0 was exceeded (Galler and Davey, 1971). Thereafter, it increased rapidly to between 60°C and 70°C. As noted above (Schulze, 1962b), in a continuous system thermophily was maintained at the high feed rate even though the pH dropped to 5.6. This discrepancy is probably a function of batch versus continuous composting.

Wiley (1956) noted that calcium carbonate had little effect on the pH profile. Golueke *et al.* (1954) found this material to moderate the initial pH decrease, but the only long-term effect compared to the control compost was a lowered nitrogen content resulting from ammonia volatilization. Addition of calcium hydroxide reduced the time to peak temperatures from 2.5 to 1.5 days and shortened the time to completion (Wiley, 1956). Schulze (1965) added lime to the waste as a pH corrective. However, there seems to be no universal requirement for deliberate pH adjustment in composting.

C. INTERACTIONS AMONG MOISTURE, AERATION, AND TEMPERATURE

1. Moisture

a. General. Water is produced metabolically during composting, and is evaporated from the materials throughout the process. Gains and losses are related to aeration and temperature, since water is a product of aerobic respiration while evaporation is a function of exposure to air currents and temperature (Wiley and Pearce, 1957). Depending on conditions within the compost (and perhaps on the weather) the water content may increase, decrease, or remain the same. During prolonged processing, composts usually tend to dry. Frequent turnings of windrows caused excessive drying (Anonymous, 1955). A turning schedule based on initial moisture content has been proposed (Golueke, 1972). Nonuniformity of moisture must be expected in unenclosed batches.

Wiley and Pearce (1957; see also Wiley, 1956) estimated the amount of water produced metabolically during laboratory-scale batch composting and losses from the system. Evaporative losses generally exceeded water production during the trials (average duration, 7 days). It was suggested that the rate of water production is a good measure of decomposition during composting.

Moisture and carbon dioxide in the exhaust increased and decreased in parallel with the temperature (Wiley and Pearce, 1957). Using estimates of water and carbon dioxide production and the VM mass balance, an equation was developed to represent the oxidation of the organic materials. The one equation encompassed all the temperature stages.

b. Optimization. Operating experience and studies based on specific criteria are in general agreement that the 50–60% moisture range is favorable for composting (Poincelot, 1972). Wiley (1955) related the initial moisture content to temperature changes in laboratory-scale batches. Stirring and aeration were set at favorable predetermined rates. Low moisture (40–53%) and high moisture (72–77%) resulted in average peak temperatures of 55°C and 44°C, respectively. The comparable value at intermediate moisture (55–69%) was 63°C. The intermediate moisture range was also most favorable as judged by VM and dry weight losses. It is probable that very different results would have been forthcoming at different aeration rates.

Utilizing temperature as the criterion for judging the optimum moisture content is complicated by the effect of water on heat capacity. Thus, the temperature of a relatively dry material of lesser biological activity may increase at a rate similar to that of a wetter mass supporting the more intensive biological activity (Bartholomew and Norman, 1953). This complication is best avoided by selecting a criterion (or criteria) of process efficiency other than temperature.

Oxygen consumption is one such criterion, and it was used in testing batches of waste incubated at 40°C (Anonymous, 1955). During the first 27 hours the samples initially at approximately 45% moisture consumed the most oxygen, but thereafter the somewhat wetter materials became more active. Based on the entire experimental period of 147 hours the material at 56% moisture proved the most active. Below 50%

moisture the end product appeared to be poorly composted, while above 60% there was evidence of anaerobiasis.

In rotating units, a moisture content of 70% resulted in clumping, with the potential for anaerobiasis (Stoller et al., 1937; Schulze, 1962b). This was corrected by adding newspaper or other conditioning materials and by rotating for brief periods only. A similar problem was encountered in turning excessively wet material with a fork (Anonymous, 1955).

Schulze (1961) found that finished compost incubated at 20°C in respirometer flasks consumed oxygen twice as fast at 60% moisture than at 50%. A slightly lesser activity at 70% moisture compared to 60% was attributed to restricted oxygen diffusion. In later studies of continuous-thermophilic composting on a laboratory scale (Schulze, 1962b) and on an operational scale (Schulze, 1965), the moisture content was adjusted so that the material in the digester remained, for the most part, within the 50% to 60% range.

High concentrations of thio alcohols were formed in poorly aerated municipal waste compost wetted to 65% moisture, causing odors (Francis, 1967). The problem was corrected by operating at 55% moisture.

For operational purposes moisture content cannot be considered apart from permeability as it affects aeration. The permeability of compost to gases is strongly influenced by the moisture content, but other factors are also of importance. Two determinations have been proposed to take into account simultaneously the water status and permeability. Wiley (1957a) suggested that the percent liquid, which includes water and lipid content, more completely represented the resistance to gas permeation than the conventional percentage of moisture, and was therefore the more useful parameter. Presumably lipids are in the liquid state at compost temperatures. Schulze (1962b) adapted the measurement of free air space from the soils literature. It is strongly influenced by the bulk weight and moisture content, and may reflect permeability to gases with reasonable accuracy. A minimum of 30% free air space appeared to be necessary for favorable composting.

2. Aeration and Temperature—Batch Operation

a. General. An oxygen gradient was noted in a solid waste windrow (Wiley and Spillane, 1961). For example, 5 days after turning and grinding (both were done together) the interstitial gas 30 cm and 60 cm in from the surface measured 7.0% and 1.0% oxygen by volume, respectively. Turning and grinding a different week-old batch increased the oxygen content at 38 cm from 10.1% to 18.6%. The corresponding temperatures immediately before and after the disturbance were 68°C and 40°C. Five hours later oxygen had decreased to the low point of 1.2% while the temperature recovered to 56°C. No attempt was made to differentiate between turning and regrinding as stimulators of oxygen consumption. In subsequent field demonstrations (Kochtitzky et al., 1969), the minimum oxygen content noted in the windrows was 0.5%. This condition apparently did not result in anaerobic symptoms.

Schulze (1958) found that compost temperature increased and decreased in parallel with the rate of oxygen uptake. In pure cultures (Cooney et al., 1968), the rate of heat evolution was found to correlate with the rate of oxygen consumption. Although the highest oxygen uptake rate attained was much greater in a well-aerated compost than in a partially anaerobic one, the cumulative uptakes were similar (Galler and Davey, 1971). The comparison included all the data collected until the composts cooled to 42°C (139 hours and 210 hours, respectively). Rates of carbon dioxide and heat production were related (Wiley, 1957b). Respiration quotients (RQ) in the range of 0.8 to 0.9 have been recorded (Wiley, 1955; Wiley and Pearce, 1957; Schulze, 1958). Bacterial plate counts followed the same pattern as oxygen uptake rates (Anonymous, 1955).

b. Optimization. Wiley and Pearce (1957) varied aeration rates to laboratory-scale units while holding other factors as constant as possible within favorable limits. The effect of aeration was judged mainly in terms of temperature. Within the range of aeration rates tested, all batches peaked between 57°C and 60°C. However, the time elapsed to the peak and the rapidity of cooling differed. With low rates (4.0–6.4 ft^3 air per day per pound initial VM) the temperature peaked on day 7, whereas this occurred on day 4 at both higher ranges. At the high rate (33–78 ft^3) there was rapid cooling and excessive drying. This material was prone to reheat upon rewetting. The intermediate flow rate (9–29 ft^3) proved most favorable, since thermophilic temperatures were maintained for a period of days, and the product appeared satisfactory.

Wiley (1956, 1957b) compared composting efficiency of batches in which the peak temperatures were moderated by cooling with the aid of a water jacket to that of noncooled controls. Efficiency was based on VM loss, carbon dioxide production, and oxygen consumption. Contrary to popular conception, by these criteria the cooled batches (peak temperatures ranged from approximately 55°C to 63°C) proved superior to the controls (peak range, 67°C to 73°C). The composts produced at the cooler temperatures tended to reheat, however. A process temperature peak of 49°C was unsatisfactory. Schulze (1958) tabulated data from three unpublished sources which indicated that oxygen uptakes were at a maximum at relatively cool temperatures (perhaps 45°C to 50°C). On the contrary, his experimental results (Schulze, 1958) showed that, over the entire range tested (33–62°C), the rate of oxygen consumption was positively related to temperature. Reconciliation of these conflicting findings does not seem presently possible, but two comments are offered in this connection. One concerns the extreme difficulty of reproducing composting conditions and the virtual certainty that significantly different experimental conditions prevailed among the investigations. The second comment, on a more fundamental level, concerns the concept of temperature optimum as applied to the batch process. Temperature optimum has no easily understood meaning when the temperature is always changing, as in batch composting. In contrast, the concept of temperature optimum for the continuous process is easily grasped.

3. Aeration and Temperature—Continuous Operation

a. General. Schulze (1926b) adjusted the air supply to his laboratory-scale unit so that oxygen comprised 5–10% of the exhaust gas. This required aeration rates of between 5.0 and 11.8 ft^3 per pound of VM per day, depending on loading. Oxygen in the exhaust was highest and compost temperature lowest just before feeding, after which these values changed in opposite directions. In the full-scale plant (Schulze, 1965), measurement of the residual oxygen in the exhaust was not possible, but the content in the interstitial gas was determined. Most of the readings fell between 3% and 12%. Since the time of that publication closer control of oxygen content has been achieved through installation of an automatic device to regulate the air supply (J. S. Coulson, Fairfield Engineering Co., Marion, Ohio, personal communication). The introduction of fresh waste caused a temporary localized depression of the temperature to a minimum of 55°C.

b. Optimization. Schulze (1962b) developed an equation to represent the relation between the rate of oxygen consumption and temperature over the range permitted by his data (approximately 35°–65°C). The results of previous batch trials (Schulze, 1958) were strongly taken into account, perhaps unjustifiably. Examination of the values from the continuous operation alone shows these to fall into two distinct patterns corresponding to the mesophilic and thermophilic ranges. The discontinuity between the groups of values comes at approximately 50°C. Apparently this information would be more accurately summarized in different equations, one for each temperature range. A priori it seems advisable to treat data representing mesophilic and thermophilic conditions separately. Nevertheless, this valuable work shows that the highest rates of oxygen consumption occurred at the highest temperatures reached through self-heating. There appears to be no reliable published data beyond 65°C.

At fixed temperatures, losses of cellulose, hemicellulose, and total weight from horse manure were slightly greater at 65°C than at 50°C for the first 9 days of a 47-day incubation (Waksman *et al.*, 1939b).

Thereafter the losses at 50°C exceeded those at 65°C. The amount of protein increased with time, probably reflecting higher rates of microbial synthesis than degradation. The lignin fraction appeared to be relatively inert. As judged by all the parameters followed, activity at 75°C was slight.

VII. Conclusion

To the microbial ecologist, the composting ecosystem is particularly fascinating because of the rapid and broadly predictable population successions and their interactions with temperature. To the practical environmentalist, the process represents a flexible and ecologically sound technology for treating solid wastes. The synthesis of these views should be rewarding on both levels.

Acknowledgments

One of us (MLM) received support from Grant EPA-T-900350, U.S. Environmental Protection Agency. We thank Dr. T. B. S. Prakasam for reading the manuscript.

References

Alexander, M. (1971). "Microbial Ecology." Wiley, New York.
Anonymous. (1953). *Univ. Calif., Berkeley, Sanit. Eng. Res. Lab., Tech. Bull.* No. 9.
Anonymous. (1955). "Preliminary Report on a Study of the Composting of Garbage and other Solid Organic Wastes." Civil Sanit. Eng. Dept., Michigan State University, Ann Arbor.
Anonymous. (1965). "Technical and Planning Aspects of Solid Wastes." Ohio Dept. of Health, USPHS, Columbus.
Anonymous. (1968). "Solid Waste Management/Composting." U.S. Dept. Health, Education and Welfare, Public Health Service, Washington, D.C.
Anonymous. (1970). "Municipal Refuse Disposal," 3rd ed. Public Administration Service, Chicago, Illinois.
Anonymous. (1971). "Composting of Municipal Solid Wastes in the United States." Publ. SW-47r., U.S. Environmental Protection Agency, Washington, D.C.
Anonymous. (1973a). *Ambio* **2**, 126.
Anonymous. (1973b). "The Terex 74-51 Composter," Rep. Terex Div., General Motors, Hudson, Ohio.
Bartholomew, W. V., and Norman, A. G. (1953). *J. Bacteriol.* **65**, 228.
Bausum, H. T., and Matney, T. S. (1965). *J. Bacteriol.* **90**, 50.
Block, S. S. (1965). *Appl. Microbiol.* **13**, 5.
Bond, R. G., and Straub, C. P. (1973). "Handbook of Environmental Control," Vol. II. CRC Press, Cleveland, Ohio.
Boussingault, J. B. (1845). "Rural Economy" (transl. by G. Law). Appleton, New York.

Bowes, P. C. (1956). *Wood (London)* **21**, 20.
Brock, T. D., and Freeze, H. (1969). *J. Bacteriol.* **98**, 289.
Browne, C. A. (1929). *U.S., Dep. Agr., Tech. Bull.* **141**.
Browne, C. A. (1933). *Science* **77**, 223.
Carlyle, R. E., and Norman, A. G. (1941). *J. Bacteriol.* **41**, 699.
Carnes, R. A., and Lossin, R. D. (1970). *Compost Sci.* **11** (5), 18.
Chang, Y. (1967). *Trans. Brit. Mycol. Soc.* **50**, 667.
Chang, Y., and Hudson, H. J. (1967). *Trans. Brit. Mycol. Soc.* **50**, 649.
Christensen, C. M., and Gordon, D. R. (1948). *Cereal Chem.* **25**, 40.
Christian, J. H. B. (1963). *Recent Advan. Food Sci.* **3**, 248–255.
Cook, H. A., Cromwell, K. L., and Wilson, H. A. (1967). *Proc. W. Va. Acad. Sci.* **39**, 107.
Cooney, C. L., Wang, D. I. C., and Mateles, R. I. (1968). *Biotechnol. Bioeng.* **11**, 269.
Cooney, D. G., and Emerson, R. (1964). "Thermophilic Fungi." Freeman, San Francisco, California.
Cross, T. (1968). *J. Appl. Bacteriol.* **31**, 36.
Cross, T., and Goodfellow, M. (1973). *In* "Actinomycetales: Characteristics and Practical Importance" (G. Sykes and F. A. Skinner, eds.), pp. 11–112. Academic Press, New York.
Davey, C. B. (1953). *Soil Sci. Soc. Amer., Proc.* **17**, 59.
Dye, M. H. (1964). *N.Z.J. Sci.* **7**, 87.
Dye, M. H., and Rothbaum, H. P. (1964). *N.Z.J. Sci.* **7**, 97.
Eastwood, D. J. (1952). *Trans. Brit. Mycol. Soc.* **35**, 215.
Edwards, O. F., and Rettger, L. F. (1937). *J. Bacteriol.* **34**, 489.
Erikson, D. (1952). *J. Gen. Microbiol.* **6**, 286.
Eweson, E. (1973). *Pollut. Eng.* **5**, 38.
Farkasdi, G. (1965). *Compost Sci.* **6** (1), 11.
Farrell, J., and Campbell, L. L. (1969). *Advan. Microbiol. Physiol.* **3**, 83.
Fergus, C. L. (1964). *Mycologia* **56**, 267.
Festenstein, G. N., Lacy, J., Skinner, F. A., Jenkins, P. A., and Pepys, J. (1965). *J. Gen. Microbiol.* **41**, 389.
Finstein, M. S., and Alexander, M. (1962). *Soil Sci.* **94**, 334.
Finstein, M. S., and Arent, D. (1974). *Compost Sci.* **15** (5), 6.
Forsyth, W. G., and Webley, D. M. (1948). *Proc. Soc. Appl. Bacteriol.* p. 34.
Francis, C. W. P. (1967). *Water Pollut. Contr.* **66**, 19.
Franz, M. (1972a). *Compost Sci.* **13** (3), 16.
Franz, M. (1972b). *Compost Sci.* **13** (6), 6.
Fuller, W. H., and Bosma, S. (1965). *Compost Sci.* **6** (2), 26.
Galler, W. S., and Davey, C. B. (1971). *In* "Livestock Waste Management and Pollution Abatement," Publ. PROC-271, pp. 159–162. Amer. Soc. Agr. Eng., St. Joseph, Michigan.
Gaskill, J. O., and Gilman, J. C. (1939). *Plant Physiol.* **14**, 31.
Gaughran, E. R. L. (1947). *Bacteriol. Rev.* **11**, 189.
Glathe, H. (1959). *Int. Res. Group Refuse Disposal, Inform. Bull.* **7**, 9.
Golueke, C. G. (1972). "Composting." Rodale Press, Emmaus, Pennsylvania.
Golueke, C. G. (1973a). *Crit. Rev. Environ. Contr.* **3**, 261.
Golueke, C. G. (1973b). *Compost Sci.* **5** (3), 5.
Golueke, C. G., and Gotaas, H. B. (1954). *Amer. J. Pub. Health* **44**, 339.
Golueke, C. G., Card, B. J., and McGauhey, P. H. (1954). *Appl. Microbiol.* **2**, 45.
Gordon, R. (1973). *In* "Handbook of Microbiology" (A. J. Laskin and H. A. Lechevalier, eds.), pp. 71–88. CRC Press, Cleveland, Ohio.

Gotaas, H. B. (1956). *World Health Organ.,* **31**.
Gregory, P. H., Lacey, M. E., Festenstein, G. N., and Skinner, F. A. (1963). *J. Gen. Microbiol.* **33**, 147.
Haldane, J. S., and Makgill, R. H. (1923). *Fuel Sci. Pract.* **ii**, 380.
Henssen, A. (1957). *Arch. Mikrobiol.* **27**, 63.
Hildebrandt, F. (1927). *Zentralbl. Bakteriol., Parasitenk., Infektionskr. Hyg., Abt.* 2 **71**, 440.
Hirano, J., Mutoh, Y., Kitamura, M., Asai, S., Nakajima, I., and Takamiya, A. (1958). *J. Gen. Appl. Microbiol.* **4**, 188.
Hortenstine, C. C., and Rothwell, D. F. (1972). *J. Environ. Qual.* **1**, 415.
Howard, A. Sir. (1940). "An Agricultural Testament." Oxford Univ. Press, London and New York.
Hoyle, D. A., and Mattingly, G. E. G. (1954). *J. Sci. Food Agr.* **5**, 54.
James, L. H., Bidwell, G. L., and McKinney, R. S. (1928). *J. Agr. Res.* **36**, 481.
Jann, G. J., Howard, D. H., and Salle, A. J. (1959). *Appl. Microbiol.* **7**, 271.
Kaila, A. (1952). *Suom. Maataloustieteellisen Seuran Julka* **78**, 1.
Kane, B. E., and Mullins, J. T. (1973). *Mycologia* **65**, 1087.
Kochtitzky, O. W., Seaman, W. K., and Wiley, J. S. (1969). *Compost Sci.* **9** (4), 5.
Krige, P. R. (1964). *J. Inst. Sewage Purif.* Part 3, p. 215.
Lacey, J. (1973). *In* "Actinomycetales: Characteristics and Practical Importance" (G. Sykes and F. A. Skinner, eds.), pp. 231–251. Academic Press, New York.
McClure, K. E., Klosterman, E. W., and Johnson, R. R. (1970). Ohio Report on Research and Developments in Agriculture Home Economics and Natural Resources. Volume 55, pp. 78–79. Ohio Agricultural Research and Development Center, Wooster, Ohio.
Malin, H. M., Jr. (1971). *Environ. Sci. Technol.* **5**, 1088.
Mason, H. G. (1969). *Compost Sci.* **10** (1–2), 26.
Mays, D. A., Terman, G. L., and Duggan, J. C. (1973). *J. Environ. Qual.* **2**, 89.
Miehe, H. (1907). "Die Selberhitzung des Heus." Fischer, Jena.
Miehe, H. (1930). *Arch. Mikrobiol.* **1**, 78.
Morris, M. L. (1975). M.S. Thesis, Rutgers University, New Brunswick, New Jersey.
Neilson, N. E., MacQuillan, M. F., and Campbell, J. J. R. (1957). *Can. J. Microbiol.* **3**, 939.
Niese, G. (1959). *Arch. Mikrobiol.* **34**, 285.
Norman, A. G. (1930). *Ann. Appl. Biol.* **17**, 575.
Norman, A. G., Richards, L. A., and Carlyle, R. E. (1941). *J. Bacteriol.* **41**, 689.
Obrist, W. (1966). *Compost Sci.* **6** (3), 27.
Oesterle, R., Rohde, G., and Rudat, K. D. (1963). *Compost Sci.* **4** (3), 19.
Olds, J. (1968). *Compost Sci.* **9**(1), 18.
Peterson, J. R., Lue-Hing, C., and Zenz, D. R. (1973). *In* "Recycling Treated Municipal Wastewater and Sludge through Forest and Cropland" (W. E. Soper and L. T. Kardos, eds.), pp. 26–37. Penn. State Univ. Press, University Park.
Peterson, M. L., and Stutzenberger, F. J. (1969). *Appl. Microbiol.* **18**, 8.
Poincelot, R. P. (1972). *Conn., Agr. Exp. Sta., New Haven, Bull.* **727**.
Poincelot, R. P., and Day, P. R. (1973). *Compost Sci.* **14** (3), 23.
Pöpel, F., and Ohnmacht, C. (1972). *Water Res.* **6**, 807.
Ramstad, P. E., and Geddes, W. F. (1942). *Minn., Agr. Exp. Sta., Tech. Bull.* No. 156.
Rose, W. W., and Mercer, W. A. (1968). "Fate of Insecticides in Composted Agricultural Waste," Progress Report Part I. West. Res. Lab., Nat. Canners Ass., Berkeley, California (cited by Golueke, 1973a).

Rothbaum, H. P. (1961). *J. Bacteriol.* **81**, 165.
Rothbaum, H. P. (1963). *J. Appl. Chem.* **13**, 291.
Rothbaum, H. P., and Dye, M. H. (1964). *N.Z.J. Sci.* **7**, 119.
Rothbaum, H. P., and Stone, H. M. (1961). *J. Bacteriol.* **81**, 172.
Scanlon, D. H., Duggan, C., and Bean, S. D. (1973). *Compost Sci.* **14** (3), 4.
Schulze, K. L. (1958). *Proc. Ind. Waste Conf.* **13**, No. 96, 541.
Schulze, K. L. (1961). *Compost Sci.* **2** (2), 32.
Schulze, K. L. (1962a). *Compost Sci.* **2** (4), 31.
Schulze, K. L. (1962b). *Appl. Microbiol.* **10**, 108.
Schulze, K. L. (1965). *Compost Sci.* **5** (3), 5.
Scott, W. J. (1957). *Advan. Food Res.* **7**, 83.
Shell, G. L., and Boyd, J. L. (1970). *Compost Sci.* **11** (3), 17.
Stoller, B. B., Smith, F. B., and Brown, P. E. (1937). *J. Amer. Soc. Agron.* **29**, 717.
Straub, H. (1950). *Ber. Abwassertech. Ver.* (cited by Schulze, 1962b).
Stutzenberger, F. J. (1972). *Appl. Microbiol.* **24**, 83.
Stutzenberger, F. J., Kaufman, A. J., and Lossin, R. D. (1970). *Can. J. Microbiol.* **16**, 553.
Tendler, M. D., and Burkholder, P. R. (1961). *Appl. Microbiol.* **9**, 394.
Terman, G. L., Soileau, J. M., and Allen, S. E. (1973). *J. Environ. Qual.* **2**, 84.
Thomsen, P. (1910). *Wiss. Meeresuntersuch., Abt. Kiel* [N.S.] **11**, 1.
Toth, S. J. (1968). *Compost Sci.* **9** (3), 27.
Vaux, W. G., Weeks, S. A., and Walukas, D. J. (1972). *Compost Sci.* **13** (2), 17.
von Klopotek, A. (1962). *Antonie van Leuwenhoek; J. Microbiol. Serol.* **28**, 141.
Waksman, S. A., Umbreit, W. W., and Cordon, T. C. (1939a). *Soil Sci.* **47**, 37.
Waksman, S. A., Cordon, T. C., and Hulpoi, N. (1939b). *Soil Sci.* **47**, 83.
Walker, I. K., and Harrison, W. J. (1960). *N.Z.J. Agr. Res.* **3**, 861.
Webley, D. M. (1947). *Proc. Soc. Appl. Bacteriol.* p. 83.
Wedberg, S. E. (1940). Ph.D. Thesis, Yale, University, New Haven, Connecticut.
Wilde, S. A. (1958). *Forest Prod. J.* **8**, 323.
Wiley, B. B., and Westerberg, S. C. (1969). *Appl. Microbiol.* **18**, 994.
Wiley, J. S. (1955). *Proc. Ind. Waste Conf.* **10**, No. 96, 306.
Wiley, J. S. (1956). *Proc. Ind. Waste Conf.* **11**, No. 96, 334.
Wiley, J. S. (1957a). *J. Sanit. Eng. Div., Amer. Soc. Civil Eng.* **83** (SA5), 1411–1.
Wiley, J. S. (1957b). *Proc. Ind. Waste Conf.* **12**, No. 96, 596.
Wiley, J. S. (1962a). *J. Water Pollut. Contr. Fed.* **34**, 80.
Wiley, J. S. (1962b). *J. Boston Soc. Civil Eng.* p. 13.
Wiley, J. S., and Pearce, G. W. (1957). *Trans. Amer. Soc. Civil Eng.* **122**, 1009.
Wiley, J. S., and Spillane, J. T. (1961). *J. Sanit. Eng. Div., Amer. Soc. Civil Eng.* **87** (SA5), 33.
Wylie, J. C. (1960). *In* "Waste Treatment" (P.C.G. Isaac, ed.), pp. 349–366. Pergamon, Oxford.

Nitrification and Denitrification Processes Related to Waste Water Treatment

D. D. FOCHT AND A. C. CHANG

Department of Soil Science and Agricultural Engineering, University of California, Riverside, California

I.	Introduction	153
II.	Biochemistry of Nitrification and Denitrification	155
	A. Nitrification	155
	B. Denitrification	157
III.	Environmental Factors Affecting Nitrification and Denitrification	161
	A. Aeration	161
	B. pH	164
	C. Temperature	166
	D. Carbon and Nitrogen Concentration	169
IV.	Comparative Waste Treatment Methods for Nitrification and Denitrification	173
	A. Activated Sludge Processes	174
	B. Trickling Filter	176
	C. Anaerobic Activated Sludge Process	177
	D. Submerged Filter	178
	E. Land Spreading	179
	F. Nitritification–Denitritification	180
V.	Summary and Conclusions	181
	References	182

I. Introduction

Biological treatment of waste water is probably the largest application of continuous culturing of microorganisms. In the initial primary treatment, colloidal suspended solids and dissolved organic matter are removed by microbial assimilation, oxidation, and precipitation. Secondary treatment of waste water has traditionally been the focal point for kinetic studies pertaining to microbial growth and the oxidation of organic matter. Reduction in the biochemical oxygen demand (BOD) of the treated effluent has thus been the ultimate goal in waste treatment. Prior to the use of the BOD test as a measure of organic strength of waste water, the appearance of nitrate in treated waste water was looked upon as a guideline for the completion of biooxidation. Although most of the effluent nitrogen leaving a domestic treatment plant seldom exceeds 50 mg/liter, approximately 85–90% is in the form of ammonium, 10–15% remains in the organic fraction, and only a small percentage is present as nitrate (McCarty and Haug, 1971).

Degradation of organic matter in waste water treatment renders most of the carbon innocuous as carbon dioxide, but ammonium, the end product of nitrogen mineralization, cannot be viewed as such. Ammonium in receiving waters can accelerate phytoplankton blooms, where phosphorus, light, or other factors are not limiting. The State of California has recently recommended that discharges into receiving waters not exceed 2 mg of N per liter to combat nutrient enrichment of surface waters. The introduction of ammonium into surface waters must also be considered in terms of the potential oxygen demand resulting from the conversion to nitrate by the ubiquitous nitrifying bacteria. The ease in which nitrate leaches through soil has also focused attention upon the potential deterioration of underground water supplies as well as surface impoundments. The U.S. Public Health Service established a standard of 10 mg of nitrate-nitrogen per liter as the maximum concentration recommended for potable water supplies on the basis of studies showing a statistical link between the incidence of methemoglobinemia (Lee, 1970) in infants and nitrate concentration in drinking waters.

Removal of nitrogen through tertiary ammonia stripping is a novel but unproved method that received considerable attention and enthusiasm as a result of the limited success of the South Tahoe, California treatment plant (Culp and Moyer, 1969). The major drawback to this method is scaling of the nitrogen removal tower with $CaCO_3$, which is brought about by the high lime concentration used to precipitate phosphates and to drive off ammonia at pH 11.0. Because the towers had to be hosed rather frequently with hydrochloric acid to prevent clogging, the N-stripping method was eventually abandoned by the South Tahoe Public Utility District. Nitrogen removal by volatilization also may have a profound environmental impact on surrounding areas. Luebs et al. (1974) have shown that half the total nitrogen from animal wastes in dairy corrals could be lost as ammonia. Significant nutrient enrichment of land and surface waters from animal waste disposal sites has been established by Hutchinson and Viets (1969). Amines constitute a small fraction of volatile nitrogen in waste treatment facilities, but they cause objectionable odors and are noticed at much lower threshold concentrations (0.2–50 ppb) than ammonia (49 ppm) (Luebs et al., 1974).

Although many bacteria can reduce molecular nitrogen to ammonium, the reverse reaction does not exist in nature. This seemingly peculiar aspect of the nitrogen cycle necessitates that ammonium (N^{3-}) must first be oxidized to nitrite (N^{3+}) or nitrate (N^{5+}) and then be subsequently reduced to produce molecular nitrogen (N^0). Thus, systems that were traditionally designed to remove organic matter from waste water are now in the process of being updated to include a tertiary step involving the removal of nitrate-nitrogen by denitrification. With 725 MGD (million gallon per day) municipal waste water treatment plants operating or under construction for nitrogen removal at various locations in the United States, more than 75% of the total waste water is scheduled to be treated by microbial nitrification and denitrification (Adams, 1973).

II. Biochemistry of Nitrification and Denitrification

A. NITRIFICATION

Early experiments on sewage purification by soil percolation led Schloesing and Müntz (1877) to the initial discovery that the production of nitrite and nitrate ions from ammonium was a biological process since it was stopped by the addition of chloroform or boiling water. Attempts to isolate the causative agents by the established procedures of Koch failed until Winogradsky (1890) realized that the oxidation of ammonium could supply energy and that the bacteria affecting this transformation would not need reduced carbon compounds and could synthesize cellular material by reduction of carbon dioxide or bicarbonate. Thus, the first concept of chemoautotrophic growth was realized. Soil perfusion columns and isolation of the causative bacteria led to the discovery that ammonium was oxidized sequentially to nitrite and nitrate by *Nitrosomonas* and *Nitrobacter*, respectively. Axenic culture studies on the nitrification process, for the most part, have focused on these bacterial genera. Other reported genera of nitrifying bacteria have not been well characterized, and many of them have been shown to be mixed cultures and to be unable to carry out nitrification as discussed in reviews by Meiklejohn (1954) and Bisset and Grace (1954).

The oxidation of ammonium to nitrite is an aerobic process, whereby oxygen serves in two ways by (1) direct incorporation of one atom into the substrate and (2) acceptance of electrons generated during electron transfer through the cytochrome system. Rees and Nason (1965) established that at least one of the atoms from molecular oxygen was incorporated into nitrite by a mixed function oxidase in *Nitrosomonas*, though the exact point in the pathway where this occurred was not known.

The initial oxidation of ammonium probably involves a two-electron transfer to form hydroxylamine inasmuch as cell suspensions of *Nitrosomonas* oxidize hydroxylamine without a lag period (Engel and Alexander, 1958) and produce large quantities of hydroxylamine from ammonium when hydrazine is used as an inhibitor (Hofman and Lees, 1952). Using ^{18}O, Verstraete and Alexander (1972) showed that the oxidation of ammonium to hydroxylamine by cells of *Arthrobacter* involved the incorporation of molecular oxygen. Although heterotrophic nitrification differs in many ways from autotrophic nitrification, it would appear that the same mechanism would exist for the oxidation of ammonium by *Nitrosomonas* since no energy is generated during this step. Verstraete (1975) suggests that molecular oxygen is probably introduced during the initial step in ammonium oxidation since oxidation of hydroxylamine has been shown to be linked to the cytochrome system.

Owing to the chemical instability of hydroxylamine and many of the proposed nitrogenous intermediates between ammonium and nitrite, it is difficult to establish whether specific reactions are chemical or biological. A logical two-electron transfer from hydroxylamine would produce a hypothetical nitroxyl, a compound whose existence has never been proven, probably because of its extreme instability. Aleem and Lees (1963) found that cell-free extracts of *Nitrosomonas*, when incubated with hydroxylamine and cytochrome c Fe^{3+}, catalyzed an oxidative condensation resulting in 2 moles of nitrite. They concluded that nitrification was thus a semicyclical process involving the participation of ferrocytochrome in the oxidation of hydroxylamine to a hypothetical nitroxyl, which condensed with nitrite to form nitrohydroxylamine as shown in Fig. 1.

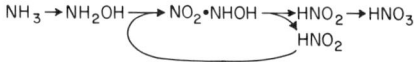

FIG. 1. Pathway of nitrification.

In the absence of oxygen, Anderson (1964) noted the formation of nitrous oxide, which was apparently generated nonenzymically. The same nonenzymic generation of nitrous oxide was also observed in resting-cell suspensions of *Nitrosomonas europeae* that were incubated for several days (Yoshida and Alexander, 1970). However, chemical production of nitrous oxide at neutral pH from hydroxylamine and nitrite has also been shown to occur with relative ease (Bothner-By and Friedman, 1952).

Oxidation of nitrite to nitrate by *Nitrobacter* is a simpler reaction that is well understood. Again the process is aerobic, although the role of oxygen is strictly as a terminal electron acceptor in reaction (1)

$$NO_2^- + XH_2 + O_2 \rightarrow NO_3^- + H_2O + X \qquad (1)$$

where X represents the species in the electron-transport chain originating with NAD and ending with cytochrome oxidase. Thus, Aleem *et al.* (1965) demonstrated conclusively that the oxygen atom in nitrate was generated from water and not from oxygen.

A recent review by Verstraete (1975) assesses the occurrence and significance of heterotrophic nitrification. Unlike autotrophic nitrification, which is linked to cellular growth and is proportional to the total cellular biomass, heterotrophic nitrification is independent of cell yields. Most of the major heterotrophic nitrification end products are formed, with few exceptions, during the stationary growth phase. Heterotrophic nitrification is common to diverse genera of bacteria, fungi, and actinomycetes. Both an organic and inorganic pathway, the latter similar to autotrophic nitrification, are recognized. Branches to the inorganic pathway may occur at each step in the organic pathway at any point from amine → oxime → C-nitroso → C-nitro to yield ammonium, hydroxylamine, nitroxyl, and nitrite/nitrate, respectively (Verstraete, 1975).

It is doubtful that significant quantities of nitrate are generated from heterotrophic nitrification in most natural systems. Autotrophic nitrification rates have been calculated to be ten times greater than heterotrophic nitrification (Verstraete, 1975). Addition of N-serve [2-chloro-6-(trichloromethyl)pyridine], which selectively inhibits autotrophic nitrification, fails to show the occurrence of significant nitrification in soil (Engel and Alexander, 1958). Nonetheless, heterotrophic nitrification may be prominent in atypical habitats having alkaline (Verstraete and Alexander, 1973) or acid pH values (Weber and Gainey, 1962; Becker and Schmidt, 1964), where autotrophic nitrification would not occur. However, organic heterotrophic nitrification may have other significant environmental implications, since many of these substituted hydroxylamines, amine oxides, nitroso and nitro compounds are toxic at concentrations of less than 1 mg/l (Verstraete, 1975).

B. Denitrification

Gayon and Dupetit (1886) noted the disappearance of nitrate and nitrite with the concomitant production of molecular nitrogen and nitrous oxide in sand columns perfused with nitrified sewage effluent. They concluded that the process was of biological origin, occurred in the absence of oxygen, and involved the reduction of nitrate to nitrite prior to gaseous formation. Weissenberg (1902) made the first assertion that reduction of nitrates and nitrites to gaseous products was brought about by aerobic bacteria that switched over to using the oxygen in nitrate when molecular oxygen was depleted. Although he was generally correct in his assertion, nitrate per se serves as a terminal electron acceptor in lieu of oxygen during the oxidation of organic matter. The coupled reduction of nitrate with the oxidation of carbonaceous compounds was recognized by earlier investigators and lead to the isolation of many species of heterotrophic, denitrifying bacteria from soil and water. Although some autotrophic bacteria can use reduced sulfur compounds (*Thiobacillus denitrificans*) or molecular hydrogen (*Micrococcus denitrificans*) as hydrogen donors, their ecological significance might appear to be minimal. Martin and Ervin (1953), however, noticed that nitrogen deficiency in citrus occurred in alkaline soils amended with sulfur, which they attributed to autotrophic denitrification.

Denitrification is distinct from assimilatory nitrate reduction and dissimilatory nitrate reduction in several ways. Assimilatory nitrate reduction involves the reduction of nitrate to ammonium for cellular synthesis, is not coupled with the respiratory electron transport chain, and is commonly found among higher green plants as well as microorganisms. Dissimilatory nitrate reduction involves the reduction of nitrate to nitrite, is coupled with the respiratory electron transport chain, and is found to occur among many species of bacteria (mostly facultative anaerobes) and among a few species of actinomycetes. Denitrification involves the reduction of nitrate to gaseous products of nitrogen (usually molecular nitrogen and nitrous oxide), is coupled with the respiratory electron transport chain and is found exclusively among the bacteria (usually aerobes). A comprehensive review of this subject has recently been written by Payne (1973).

Inasmuch as assimilatory nitrate reduction commonly occurs in the presence of oxygen, it would follow that assimilatory and dissimilatory nitrate reduction would be mediated by different enzymes. Payne (1973) pointed out that all denitrifying and nitrate respiring bacteria contain the dissimilatory reductase although not all appear to contain the assimilatory reductase despite their ability to utilize nitrate for nutritional purposes. The two reductases of *Aerobacter aerogenes* have been shown to have similar kinetic properties and pH optima (Van't Riet *et al.*, 1972), and no immunological distinction or separation between the two reductases of *E. coli* has been shown (Murray and Sanwal, 1963). Thus, these studies would suggest formation of a similar protein which combines in different ways depending on the regulatory control of the cell. Numerous studies by Pinchonoty's colleagues (see Payne, 1973, for a complete citation) have indicated that assimilatory nitrate reductase is a soluble enzyme that is inhibited by chlorate, cyanide, and azide and is not inhibited by oxygen: by contrast, the dissimilatory nitrate reductase is particle bound, inhibition by azide is competitive and reversible, and chlorate is reduced. Prakash and Sadana (1973) showed that dissimilatory nitrate reduction in *Achromobacter fischeri* resulted in the production of ammonium, although no energy was generated through reduction of nitrite. Inasmuch as nitrate increased cell yields under reduced oxygen tension while nitrite was shown to be toxic, they concluded that reduction of nitrite through the assimilatory pathway provided a detoxifying mechanism to the cells.

Beijerinck and Minkman (1910) were the first to suggest several intermediates to account for the successive reduction of nitrate to molecular nitrogen, though Kluyver and Donker (1926) proposed the first general scheme which followed a two-electron addition per substrate in the following order: nitrate, nitrite, dimerization of two nitroxyls to form hyponitrate, nitrous oxide, molecular nitrogen. They suggested that assimilatory nitrate reduction and denitrification diverged from hyponitrite and that hydroxylamine was the intermediate between hyponitrite and ammonium. Experiments with *Pseudomonas aeruginosa* (Allen and van Niel, 1952) and *Micrococcus denitrificans* (Kluyver and Verhoeven, 1954) were unable to show that hyponitrite was reduced. Allen and van Niel (1952) proposed nitramide as an intermediate and showed that it was rapidly converted to molecular nitrogen, although Kluyver and Verhoeven (1954) showed that nitramide was decomposed spontaneously in the absence of cells and dismissed this as an intermediate. Campbell and Lees (1967) simply postulated that assimilatory nitrate reduction was the reverse of denitrification and thereby suggested nitrohydroxylamine as the likely nitrogen dimer, from where the assimilatory and dissimilatory pathways diverged. Their scheme showed nitrous oxide to be direct precursor en route to molecular nitrogen and nitric oxide to be off the main pathway.

Although nitrous oxide has been observed in numerous denitrification studies since Gayon and Dupetit (1886), its role as a direct, rather than indirect, precursor to molecular nitrogen was dismissed on grounds that denitrifying cell suspensions presumably reduced nitrous oxide only after a lag period and that cells incubated with sodium azide or cyanide produced only molecular nitrogen from nitrite, yet would not reduce nitrous oxide (Allen and van Niel, 1952; Sacks and Barker, 1952). Objections to the first point (Focht, 1974) have been made on grounds that negative gas readings probably reflect diffusion from the gaseous phase into the solute and would account for a lag period if the rate of nitrous oxide reduction were concentration dependent. The generation of molecular nitrogen from nitrite is a point worth considering since experimental evidence has shown that reduction of nitrite and nitric oxide does not involve ATP phosphorylation and is mediated by constituents associated with the soluble fraction of the cell, whereas reduction of nitrous oxide does generate ATP phosphorylation and is linked to particulate cytochromes (Cox et al., 1971; Payne et al., 1971; Payne, 1973). Thus, it seems reasonable to conclude that nitrous oxide is a direct precursor to molecular nitrogen. The apparent anomaly in studies involving additions of high nitrite concentrations to heavy cell suspensions of bacteria may be resolved by noting that high nitrite concentrations repress the formation of both nitrous oxide and molecular nitrogen (Payne and Riley, 1969), which results in the accumulation of nitric oxide. Many of these earlier studies (Sacks and Barker, 1952; Allen and van Niel,

1952) did not use gas chromatography or a definitive method for identifying gases—particularly nitric oxide and molecular nitrogen, which have similar solubilities and vapor pressures.

Payne (1973) concluded that denitrification simply proceeded from $NO_3^- \to NO_2^- \to NO \to N_2O \to N_2$. The only unanswered questions in this scheme are (1) where the pathway diverges to assimilatory nitrate reduction and (2) the role of nitric oxide—a gas not usually observed at neutral pH values with intact cells. The assimilatory nitrate reduction pathway proposed by Payne (1973) is essentially the reverse of nitrification (i.e., $NO_3 \to NO_2 \to X \to NH_2OH \to NH_3$) and is similar to that proposed by Campbell and Lees (1967), where the unknown intermediate X is nitrohydroxylamine. The following pathway is proposed in Fig. 2, which incorporates concepts of Campbell and Lees (1967) and Payne (1973). The pathway thus considers each reductive step to be a two-electron transfer.

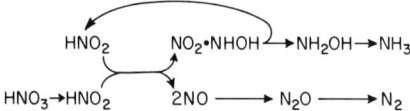

Fig. 2. Pathway of denitrification and of assimilatory and dissimilatory nitrate reduction.

Cox et al. (1971) and Cox and Payne (1973) isolated two soluble fractions of c-type cytochromes from P. perfectomarinus, which were involved exclusively in the reduction of nitrite and nitric oxide. Electron paramagnetic resonance (EPR) measurements revealed the formation of a heme–nitric oxide complex with a cytochrome c_{548} fraction, which eventually released nitric oxide. The second fraction, however, indicated formation of a different heme–nitric oxide complex that was more tightly bound and eventually released nitrous oxide. These studies are in agreement with those of Gordy and Rexroad (1961), who showed that two molecules of nitric oxide bind very strongly with the iron atom of hemoglobin. Thus, if these two molecules are reduced on the heme moiety to nitrous oxide, it is not too likely that free nitric oxide would be found in intact cell suspensions or in soil. Only where further reduction of nitric oxide is prohibitive (e.g., inactivation of nitric oxide reductase or repression of intact cells by high nitrite concentration) will it thus be released into the gaseous phase.

The branch point in the electron transport chain to nitrate or oxygen occurs at cytochrome c (Payne, 1973). Electron transport to nitrate involves the participation of a molybdenum–iron–cytochrome c complex (Fewson and Nicholas, 1961; Forget and Dervartanian, 1972), whereas transport to oxygen proceeds through cytochrome a_3 and does not require molybdenum. Cells grown anaerobically with nitrate produce more cytochrome c than when grown aerobically (Downey and Kiszkiss, 1969; Sapshead and Wimpenny, 1972; Daniel and Appleby, 1972).

III. Environmental Factors Affecting Nitrification and Denitrification

Because the rate at which a substrate will be metabolized is inherently dependent upon a number of environmental parameters, it is difficult to compare rates derived from different experimental conditions in a quantitative manner. Although the direction in which a specific physical condition influences a biochemical process may be known, there are frequently interactions among two or more of these variables. For example, the oxidation–reduction potential (E_h), is a factor that acts in opposite ways in denitrification and nitrification processes, yet is a direct function of pH and temperature by the Nernst equation. Biochemical parameters, such as organic matter concentration, may also have several functions by lowering the oxygen tension through increased respiratory activity, by increasing the biomass, and/or by serving as a reductant during the denitrification process. Thus, the term "maximal" or "optimal" rates must be used judiciously when referring to a given set of environmental conditions. The purpose of the section which follows is to characterize the effects of these variables upon empirical rate constants.

A. Aeration

In considering the aeration status of the environment, it is obvious that both nitrification and denitrification could not occur optimally in the same locale at the same time. However, this does not rule out the occurrence of both processes within close proximity of each other, albeit at rates less than maximal. The cyclic regeneration and sloughing of the zoogloeal film on the rocks of trickling filters is an example of the occurrence of oxidative and highly anoxic environments existing in close proximity, where the oxygen diffusion to the inside of the film becomes reduced as the biomass and oxygen demand near the outside increase. Wuhrmann (1964) has reviewed the conditions whereby anoxic environments can occur in microsites as small as the flocs in an activated sludge tank.

The spatial geometry of the environment is very important as it determines whether the rate of microbial respiration may exceed the rate of oxygen diffusion. Wuhrmann (1963a) showed that the net diffusion of oxygen from the outermost to innermost cell in a fixed volume decreased for the following geometries: sphere, cylinder, two-faced film, and one-faced film. Using a respiration rate of 10^{-4} mg O_2/cm^3 and a diffusion coefficient of 5×10^{-6} cm^2/sec, a dissolved oxygen concentration of 2 mg/liter on the outside would be reduced to a "critical level" of 0.1 mg/liter at the innermost part of a sphere having a 500-μm diameter. Yet it would require only a 145 μm diameter in a one-faced film to effect the same critical oxygen tension. Since flocs in activated sludge and films in trickling filters generally exceed these dimensions, it can be concluded that anaerobic processes are operative in what otherwise appears to be an "aerobic" environment.

Unaccounted-for losses of nitrogen in seemingly well-aerated soils have usually been ascribed to denitrification (Allison, 1955) though the gaseous losses have never been measured directly. The microsite concept (Greenwood, 1961) thus holds for soil as well. Starr et al. (1974) observed that concurrent nitrification–denitrification occurred in soil columns continuously perfused with ammonium and maintained at an average overall water content of 85–90% saturation. A comprehensive review of the literature by Wesseling and van Wijk (1957) showed that oxygen diffusion in soil became critical at about 85–90% saturation. Pilot and Patrick (1972) have shown that denitrification ceases at about 20–30 cm tension. Studies on the atmosphere of soils receiving animal wastes showed that nitrous oxide concentrations were significantly above ambient levels at tension readings less than 25 cm, indicating the cessation of denitrification at higher tensions (Focht et al., 1974). These "critical" tensions correspond approximately to about 85–90% saturation, depending on the soil. Bremner and Shaw (1958) showed that denitrification increased as the soils were ponded beyond their water-holding capacity. Presumably, the increased thickness of the aqueous layer resulted in greater reduction of oxygen diffusion.

The critical factor governing denitrification is the dissolved oxygen concentration, not the atmosphere composition. Despite this fact, many investigators still continue to make injudicious use of the terms aerobic or anaerobic on the basis of preconceived notions regarding the atmosphere in which the sample is incubated. Studies by Meek et al. (1969; personal communication, 1974) showed that gaseous oxygen concentrations had no correlations whatever to the reduction of dissolved oxygen, E_h, and nitrates in soils treated with animal waste. Concurrent high concentrations of nitrous oxide were observed in these same profiles despite the fact that the gaseous oxygen concentration never fell below 14% and remained closer to 17% throughout the study (Focht et al., 1974). Pure culture studies (Sacks and Barker, 1949) have shown that denitrification still occurred at ambient oxygen concentrations, although the rate was reduced to one-fourth that occurring in the absence of oxygen. The geometrical importance of gaseous to aqueous diffusion was shown by Collins (1955), who found that denitrification occurred with pure cultures of P. aeruginosa under apparent "aerobic" conditions, and that the shape of the culture flask was directly related to this.

In a thorough review of this subject, Painter (1970) concluded, with sufficient justification and documentation, that many studies, assumed to be conducted under aerated conditions, must be viewed as equivocal in lieu of soluble oxygen data.

The critical oxygen concentration of 0.1 mg/liter proposed by Wuhrmann (1964) is close to that observed when utilization of oxygen shifts from an apparent zero-order to an apparent first-order reaction at 0.13 mg/liter (Chance, 1957). The independence of oxygen concentration on respiration rates, except at very low concentrations, is apparently due to the small Michaelis constant ($K_m \cong 10^{-6}$ M) for respiration rates by microorganisms (Greenwood, 1961; Wimpenny, 1969; Painter, 1970). Thus, the rate-limiting step in the electron transport chain is not in the terminal oxidase, but usually at the primary dehydrogenase level (i.e., initial reduction of organic substrate). However, White and Sinclair (1971) have suggested that the rate-limiting step in the electron transport chain is at the branch point to the terminal electron acceptor by noting that the critical oxygen concentration is increased when cellular levels of cytochrome c are elevated by previous anaerobic induction with nitrate or nitrite as electron acceptor. Harrison (1973) reviewed the literature on the kinetics of oxygen consumption and suggested caution in the use of Michaelis–Menten kinetics to describe rate processes related to oxygen tension where enzyme concentrations of a branched-chain electron transport system are not constant.

A critical oxygen concentration of 0.2 mg/liter, above which denitrification did not occur, was reported in cultures of *Pseudomonas* (Skerman and MacRae, 1957) and in sewage (Dawson and Murphy, 1973). Denitrification in oceans has been observed at dissolved oxygen concentrations of 0.16 mg/liter (Richards and Broenkow, 1971) and 0.70 mg/liter (Goering and Cline, 1970). Goering (1968) found that the rate of denitrification in the ocean was approximately halved from 0.34 to 0.19 mg of N per liter per day when the dissolved oxygen concentration increased from 0.029 to 0.27 mg/liter. Wheatland et al. (1959) noted a similar half reduction in the rate of denitrification at 0.20 mg/liter as compared with the anaerobic rate in sewage. They also reported the occurrence of denitrification at 2 mg/liter, though the rate was reduced by 90%.

The aeration status may be expressed in terms of the oxidation–reduction potential (E_h) when dissolved oxygen concentrations near the critical level are difficult to measure with accuracy. Lower cost, ease of installation, and the occurrence of wetting and drying cycles in soils have made redox electrodes (usually platinum) the favored choice of soil scientists. Although the NO_3^-/NO_2^- couple determined under ideal conditions is 421 mV (Latimer, 1952), nitrate appears to be reduced at potentials between 300 and 350 mV in soil (Patrick, 1961; Meek et al., 1969; Pearsall and Mortimer, 1939; Bailey and Beauchamp, 1973a) and in culture (Kefauver and Allison, 1957). Cytochrome c, the point at which electron flow diverges to nitrate, is slightly below this with a range of 250–350 mV in *P. denitrificans* (Kamen and Vernon, 1955; Iwasaki, 1960), *P. aeruginosa* (Kamen and Takeda, 1956), *Micrococcus denitrificans* (Kamen and Vernon, 1955), and *Thiobacillus denitrificans* (Aubert et al., 1959).

The aeration status is important not only in influencing the rate of denitrification, but also in determining whether any intermediate products might accumulate. The observations that nitrite is reduced at about 200 mV, which is a lower potential than nitrate (Kefauver and Allison, 1957; Bailey and Beauchamp, 1973b) or many of the c-type cytochromes, probably explains why nitrite is commonly observed as a transient intermediate and is not reduced until most of the nitrite has disappeared in sewage (Dawson and Murphy, 1973), soils (Cooper and Smith, 1963; Bremner and Shaw, 1958; Cady and Bartholemew, 1960), lake sediments (Goering and Dugdale, 1966), and oceans (Richards and Broenkow, 1971; Carlucci and Schubert, 1969; Goering and Cline, 1970; Goering, 1968). However, most studies with cell suspensions (Sacks and Barker, 1949; Allen and van Niel, 1952), sewage (Prakasam and Loehr, 1972), and soil (Nommik, 1956; Bailey and Beauchamp, 1973a) have shown that nitrite, when added as substrate, is always reduced faster than nitrate, whether the two of them are incubated together or separately, under anaerobic conditions. This apparent anomaly can be resolved when it is remembered that high nitrite concentrations inhibit nitrate reduction (Payne, 1973). Consequently, where nitrite exists in low concentrations as a transient intermediate, redox potential would be a factor determining whether or not nitrite would be reduced.

Since microbial respiration is generally independent of oxygen concentration beyond the "critical" point, it might be expected that nitrification would respond in the same way. Schoberl and Engel (1964), however, report relatively high Michaelis constants of 0.5 and 1.0 mg/liter for oxygen consumption by *Nitrosomonas* and *Nitrobacter*, respectively. Loveless and Painter (1968) report a similar value of 0.3 mg/liter for *Nitrosomonas*, while Boon and Laudelout (1962) report a slightly lower value of 0.5 mg/liter for *Nitrobacter*. Wuhrmann (1963b) has suggested 2.0 mg/liter as the minimum level, above which nitrification would be constant based on experiments in treatment plants, showing reduced nitrification rates at 1 mg/liter.

B. pH

Both nitrification and denitrification processes occur optimally at a neutral to slightly alkaline pH. Though heterotrophic nitrification by fungi and acidophilic bacteria have been demonstrated, one can only speculate on the ecological significance. Verstraete (1975) suggested that heterotrophic nitrification may be significant in peat and acid forest soils. Under such conditions (pH 4) any gaseous losses of nitrogen that might occur would be through chemical decomposition of nitrous acid to nitric oxide or to nitrogen by the Van Slyke reaction.

Early studies by Meyerhof (1916a,b, 1917) showed that the pH optima of *Nitrosomonas* and *Nitrobacter* ranged between 8.5 and 8.8, and the growth ranges were between 7.6 and 9.4 and 5.7 and 10.2, respectively. The limiting-growth range of *Nitrosomonas* at pH 7.6 has been discredited in the light of further studies, which show that nitrification still occurs at pH 5.0 in soil and sewage (Meiklejohn, 1954; Heubült, 1929; Prakasam and Loehr, 1972). There are apparent differences among strains of *Nitrosomonas* regarding pH optima as witnessed by narrow and wide flat-topped curves reported by different investigators (Loveless and Painter, 1968; Hofman and Lees, 1953; Engel and Alexander, 1958). Nevertheless, the optimum pH in all cases is above 7.0. Although *Nitrobacter* can grow in pure culture up to pH 10.2 when supplied with nitrite, it is rather unlikely that growth in soil or highly nitrogenous waste waters would be prolific since free ammonia is toxic to the bacterium. Although Prakasam and Loehr (1972) found that nitrification was unaffected up to pH 11.0, they observed that a free ammonia concentration greater than 0.02 ppm N inhibited further oxidation of nitrite. Broadbent et al. (1957) observed the accumulation of nitrite and further inhibition of oxidation at pH 8.5 in several soils; this inhibition was more marked with increased concentrations of ammoniacal nitrogen and decreased clay content. Thus, the effect of elevated pH on the nitrification process must be considered as it relates to the concentration of free ammonia in solution ($pK_a = 9.24$).

Denitrification is less influenced by increased alkalinity than nitrification. Dawson and Murphy (1972) have shown that denitrification rates give parabolic curves as a function of pH with a peak at 7.0. The rates at pH 6.0 and pH 8.0 were approximately halved. However, Nommik (1956), Wiljer and Delwiche (1954), and Bremner and Shaw (1958) have shown that the rate of denitrification increases linearly from pH 4, levels off between pH 7 and 8, and declines, though not ceasing, to pH 9.5. Neutral to slightly alkaline pH ranges not only effect faster rates of denitrification, but also the completion to N_2. Proportionally less nitrous oxide is observed in soil as the pH rises (Nommik, 1956; Wiljer and Delwiche, 1954; Hauck and Melsted, 1956). This cannot be explained on increases in solubility of nitrous oxide since it diminishes with increasing pH; thus, the rate of nitrous oxide reduction apparently increases with pH faster than its rate of formation.

Nitric oxide also is a gaseous product of denitrification that is usually found under acid conditions (Nommik, 1956; Wiljer and Delwiche, 1954; Bollag et al., 1973; Cady and Bartholemew, 1960). Questions have been raised as to whether it is formed chemically from decomposition of nitrous acid or biochemically. Bollag et al. (1973) concluded that formation of nitric oxide in acid soils was largely chemical since sterilized soils evolved as much nitric oxide as controls upon addition of nitrite.

C. Temperature

Temperature is undoubtedly the most difficult environmental parameter to control in waste treatment facilities and causes the greatest problem during the winter months. In most environments, nitrification and denitrification proceed at suboptimal temperatures. The optimal temperature for nitrification ranges between 30° and 36°C (Meiklejohn, 1954; Buswell et al., 1954; Deppe and Engel, 1960). Laudelout and van Tichelen (1960) reported an optimal nitrification temperature of 42°C with growth diminishing but continuing at 49°C. They also found that oxygen consumption rates were maximal at 49°C. Denitrification, by contrast, is optimal between 65° and 75°C and ceases at 85°C (Nommik, 1956). The high temperature optimum is presumably due to the exclusive predominance of themophilic species of *Bacillus*. Since nitrification is affected by highly specialized bacteria of extremely narrow species diversity, it is not surprising that the process does not encompass wider temperature ranges in contrast to denitrification.

Kinetic constants usually conform to the Arrhenius equation for nitrification though Buswell et al. (1954) did not obtain a linear plot of log K vs. T^{-1}. Dawson and Murphy (1972) showed that denitrification conformed to Arrhenius kinetics from 3 to 28°C although most other studies have shown that the rate is affected proportionally more below the 10–15°C range (Nommik, 1956; Bremner and Shaw, 1958; McCarty et al., 1969; Bailey and Beauchamp, 1973a). The sharp break in the temperature curve at lower temperatures is generally thought to be due to the effect of physical factors such as solubility or diffusion becoming more pronounced or to one of several reactions exerting a greater influence at a respective temperature range (Ingraham, 1962). The quotient (Q_{10}) of K_T/K_{T-10} enables one to express the temperature (T) as a direct, rather than reciprocal, function of the rate constant (K) and to partition sections of a nonlinear curve for selected temperature ranges. Table I lists Q_{10} values reported in the literature or calculated from tables and graphs where they are not given. Although there is considerable range among the tabulated Q_{10} values, certain distinct patterns emerge. Lower temperature ranges for both nitrification and denitrifica-

TABLE I
Temperature Coefficients for Nitrification and Denitrification

Reaction and initial substrate	Maximum rate, per day	Q_{10}	System	Determination	Reference
Denitrification, nitrate	4.3 μg N/ml/mg	3.0	Mixed cultures	Nitrate loss	Dawson and Murphy (1972b)
	77 μg N/gm	1.6, 16[a]	Soil	Gaseous N	Nommik (1956); Focht, (1973)
	115 μg N/gm	3.0, 16[b]	Soil #1	Nitrate loss	Bremner and Shaw (1958)
	125 μg N/gm	3.2	Soil #4		
	310 μg N/gm	1.4	Manhattan soil	Nitrate loss	Cooper and Smith (1963)
	520 μg N/gm	2.5	Millville soil		
	480 μg N/gm	2.9	Yolo soil		
	210 μg N/gm	2.5	Soil	Nitrate loss	Bailey and Beauchamp (1973b)
	238 μg N/gm	3.6	Soil	Gaseous N	Mulbarger (1971)
	600 μg N/MLVSS	2.6	Activated sludge	Nitrate loss	Bailey and Beauchamp (1973b)
Denitrification, nitrite	275 μg N/gm	2.4, 5.0[c]	Soil	Nitrite loss	
	326 μg N/gm	3.2	Soil	Gaseous N	Nommik (1956); Focht, (1973)
N_2O reduction, nitrate	55 μg N/gm	1.4	Soil	Gaseous N	Cooper and Smith (1963)
	480 μg N/gm	1.8	Manhattan soil	Gaseous N	
	900 μg N/gm	2.7	Millville soil		
	520 μg N/gm	2.0	Yolo soil		
Nitrification, ammonium	2.2	1.7, 8.0[d]	Nitrosomonas	Growth rate constant	Buswell et al. (1954)
	1.9	1.9, 8.0[d]	Nitrosomonas	Nitrite rate constant	
	2.0	2.7	River water	Growth rate constant	Knowles et al. (1965)
	180 μg/gm MLVSS[e]	2.3	Activated sludge	Loss of Kjeldahl N	Mulbarger (1971)
Nitrification, nitrite	2.0	1.9	River water	Growth rate constant	Knowles et al. (1965)
	1.5 ml O_2	1.9	Nitrobacter	Oxygen uptake	Laudelout and van Tichelen, (1960)

[a] 3–21°C.
[b] 2–12°C.
[c] 5–15°C.
[d] 10–22°C.
[e] Mixed liquid volume of suspended solids.

tion have higher Q_{10} values than temperature ranges above 15°C, for reasons discussed earlier. Higher Q_{10} values for denitrification occur in all but one case (Yolo soil; Cooper and Smith, 1963) when the rates are derived from gaseous product formation than when derived from substrate disappearance. Most likely, this is due to the decreased solubility of nitrous oxide and molecular nitrogen with increasing temperature. Although denitrification rates are generally higher with nitrite than with nitrate as substrate, there do not appear to be any significant differences between Q_{10} values.

Changes in temperature may also influence the composition of the products formed during denitrification if the distinct reactions have different Q_{10} as determined under the same experimental conditions. Although the relative proportions of nitrous oxide and molecular nitrogen produced during denitrification vary slightly, nitric oxide is detected in greater quantities at lower temperatures in soil (Bailey and Beauchamp, 1973a; Nommik, 1956). Bailey and Beauchamp (1973a) observed no reduction of nitrate and no formation of gaseous products in soil after 22 days at 5°C. However, they did detect nitric oxide as the only gas formed when soils were incubated at 5°C with nitrite. They concluded that reduction of nitrate ceased at 5°C and that the evolution of nitric oxide from nitrite was produced by chemodenitrification. Inasmuch as the pH of their soil (7.1) was rather high for chemodenitrification to occur, and since much greater quantities of nitric oxide were evolved at higher temperatures from nitrite or nitrate, their conclusions regarding chemodenitrification are equivocal in lieu of controls using sterilized soils. What appears more probable is that nitrite inhibited further reduction of nitric oxide much like that observed in pure culture, as discussed in Section II,B. Most studies, however, have shown that denitrification occurs at temperatures below 5°C when nitrate is used as substrate (Nommik, 1956; Bremner and Shaw, 1958; Mulbarger, 1971; Dawson and Murphy, 1972).

Novak (1974) considered the effects of temperature upon the Michaelis constant K_m and the maximum rate constant V_{max} and modified the standard Michaelis–Menten equation to consider exponential coefficients for both K_m and V_{max}. He also showed that Q_{10} was influenced by substrate concentration—particularly at the lower levels. Q_{10} was inversely and directly proportional to substrate concentration under anaerobic and aerobic conditions, respectively. When plotted as a function of substrate concentration, Q_{10} was almost linear for complex waste from substrate concentrations approaching zero and fell roughly along the same curve where acetic concentrations were greater than 1000 mg/liter. The change in Q_{10} for acetic acid concentrations between 0 and 1000 mg/liter was exponential. Under aerobic conditions, there was relatively little effect of glucose concentration from 0 to 500 mg/liter on Q_{10}. Synthetic waste concentrations from 20 to 200 mg/liter, however, effectively tripled the Q_{10}.

D. Carbon and Nitrogen Concentration

1. Kinetics

The kinetics of nitrification were applied to waste water treatment by Downing *et al.* (1964), who used the Michaelis–Menten equation to show that continuous nitrification could occur in activated sludge if the residence time was sufficient to maintain the slower growing nitrifiers. Although a growth rate constant of 2.2 day^{-1} (8 hour generation time) has been reported in pure cultures for *Nitrosomonas* (Skinner and Walker, 1961), rates less than half this are normally observed in sewage and culture, as seen in Table I. the Michaelis constant for oxidation of ammonium generally ranges between 1 and 10 mg of N per liter for temperatures between 20° and 30°C (Hofman and Lees, 1953; Ulken, 1963; Loveless and Painter, 1968), and K_m for nitrite oxidation generally falls between 5 and 9 mg/liter (Lees and Simpson, 1957; Gould and Lees, 1960; Laudelout and van Tichelen, 1960; Ulken, 1963). Since K_m is a function of temperature (log K_m vs. T usually gives a straight line), it is not unusual that lower values of 0.2 mg of N per liter at 8°C (Knowles *et al.*, 1965) and higher values of 29 mg of N per liter at 49°C (Laudelout and van Tichelen, 1960) have been observed. Thus good agreement between reports on the rate constants has been obtained in pure culture or mixed culture systems, though rates in activated sludge are generally lower. McLaren (1971) studied the growth of nitrifiers in soil and suggested that rate constants determined for autotrophic bacteria should have a more uniform meaning on the basis of a narrow species diversity evolving from competitive exclusion for a highly specialized niche.

The kinetics for denitrification are far more complex than nitrification and have not been well developed. Because reduction of nitrate is coupled with the oxidation of a carbonaceous substrate, a kinetic approach must involve at least a dual substrate–enzyme(s) complex. Such equations have been developed (Bray and White, 1966) for idealized systems, but denitrification is far more complex because the nitrogenous intermediates presumably have different saturation constants, they may be competitively inhibited at specific redox potentials, and the association complex of reductant and oxidant may involve one or several enzymes. Unfortunately, most studies have not even considered whether Michaelis–Menten kinetics are applicable and have taken a grosser, simplistic outlook by attempting to describe denitrification as a zero or first-order reaction. Investigation with soil show the rate to be independent of nitrate concentration (Wiljer and Delwiche, 1954; Nommik, 1956; Bremner and Shaw, 1958; Cooper and Smith, 1963), but this is apparently due to the use of high nitrate solution concentrations (usually 100 mg of N per liter or greater), which appear to be much lower when considered on a dry weight basis. Thus, apparent first-order kinetics were reported by Stanford et al. (1975) in soil when solution concentrations lower than 32 ppm NO_3-N were used. Bowman and Focht (1974) showed that the apparent zero-order and first-order states were merely the extreme ends of the standard Michaelis–Menten curve in describing denitrification—providing the concentration of reductant or oxidant, when held constant while the other was varied, was not limiting. Where the reaction followed Michaelis–Menten kinetics, V_{max} = 150 mg of N per milliliter per day and K_m = 170 mg of N per milliliter for a desert soil supplemented with glucose. The Michaelis constant was considerably higher than that reported for dissimilatory nitrate reductases of *P. aeruginosa* (Fewson and Nicholas, 1961) and *E. coli* (Nason, 1962), which were 0.22 and 7 mg of N per liter, respectively.

Apparent discrepancies between reported zero-order (Wuhrmann, 1963a; Dawson and Murphy, 1973; and first-order (Johnson, 1968; Balakrishnan and Eckenfelder, 1969; Mulbarger, 1971) kinetics observed in waste water treatment are probably explained by whether the system is carbon or nitrogen limiting. Secondary-treated waste waters are usually carbon-limiting so that addition of more nitrate should not increase the rate of denitrification. Thus, Johnson (1968) and Balakrishnan and Eckenfelder (1969) showed that the rate was dependent on both nitrate and carbon concentration. Wuhrmann (1964) has pointed out that there are generally sufficient nutrients including endogenous carbon in sewage biomass to effectively remove the nitrate introduced during the holding period, but it is generally agreed that the rate of denitrification is increased by addition of exogenous carbon in carbon-limiting systems.

Although the stoichiometric carbon–nitrogen relationship depends upon the electrons supplied per mole of carbonaceous substrate, the existing C/N ratio of the sewage biomass and the endogenous organic material is relatively uniform. The rate of denitrification generally proceeds maximally between a C/N ratio of 2 to 3 (Dawson and Murphy, 1973; Wuhrmann and Mechsner, 1973). Surprisingly, this same ratio has also been found to apply to soils, which have much higher C/N ratios than sewage, when exogenous carbon was added (Bremner and Shaw, 1958; Bowman and Focht, 1974). This would suggest that "available" carbon in soil is limiting and that, like sewage, the bulk of it is used for respiratory rather than assimilatory purposes. Increasing the C/N ratio by additions of methanol or other exogenous substrates beyond that needed for denitrification does not significantly reduce the effluent nitrate concentration and increases the effluent BOD as Wuhrmann and Mechsner (1973) have shown (Fig. 3). Thus, additions of exogenous carbon to waste

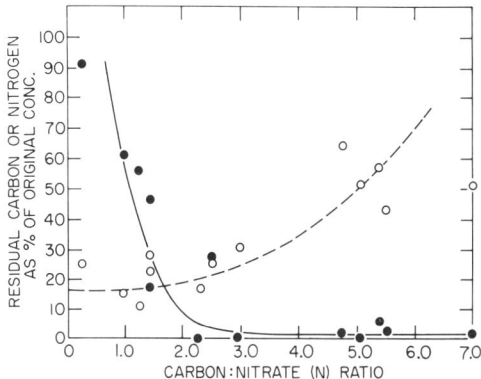

FIG. 3. Effect of ratio of carbon (- - -) nitrogen (———) on removal of nitrate. From Wuhrmann and Mechsner (1972).

waters that are carbon-limiting should not exceed the dissimilatory nitrogen demand.

Retention times required for nitrogen removal in carbon-limiting waste waters can thus be shortened considerably by addition of an energy source. Most studies show that methanol provides the cheapest and most efficient energy source and effects the least amount of carbon for assimilatory purposes (Mulbarger, 1971; McCarty and Haug, 1971; Barth et al., 1968; Sperl and Hoare, 1971). Molasses, humus (Finsen and Sampson, 1959), cellulose (Skinner, 1972), and sugars from bakery wastes (Adams et al., 1970) have also been used but with considerably less efficiency in terms of larger amounts, particularly sugars, being assimilated into biomass, which results in a higher effluent BOD. High sugar concentrations and/or high C/N ratios appear to favor fungal growth and inhibit denitrifiers in soil (Bowman and Focht, 1974). Methane has been used, though with limited success, because of its slower oxidation in comparison to methanol (Pretorius, 1973). Although the use of by-passed primary effluent has been found to remove most of the nitrates, the resulting effluent is higher in BOD and organic nitrogen (Mulbarger, 1971; Johnson and Schroepfer, 1964). Removal of nitrate in columns of soil poor in available carbon is greatly accelerated by the incorporation of sulfur as an energy source favorable for the growth of *Thiobacillus denitrificans* (Mann et al., 1972).

2. Inhibition

Substrate inhibition to both nitrifying bacteria was first observed by Meyerhof (1916a,b, 1917), who showed that optimal oxygen consumption rates for *Nitrosomonas* and *Nitrobacter* declined at respective ammonium N and nitrite N concentrations beyond 60 and 350 mg/liter. Respiration of *Nitrosomonas* was less than 10% at an ammonium N concentration of 1400 mg of N per liter, while respiration of *Nitrobacter* was less severely inhibited to about 20% of maximum at nitrite N concentrations as high as 0.5%. Boon and Laudelout (1962) showed that nitrite oxidation was maximal at 280 mg of N per liter and attributed substrate repression at high concentrations to undissociated nitrous acid. Similar conclusions were reached by Prakasam and Loehr (1972), who showed that nitrous acid, not nitrite per se, inhibited oxidation of nitrite in activated sludge units.

End-product inhibition has also been demonstrated with both *Nitrosomonas* and *Nitrobacter*. Nitrite is apparently more toxic to *Nitrosomonas* during the lag phase of growth (500 mg of N per liter) than the log phase (2500 mg of N per liter) (Painter, 1970). Nitrite concentrations of 1400 mg and 4700 mg of N per liter are required to effect a respective 26% and 100% inhibition of oxygen uptake by resting cell suspensions of *Nitrosomonas* (Meyerhof, 1916a,b). Boon and Laudelout (1962) showed that nitrite oxidation by *Nitrobacter* was noncompetitively inhibited by nitrate, and Gould and Lees (1960) showed that maximal rates of oxidation could be maintained if nitrate was continually removed by dialysis. End-product inhibition noted in pure culture studies apparently accounts for the observations of Prakasam and Loehr (1972), which showed that oxidation of ammonium with highly nitrogenous chicken manure usually ceased at about 80% completion. Following removal of nitrite and nitrate by denitrification, the residual ammonium was then completely oxidized upon subsequent aeration.

In an atmosphere containing 78% N_2, it does not appear likely that end-product inhibition is of any consequence with denitrification. No inhibition by nitrous oxide has been observed; in fact, atmospheres containing 100% nitrous oxide greatly reduce the lag period observed in denitrification studies (R. L. Van der Staay and D. D. Focht, unpublished data). Repression of both nitrate and nitrous oxide reduction by nitrite has been discussed in the previous sections.

Though nitrate does not appear to have any inhibitory effects upon denitrifying bacteria at concentrations exceeding 1%, Verhoeven (1950) noted that *Bacillus* was predominant at 5–12% KNO_3 concentrations while *Pseudomonas* was predominant at 1–2% concentrations. It is not known whether this selection process is due to toxicity of nitrate at higher concentrations or to the competitive abilities of the two genera at different substrate concentrations. For example, the organism (e.g., *Pseudomonas*) with a smaller K_m would be favored at lower substrate concentrations.

Organic nitrogen compounds, such as peptone, aniline, guanidines, urethanes, ureas, and pyridines, have been shown to have inhibitory effects upon the nitrifiers. Most studies have shown no inhibitory effect from glucose or other carbohydrates (see reviews by Meiklejohn, 1954; Lees, 1954; Quastel, 1965; Painter, 1970). Yet, since the discovery of the autotrophic mode of life by Winogradsky (1890), many erroneous contentions have arisen that organic matter per se is in some way "toxic" to the nitrifiers and other chemolithotrophic bacteria. Hopefully, reviews by Meiklejohn (1954) and, more succinctly, Rittenberg (1969) should lay this concept to rest. In noting that most of the inhibitory nitrogen compounds are good chelants, Lees (1952) suggested that divalent ions essential for nitrification may be unavailable. However, Painter has cautioned that reduction of nitrification by addition of sludge might be due to oxygen limitation brought on by heterotrophic activity. Tomlinson *et al.* (1966) noted that nitrification was inhibited by 75% in pure culture by respective concentrations per liter of 4 mg of Cu^{2+} and 0.8 mg of Hg^{2+}. In activated sludge, 150 mg of Cu^{2+} and 90 mg of Hg^{2+} per liter were required to achieve the same level of inhibition. The formidable concentration of detergent sequestering agents and amino nitrogen compounds in domestic waste waters is obviously a factor in reducing the toxicity of heavy metal ions to the nitrifiers and other bacteria as well. Whether or not the chelation mechanism explains the reduced growth of the nitrifiers has not been established with certainty.

IV. Comparative Waste Treatment Methods for Nitrification and Dentrification

Biological nitrification and denitrification has been considered the most favorable process for adaptation in existing waste water treatment systems. Since treatment processes were usually designed for suspended solids and dissolved organic matter removal, the removal of nitrogen presumably by nitrification and denitrification has been more or less incidental and has not been always consistent (Barth *et al.*, 1966; Johnson and Schroepfer, 1964). Upgrading treatment for nitrogen removal requires modification of the existing operation or the incorporation of additional unit processes. In light of the increasing demand of nitrogen removal from waste water, many different attempts have been made to achieve this goal.

A. ACTIVATED SLUDGE PROCESSES

1. Combined Carbonaceous and Nitrogeneous Oxidation

In a combined carbonaceous and nitrogeneous reactor, BOD removal and nitrification are accomplished simultaneously. The rate and degree of nitrification at any given time is dependent largely on the density of nitrifying bacteria, which constitute a relatively small fraction of the sludge mass (Wuhrmann, 1964). Owing to a much slower growth rate, a high level of nitrification can be maintained only when the growth rate of nitrifiers is fast enough to replace the loss through sludge wasting. Downing et al. (1964) and Jenkins and Garrison (1968) outlined the basic operating condition that are required to achieve this objective, so that washout will not occur.

Load factor, which is both a function of (hydraulic) detention time and mixed liquid suspended solids concentration (MLSS), is the governing parameter of nitrification in this case. Since both carbon and nitrogen oxidation are a function of the load factor, it is essential that the load factor of a combined reactor be set in the narrow range where both reactions would be operated near their optimum, as seen in Fig. 4 (Mechalas et al., 1970). It is generally agreed that the load factor should not exceed 0.3–0.4 kg BOD per kilogram of MLSS per day (Balakrishnan and Eckenfelder, 1970; Johnson and Schroepfer, 1964). For effluent from primary settling, where BOD input cannot be controlled, the lowering of the load factor can be achieved only by operating at either a lower flow rate, (i.e., higher hydraulic detention time) or at a high MLSS. A high MLSS in the reactor is impractical and difficult to maintain, so that the load factor is adjusted by manipulating the flow rate. The hydraulic retention time that would accommodate the optimal loading varies from less than 1 hour (Wuhrmann, 1964) to 8 hours (Downing et al., 1964). Most reports suggested 4–6 hours as the most suitable retention time (Balakrishnan and Eckenfelder, 1970). When the reactor is operated at the combined oxidation mode, the sludge yield is lowered. For a given MLSS, the percentage of sludge recycled will be increased, which increases the sludge age. Although there are systems where a sludge age of 15–20 days has been maintained, concomitant with approximately 50% completion of nitrification (Beckman et al., 1972), a minimum of 3–4 days is usually required to ensure successful nitrification (Balakrishnan and Eckenfelder, 1970; Wuhrmann, 1964).

A typical problem with this type of reactor is the maintenance of large viable suspended solids through sludge recycling when settling characteristics of the sludge are poor. A poor liquid–solid separation also affects the overall treatment efficiency. It also has been well estab-

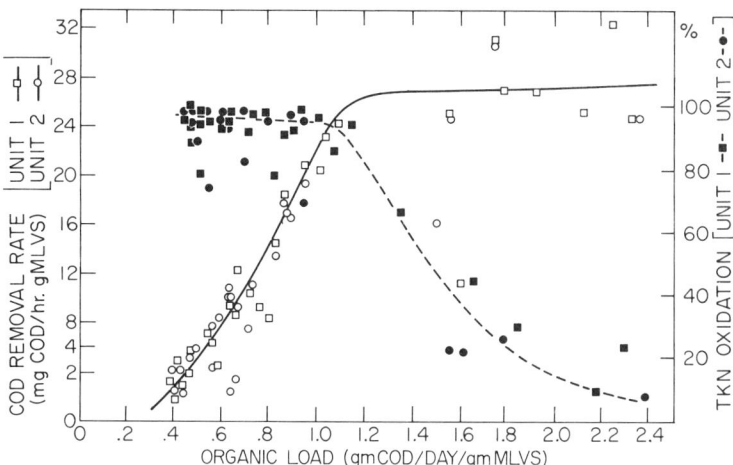

FIG. 4. Carbonaceous and nitrogenous oxidation vs. organic loading. MLVS, mixed liquor volatile suspended solids; COD, chemical oxygen demand; TKN, total Kjeldahl nitrogen. From Mechalas et al. (1970).

lished that treatment plants seldom can accomplish complete nitrification on a year-round basis owing to seasonal change of ambient temperature (Horstkoote et al., 1974; Beckman et al., 1972; Barth et al., 1966). When denitrification is desired, a combined oxidation process will be followed by a denitrification process to reduce oxidized nitrogen.

2. Separate Carbon and Nitrogen Oxidation

An attempt to remedy the shortcoming of combined activated sludge is to separate nitrification from carbon oxidation so that each stage can function closer to its optimum. This minimizes washout with the nitrification stage, and the process can be operated successfully at shorter detention time, lower MLSS, and sludge age. Pilot-plant testing of such a system resulted in a high degree of success (Rimer and Woodward, 1972). However, the initial development of nitrifying flora in the second stage was very slow at temperatures below 10°C, although, once established, the population could be maintained. Temperature again turned out to be the rate-limiting factor, and nitrate conversion was seriously affected at winter months. Mulbarger (1971) compared the performance of several possible modifications of the activated sludge process and concluded the two-stage nitrification had the most consistent, satisfactory results. With two reactors, two settling basins, and separate sludge return, costs of construction and operation would be considerably higher. In terms of overall nitrogen removal, he found that a three-stage system, which encompassed separate carbon removal, nitrification, and denitrification units gave the best results.

B. Trickling Filter

In a trickling filter, carbon oxidation and nitrification can also occur simultaneously. When a conventional rock filter is used, the organic loading should not exceed 0.19 kg BOD per day per cubic meter. The most significant innovation in trickling filter technology in recent years has been the use of synthetic media in place of rock. The light-weight synthetic media have higher specific surface to support greater slime growth and permit higher loading per unit volume. When synthetic media are used, the optimum loading can increase to 0.39 kg BOD per day per cubic meter, owing obviously to greater specific surface area per volume of the synthetic media (87.1–112.9 m^{-1} vs 38.7–58.1 m^{-1}) (Stenquist et al., 1974).

Carbon oxidation and nitrification of waste water can also be accomplished by using the fixed-film rotating biological contactor, which was developed approximately 30 years ago (Antonie et al., 1974). The active biomass is attached on the surface of a set of closely spaced thin disks, partially submerged in a channel where waste water flows through. The rotation of disks facilitate biomass contact with the atmosphere and the waste water. A series of 6–10 sets of disks are arranged in sequence to form the contact tank. Nitrification occurs at the later stage of treatment where BOD is reduced below 10 mg/liter, but the reaction is not always complete, and nitrite may accumulate. Statistically, the degree of nitrification is a function of load factor, total surface area, and disk rotating speed (Weng and Molof, 1974).

More often, the trickling filter has been used to improve secondary treatment effluent for additional BOD removal and nitrification, so that it functions somewhat analogously to the second stage of the two-stage system (Duddles et al., 1974). Waste water entering the filter thus has a much reduced organic load. The sludge yield in filters used solely for nitrification is small, and the amount of suspended soils in the effluent does not require a settling basin. In turn, hydraulic loading becomes a more important factor that limits the operation. The degree of nitrification depends upon the actual contact time. With high surface loading, water may trickle through the entire path before completing the oxidation. When synthetic filter media are used, the hydraulic loading rate may be increased to 1.43 liters/second per square meter, while organic loading still would not exceed 0.13 kg BOD per day per cubic meter. The overall degree of nitrification in this case is about 63%. If hydraulic loading is lowered, greater nitrate conversion (better than 90%) can be achieved. Hydraulic loading is also reduced to accommodate filters operating at low temperature.

C. ANAEROBIC ACTIVATED SLUDGE PROCESS

When denitrification of nitrified waste water effluent is desired, the activated sludge process is operated anaerobically. Johnson (1972) maintained that the kinetics developed for the mixed, aerobic activated-sludge reactor could be used to describe denitrification with slight modification. He considered nitrate as a dissolved oxygen equivalent. In effect, this assumes that the kinetic constants for oxygen consumption (K_m and V_{max}) are the same as for denitrification. Mulbarger (1971) found that the measured rates of respiration and denitrification were very close when an equivalent mass ratio (O_2/NO_3^-) of 2.86 was used: respective oxygen and equivalent denitrification rates of 20 ± 10 mg and 25 ± 9 mg O_2/hour per gram of MLVSS were obtained. The results agree reasonably well with experimental data of Stensel (1973). Moore and Schroeder (1971a,b) investigated effects of nitrate feed rate and cell residence time on complete, mixed continuous-flow reactors operated at the steady state. They concluded that denitrification processes can be operated at near-maximum unit removal rates and still obtain acceptable nitrogen conversion (less than 2 mg of NO_3^- N per liter in the denitrified effluent). Stensel et al. (1973) also indicated that cell retention time in the reactor will depend on the organic carbon requirement and nitrate removal efficiency.

In batch experiments, it has been demonstrated that the stored intracellular food reserves of the sludge mass could be used as a source of oxidizable organic substrate for denitrification (Wuhrmann, 1964). Attempts to put it to a full-scale, continuous operation have yet to be made. Undoubtedly, many operating difficulties, such as switching the activated sludge from an aerobic to anaerobic mode and reversing the flow, must be overcome before the system can be operated successfully. In common practice, an external source of organic carbon is supplied. In sewage treatment, a readily available source of substrate would be the unnitrified primary effluent. Johnson and Schroepfer (1964) diverted a portion of the primary effluent to the nitrified effluent in the anaerobic activated sludge reactor. While the nitrate was denitrified, unoxidized nitrogen in the primary effluent went through the system unaffected and reduced the overall nitrogen removal efficiency. There was also an increase of BOD in the effluent. McCarty et al. (1969) evaluated the use of alternate carbon sources in terms of their consumption ratio (the ratio of total amount of an organic chemical consumed during denitrification to the stoichiometric requirement for denitrification). Thus, a lower ratio would mean more efficient use of the substrate for denitrification. Methanol was found to produce the lowest consumption ratio. An empirical equation (2) was developed to estimate the methanol requirement at 20°C.

$$C_m = 2.47 N_o + 1.53 N_i \text{ to } 0.87 D_o \tag{2}$$

where C_m = required methanol concentration (mg/liter), N_o = initial nitrate nitrogen concentration (mg/liter), N_i = initial nitrite nitrogen concentration (mg/liter), D_o = initial dissolved oxygen concentration (mg/liter). In operation, a commonly used parameter is the methanol/nitrate ratio (M/N), which indicates units of methanol required for each unit of nitrate reduced. Under normal operating conditions, the M/N ratio ranged between 2 and 4 (Gulf South Research Institute, 1970).

D. Submerged Filter

Compared to the activated sludge process, trickling filters are more efficient in retaining large populations of viable biomass. There is, however, little control over the retention time, which is a crucial parameter in determining the degree of conversion. Submerged filters are developed to remedy shortcomings of both activated sludge and trickling filters. In the reactor, bacteria are effectively retained on the surface of the filter packing. Waste water flows upward through voids in the media and exits from the top of the filter. The hydraulic retention time can be properly controlled by regulating the inflow. Depending on whether oxygen or methanol is supplied, this system can function for both nitrification or denitrification.

During nitrification, the filter not only converts ammonium to nitrate, but removes residual BOD and suspended solids from the secondary effluent (McHarness and McCarty, 1973). A filter filled with a 6 meter depth of 2.5 cm size rock can achieve 90% conversion in a retention time of less than 1 hour.

When carbonaceous substrate is supplied, the submerged filter is also effective for denitrification. The first attempt was made by Finsen and Sampson (1959), who passed waste water through a column filled with 2.54 cm in diameter glass marble, using molasses as the substrate. Better than half-fold reduction of nitrates was observed within 37 minutes after the waste water had entered the unit, and almost complete denitrification occurred in less than 1 hour. In other studies (Tamblyn and Sword, 1969), material of various particle sizes and adsorption characteristics were used to determine the most suitable media supporting bacterial growth. In terms of overall nitrogen removal, filters using media of high adsorption capacity or high specific surface area did not appear to have any distinct advantage over commonly used rocks. Fine material, such as sand and activated carbon, required a high hydraulic head, subsequently causing short-circuiting and displacement of the media. Round rock aggregates with diameters ranging from 1.91 to 3.81 cm were the most acceptable. In long-term operations, clogging and associated problems appeared to be the major difficulty.

E. Land Spreading

Land spreading of waste-water effluent is a successful means of nitrogen removal where sufficient land is available and where infiltration rates are not impeded. Kardos (1967) calculated that 15.6–39.4 ha were needed for disposal of 1 MLD (million liters per day) in Pennsylvania (or 156–394 acres/MGD). This is in agreement with studies conducted in Wisconsin, which show that 13.8 ha/MLD are required (Fetter and Holzmacher, 1974). Bouwer et al. (1974) found that only 0.37 ha/MLD were needed in the Flushing Meadows project conducted near Phoenix, Arizona. Consequently, the applicability of this method to large urban areas will depend on the climate and on the proximity to available land disposal sites.

Clogging of the soil pores is frequently a problem with continuous application as discussed by Fetter and Holzmacher (1974). This problem has been minimized with the Flushing Meadows Project by use of alternate wetting and drying cycles, which facilitates mineralization of organic matter that frequently contributes to sealing. Bouwer (1970) found that short cycles of 2 days wet and 3 days dry resulted in complete conversion to nitrate, while longer cycles of 14 days wet and 7 days dry removed 90% of the total nitrogen applied—presumably through denitrification. When waste water containing 25 mg of N per liter was added to soil columns at 10 days wet, 10 days dry intervals, Lance (1972) found that 67% of the total nitrogen was removed by denitrification, although nitrate peaks at 5, 15, and 25 days were present in the effluent, which averaged about 50 mg of N per liter. Five-day intervals between the peaks were virtually devoid of nitrate. He concluded that, by proper timing, the effluent high in nitrogen could be used for agricultural recharge, and the effluent low in nitrogen could be used for recreational purposes. Lance and Whistler (1972) found that nitrogen removal could be increased to 80% by reducing the infiltration rate by half. They also observed the same increase in nitrogen removal by mixing the recharged water, which was high in nitrate, with secondary effluent. They concluded that denitrification was limited by available carbon in the soil.

In a field study conducted over a three-year period from 1969 to 1972, Bouwer et al. (1974) found that 30% of the applied nitrogen was lost through denitrification and that a nitrogen load of 28,000 kg/ha per year could be maintained. Using a 15 days wet, 15 days dry cycle, they observed monthly nitrate peaks, which ranged between 5 and 50 mg of N per liter in one well. They concluded that land disposal was a much cheaper method than tertiary sewage treatment to produce a renovated water of the same quality.

F. NITRITIFICATION–DENITRITIFICATION

Nitritification and nitratification are defined as the respective oxidation of ammonium to nitrite and of nitrite to nitrate; denitritification and denitratification are defined as the respective reduction of nitrite and nitrate to gaseous products (Prakasam and Loehr, 1972). These terms have been used since neither nitrification nor denitrification make any distinction between specific involvement of nitrate or nitrite. The authors studied the treatment of poultry wastes and noticed that a solids retention time greater than 2 days converted about half of the total Kjeldahl N to nitrite and nitrate. Since poultry wastes are much higher in ammoniacal nitrogen than domestic wastes, it was not unusual to observe concentrations ranging from 30 to 645 mg of N per liter in the spent, mixed liquors. After removal of nitrate and nitrite by denitrification, the residual ammonium could be removed by another nitrification–denitrification cycle, which led Prakasam and Loehr to suggest that feedback inhibition by nitrate or nitrite was limiting complete oxidation of ammonium. They found that removal of ammoniacal nitrogen could be accelerated by bypassing the nitratification step. Thus, they suggested that loading rates could be increased to 0.9 lb COD per kilogram of MLVSS per day.

Nitritification–denitritification has considerable merit over the complete oxidation to nitrate and the subsequent reduction on several grounds. First of all, toxicity of ammonia to *Nitrobacter* does not have to be considered, so that loading rates of particularly high ammonium nitrogen concentration can be considered strictly in terms of kinetics of nitritification. Second, the overall process results in less energy expenditure and would consequently reduce the methanol requirement or optimal C/N ratio needed for effective denitritification. Third, denitrification is more rapid than denitratification, as discussed previously. Nitrite can also be removed chemically by acidification and addition of urea to yield molecular nitrogen.

Prakasam and Loehr (1972), however, did not indicate whether or not this procedure could lead to the formation of highly carcinogenic N-nitrosamines, which are formed chemically and biochemically by condensation of a secondary amine with nitrite. Dimethylnitrosamine has been produced in laboratory studies with sewage and lake water treated with secondary amines and nitrite (Ayanaba et al., 1973; Ayanaba and Alexander, 1974). Batch studies by Voets et al. (1975) have failed to detect the presence of nitrosamines during nitritification and denitritification. They concluded, nevertheless, that regular analysis for nitrosamines be conducted to control effluent quality.

V. Summary and Conclusions

Nitrification followed by subsequent denitrification is the easiest and most economical method currently available for removal of nitrogen from waste waters. The rate-limiting step in the overall process is usually the nitrification step, although this can be accelerated by the use of separate systems for carbon oxidation and nitrification. Denitrification is rapid since it is a respiratory process, yet the rate is frequently limited by the amount of available carbonaceous substrate needed for reduction of nitrate.

The kinetics for nitrification are relatively straightforward and well developed by comparison to denitrification. Even in pure culture systems, "apparent" Michaelis constants are difficult to determine because two substrates (reductant and oxidant) are involved with one, or more, enzyme(s). Yet, the respective K_m values for oxidant and reductant would be important selective mechanisms—particularly at low concentrations—in determining the climax community. Different reported values for the "critical oxygen concentration," above which denitrification does not occur, may reflect the selection of different bacterial species, which have different affinity constants for oxygen or different amounts of cytochrome c. Since it is not known how great this variation is among different species, it may be premature to assume that mixed culture systems will adhere to standard kinetic equations.

Low temperatures (below 15°C) cause the greatest problem in waste treatment of nitrogen and affect the reduction of nitrate to nitrite most drastically. The pronounced effect of low substrate (reductant) concentrations uon Q_{10} in mixed culture systems is not well understood either. Whether these phenomena are the result of selection of different bacterial types in response to temperature or substrate concentrations or of physical factors, such as increased oxygen solubility, is not known.

Nitritification and denitritification represent an improvement over the general nitrification–denitrification procedure in terms of simplicity and time. Toxicity of ammonia to *Nitrobacter* does not have to be considered, so that loading rates can be increased, and the kinetics can be simplified by eliminating the nitratification step. Less energy is required for reduction of nitrite, and nitrite is reduced faster than nitrate—particularly at temperatures near freezing. More research on these methods should be focused not only on the kinetics, but on careful monitoring to ensure that nitrosamine formation will not pose any environmental problems.

Land spreading of secondary effluent appears to be the best long-range method for recycling nutrients to croplands or forage grass in lieu of problems associated with toxic trace metals or salinity. The greatest problem associated with this method is the high nitrate peaks that result with alternate wet and dry cycles. Thus, this system must be designed to (1) remove sufficient nitrate by denitrification or to (2) recover the nitrate-laden waters before they mix with the groundwater.

REFERENCES

Adams, C. E. (1973). *Environ. Sci. Technol.* 7, 696–701.
Adams, C. E., Krenkel, P. A., and Bingham, E. C. (1970). In "Advances in Water Pollution Research" (S. H. Jenkins, ed.), pp. 1–13/1 to 1–13/13. Pergamon, Oxford.
Aleem, M. I. H., and Lees, H. (1963). *Can. J. Biochem. Physiol.* 41, 763–778.
Aleem, M. I. H., Hock, G. E., and Vanner, J. E. (1965). *Proc. Nat. Acad. Sci. U.S.* 54, 869–873.
Allen, M. B., and van Niel, C. B. (1952). *J. Bacteriol.* 64, 397–412.
Allison, F. E. (1955). *Advan. Agron.* 7, 213–250.
Anderson, J. H. (1964). *Biochem. J.* 91, 8–17.
Antonie, R. L., Kluge, D. L., and Mielke, J. H. (1974). *J. Water Pollut. Contr. Fed.* 46, 498–511.
Aubert, J. P., Millet, J., and Milhaud, G. (1959). *Ann. Inst. Pasteur, Paris* 96, 559–576.
Ayanaba, A., and Alexander, M. (1974). *J. Environ. Qual.* 3, 83–89.
Ayanaba, A., Verstraete, W., and Alexander, M. (1973). *J. Nat. Cancer Inst.* 50, 811–813.
Bailey, L. D., and Beauchamp, E. G. (1973a). *Can. J. Soil. Sci.* 53, 213–218.
Bailey, L. D., and Beauchamp, E. G. (1973b). *Can. J. Soil Sci.* 53, 219–230.
Balakrishnan, S., and Eckenfelder, W. W. (1969). *Water Res.* 3, 177–188.
Balakrishnan, S., and Eckenfelder, W. W. (1970). *J. Sanit. Eng. Div., Amer. Soc. Civil Eng.* 96, SA2, 501–512.
Barth, E. F., Mulbarger, M., Talotto, B. V., and Ettinger, M. B. (1966). *J. Water Pollut. Contr. Fed.* 38, 1200–1219.
Barth, E. F., Brenner, R. C., and Lewis, R. F. (1968). *J. Water Pollut. Contr. Fed.* 40, 2040–2054.
Becker, G. F., and Schmidt, E. L. (1964). *Arch. Mikrobiol.* 49, 167–175.
Beckman, W. J., Avendt, R. J., Mulligan, T. J., and Kehrberger, G. J. (1972). *J. Water Pollut. Contr. Fed.* 44, 1916–1930.
Beijerinck, M. W., and Minkman, D. C. J. (1910). *Zentralbl. Bakteriol., Parasitenk., Infektionskr. Hyg., Abt.* 2 25, 30–63.
Bisset, K. A., and Grace, J. B. (1954). In "Autotrophic Micro-organisms" (B. A.

Fry and J. L. Peel, eds.), pp. 28–53. Cambridge Univ. Press, London and New York.
Bollag, J. M., Drzymala, S., and Kardos, L. T. (1973). *Soil Sci.* 116, 44–50.
Boon, B., and Laudelout, H. (1962). *Biochem. J.* 85, 440–447.
Bothner-By, A. A., and Friedman, L. (1952). *J. Chem. Phys.* 20, 459–462.
Bouwer, H. (1970). *J. Sanit. Eng. Div., Amer. Soc. Civil Eng.* 96, SA1, 59–74.
Bouwer, H., Lance, J. C., and Rigs, M. S. (1974). *J. Water Pollut. Contr. Fed.* 46, 844–859.
Bowman, R. A., and Focht, D. D. (1974). *Soil Biol. & Biochem.* 6, 297–301.
Bray, H. G., and White, K. (1966). "Kinetics and Thermodynamics in Biochemistry," 2nd ed. Academic Press, New York.
Bremner, J. M., and Shaw, K. (1958). *J. Agr. Sci.* 51, 40–52.
Broadbent, F. E., Taylor, N. B., and Hill, G. N. (1957). *Hilgardia* 27, 247–267.
Buswell, A. M., Shiota, T., Lawrence, N., and Meter, I. V. (1954). *Appl. Microbiol.* 2, 21–25.
Cady, F. B., and Bartholemew, W. V. (1960). *Soil Sci. Soc. Amer., Proc.* 24, 477–482.
Campbell, N. E. R., and Lees, H. (1967). In "Soil Biochemistry" (A. D. McLaren and G. H. Peterson, eds.), Vol. 1, pp. 194–215. Dekker, New York.
Carlucci, A. F., and Schubert, H. R. (1969). *Limnol. Oceanogr.* 14, 187–193.
Chance, B. (1957). *Fed. Proc., Fed. Amer. Soc. Exp. Biol.* 16, 671–680.
Collins, F. M. (1955). *Nature (London)* 175, 173–174.
Cooper, G. S., and Smith, R. L. (1963). *Soil Sci. Soc. Amer., Proc.* 27, 659–662.
Cox, C. D., and Payne, W. J. (1973). *Can. J. Microbiol.* 19, 861–872.
Cox, C. D., Payne, W. J., and Dervartanian, D. V. (1971). *Biochim. Biophys. Acta* 253, 290–294.
Culp, R. L., and Moyer, H. E. (1969). *Civil Env. (New York)* 39, 38–42.
Daniel, R. M., and Appleby, C. A. (1972). *Biochim. Biophys. Acta* 275, 347–354.
Dawson, R. W., and Murphy, K. L. (1972). *Water Res.* 6, 71–83.
Dawson, R. W., and Murphy, K. L. (1973). In "Advances in Water Pollution Research" (S. H. Jenkins, ed), pp. 671–680. Pergamon, Oxford.
Deppe, K., and Engel, H. (1960). *Zentralbl. Bakteriol., Parasitenk., Infektionskr. Hyg., Abt. 2* 113, 561–568.
Downey, R. J., and Kiszkiss, D. F. (1969). *Microbios* 2, 145–153.
Downing, A. L., Painter, H. A., and Knowles, G. (1964). *Inst. Sewage Purif., J. Proc.* pp. 130–153.
Duddles, G. A., Richardson, S. E., and Barth, E. F. (1974). *J. Water Pollut. Contr. Fed.* 46, 937–946.
Engel, M. S., and Alexander, M. (1958). *J. Bacteriol.* 76, 217–222.
Fetter, C. W., and Holzmacher, R. G. (1974). *J. Water Pollut. Contr. Fed.* 46, 260–270.
Fewson, C. A., and Nicholas, D. J. D. (1961). *Biochim. Biophys. Acta* 49, 335–349.
Finsen, P. O., and Sampson, D. (1959). *Water Waste Treat. J.* 7, 298–300.
Focht, D. D. (1974). *Soil Sci.* 118, 173–179.
Focht, D. D., Fetter, N. R., Lonkerd, W., and Stolzy, L. H. (1974). *Proc. 2nd Annu. NSF-RANN Trace Contaminants Conf.*
Forget, P., and Dervartanian, D. V. (1972). *Biochem. Biophys. Acta* 256, 600–606.
Gayon, V., and Dupetit, G. (1886). *Mem. Soc. Bordeaux, Ser. 3* 2, 200–204.
Goering, J. J. (1968). *Deep-Sea Res.* 15, 157–164.
Goering, J. J., and Cline, J. D. (1970). *Limnol. Oceanogr.* 15, 306–308.
Goering, J. J., and Dugdale, V. A. (1966). *Limnol. Oceanogr.* 11, 113–117.
Gordy, W., and Rexroad, H. N. (1961). *Free Radicals Biol. System., Proc. Symp., 1960* pp. 263–277.

Gould, G. W., and Lees, H. (1960). *Can. J. Microbiol.* **6**, 299–307.
Greenwood, D. J. (1961). *Plant Soil.* **14**, 360–376.
Gulf South Research Institute. (1970). *Water Pollut. Contr. Res. Ser.* **17010 DHT091/70**.
Harrison, D. E. F. (1973). *CRC Crit. Rev. Microbiol.* **2**, 185–228.
Hauck, R. D., and Melsted, S. W. (1956). *Soil Sci. Sco. Amer., Proc.* **20**, 361–364.
Heubült, J. (1929). *Planta* **8**, 398–402.
Hofman, T., and Lees, H. (1952). *Biochem. J.* **52**, 140–142.
Hofman, T., and Lees, H. (1953). *Biochem. J.* **54**, 579–583.
Horstkotte, G. A., Niles, D. G., Parker, D. S., and Caldwell, D. H. (1974). *J. Water Pollut. Contr. Fed.* **46**, 181–197.
Hutchinson, G. L., and Viets, F. G. (1969). *Science* **166**, 514–515.
Ingraham, J. L. (1962). In "The Bacteria" (I. C. Gunsalus and R. Y. Stanier, eds.), Vol. 4, pp. 265–296. Academic Press, New York.
Iwasaki, H. (1960). *J. Biochem. (Tokyo)* **47**, 174–184.
Jenkins, D., and Garrison, W. E. (1968). *J. Water Pollut. Contr. Fed.* **40**, 1905–1919.
Johnson, W. K. (1968). In "Advances in Water Quality Improvement" (E. F. Gloyna and W. W. Eckenfelder, eds.), pp. 178–189. Univ. of Texas Press, Austin.
Johnson, W. K. (1972). *J. Sanit. Eng. Div., Amer. Soc. Civil Eng.* **98**, SA4, 623–634.
Johnson, W. K., and Schroepfer, G. J. (1964). *J. Water Pollut. Contr. Fed.* **36**, 1015–1036.
Kamen, M. D., and Takeda, Y. (1956). *Biochim. Biophys. Acta* **21**, 518–523.
Kamen, M. D., and Vernon, L. P. (1955). *Biochim. Biophys. Acta* **17**, 10–22.
Kardos, L. T. (1967). In "Agriculture and the Quality of our Environment" (N. C. Brady, ed.), Publ. No. 85, pp. 241–250. Amer. Assoc. Advan. Sci., Washington, D.C.
Kefauver, M., and Allison, F. E. (1957). *J. Bacteriol.* **73**, 8–14.
Kluyver, A. J., and Donker, H. J. L. (1926). *Chem. Zelle Gewebe* **13**, 134–190.
Kluyver, A. J., and Verhoeven, W. (1954). *Antonie van Leeuwenhoek; J. Microbiol. Serol.* **20**, 241–262.
Knowles, G., Downing, A. L., and Barrett, M. J. (1965). *J. Gen. Microbiol.* **38**, 263–278.
Lance, J. C. (1972). *J. Water Pollut. Contr. Fed.* **44**, 1352–1361.
Lance, J. C., and Whistler, R. D. (1972). *J. Environ. Qual.* **1**, 180–186.
Latimer, W. M. (1952). "The Oxidation States of the Elements and their Potentials in Aqueous Solutions." Prentice-Hall, Englewood Cliffs, New Jersey.
Laudelout, H., and van Tichelen, L. (1960). *J. Bacteriol.* **79**, 39–42.
Lee, D. H. K. (1970). *Environ. Res.* **3**, 484–511.
Lees, H. (1952). *Biochem. J.* **52**, 134–139.
Lees, H. (1954). In "Autotrophic Micro-organisms" (B. A. Fry and J. L. Peel, eds.), pp. 84–98. Cambridge Univ. Press, London and New York.
Lees, H., and Simpson, J. R. (1957). *Biochem. J.* **65**, 297–305.
Loveless, J. E., and Painter, H. A. (1968). *J. Gen. Microbiol.* **52**, 1–14.
Luebs, R. E., Davis, K. R., and Laag, A. E. (1974). *J. Environ. Qual.* **3**, 265–269.
McCarty, P. L., and Haug, R. T. (1971). In "Microbial Aspects of Pollution" (G. Sykes and F. A. Skinner, eds.), pp. 215–232. Academic Press, New York.
McCarty, P. L., Beck, L., and St. Amant, P. (1969). *Proc. Waste Conf.* **24**, 1271–1285.
McHarness, D. D., and McCarty, P. L. (1973). *Environ. Protect. Tech. Ser.* **EPA-R2-73-158**.
McLaren, A. D. (1971). *Soil Sci. Soc. Amer., Proc.* **35**, 91–95.
Mann, L. D., Focht, D. D., Joseph, H. A., and Stolzy, L. H. (1972). *J. Environ. Qual.* **1**, 329–332.

Martin, J. P., and Ervin, J. O. (1953). *Calif. Citrogr.* **39**, 38, 54–56.
Mechalas, B. J., Allen, P. M., and Matyskiella, W. W. (1970). *Water Pollut. Contr. Res. Ser.* **17010 DRD07/70.**
Meek, B. D., Grass, L. B., and MacKenzie, A. J. (1969). *Soil Sci. Soc. Amer., Proc.* **33**, 575–578.
Meiklejohn, J. (1954). *In* "Autotrophic Micro-organisms" (B. A. Fry and J. L. Peel, eds.), pp. 68–83. Cambridge Univ. Press, London and New York.
Meyerhof, O. (1916a). *Pfluegers Arch. Gesamte Physiol. Menschen Tiere* **164**, 353–427.
Meyerhof, O. (1916b). *Pfluegers Arch. Gesamte Physiol. Menschen Tiere* **165**, 229–284.
Meyerhof, O. (1917). *Pfluegers Arch. Gesamte Physiol. Menschen Tiere* **166**, 240–280.
Moore, S. F., and Schroeder, E. D. (1971a). *Water Res.* **5**, 445–452.
Moore, S. F., and Schroeder, E. D. (1971b). *Water Res.* **5**, 685–694.
Mulbarger, M. C. (1971). *J. Water Pollut. Contr. Fed.* **43**, 2059–2070.
Murray, E. D., and Sanwal, B. D. (1963). *Can. J. Microbiol.* **9**, 781–790.
Nason, A. (1962). *Bacteriol. Rev.* **26**, 16–41.
Nommik, H. (1956). *Acta Agr. Scand.* **6**, 195–228.
Novak, J. T. (1974). *J. Water Pollut. Contr. Fed.* **46**, 1984–1994.
Painter, H. A. (1970). *Water Res.* **4**, 393–450.
Patrick, W. H. (1961). *Proc. Int. Congr. Soil Sci., 7th, 1960* pp. 494–500.
Payne, W. J. (1973). *Bacteriol. Rev.* **37**, 409–452.
Payne, W. J., and Riley, P. S. (1969). *Proc. Soc. Exp. Biol. Med.* **132**, 258–260.
Payne, W. J., Riley, P. S., and Cox, C. D. (1971). *J. Bacteriol.* **106**, 356–361.
Pearsall, W. H., and Mortimer, C. H. (1939). *J. Ecol.* **27**, 483–501.
Pilot, L., and Patrick, W. H. (1972). *Soil Sci.* **114**, 312–316.
Prakasam, T. B. S., and Loehr, R. C. (1972). *Water Res.* **6**, 859–869.
Prakash, O., and Sadana, J. C. (1973). *Can. J. Microbiol.* **19**, 15–25.
Pretorius, W. A. (1973). *In* "Advances in Water Pollution Research" (S. H. Jenkins, ed.), pp. 685–693. Pergamon, Oxford.
Quastel, J. H. (1965). *Annu. Rev. Plant Physiol.* **16**, 217–240.
Richards, F. A., and Broenkow, W. W. (1971). *Limnol. Oceanogr.* **16**, 758–765.
Rimer, A. E., and Woodward, R. L. (1972). *J. Water Pollut. Contr. Fed.* **44**, 101–116.
Rittenberg, S. C. (1969). *Advan. Microbial Physiol.* **3**, 159–196.
Rees, M., and Nason, S. (1965). *Biochim. Biophys. Acta* **113**, 398–401.
Sacks, L. E., and Barker, H. A. (1949). *J. Bacteriol.* **58**, 11–22.
Sacks, L. E., and Barker, H. A. (1952). *J. Bacteriol.* **64**, 247–252.
Sapshead, L. M., and Wimpenny, J. M. T. (1972). *Biochim. Biophys. Acta* **267**, 388–397.
Schloesing, J. J. T., and Müntz, A. (1877). *C. R. Acad. Sci.* **84**, 301–303.
Schoberl, D., and Engel, H. (1964). *Arch. Mikrobiol.* **48**, 393–400.
Skerman, V. B. D., and MacRae, J. C. (1957). *Can. J. Microbiol.* **3**, 215–230.
Skinner, F. A. (1972). *J. Appl. Bacteriol.* **35**, 453–462.
Skinner, F. A., and Walker, N. (1961). *Arch. Mikrobiol.* **38**, 339–349.
Sperl, G. T., and Hoare, D. S. (1971). *J. Bacteriol.* **108**, 733–736.
Stanford, G., Vander Pol, R. A., and Dzienzia, S. (1975). *Soil Sci. Soc. Amer., Proc.* **39**, 284–289.
Starr, J. L., Broadbent, F. E., and Nielsen, D. R. (1974). *Soil Sci. Soc. Amer., Proc.* **38**, 283–289.
Stenquist, R. J., Parker, D. S., and Dosh, T. J. (1974). *J. Water Pollut. Contr. Fed.* **46**, 2327–2339.
Stensel, H. D. (1973). *J. Sanit. Eng. Div., Amer. Soc. Civil Eng.* **99**, EE3, 388–390.

Stensel, H. D., Loehr, R. C., and Lawrence, A. W. (1973). *J. Water Pollut. Contr. Fed.* **45,** 249–261.
Tamblyn, T. A., and Sword, B. R. (1969). *Proc. Ind. Waste Conf.* **24,** 1135–1150.
Tomlinson, T. G., Boon, A. G., and Trotman, C. N. A. (1966). *J. Appl. Bacteriol.* **29,** 266–291.
Ulken, A. (1963). *Arch. Hydrobiol.* **59,** 486–501.
Van't Riet, J., Knook, D. L., and Planta, R. J. (1972). *FEBS (Fed. Eur. Biochem. Soc.) Lett.* **23,** 44–46.
Verhoeven, W. (1950). *Antonie van Leeuwenhoek; J. Microbiol. Serol.* **16,** 269–281.
Verstraete, W. (1975). *Proc. Acad. Sci. USSR, Biol. Sci. Sect.* (in press).
Verstraete, W., and Alexander, M. (1972). *Biochim. Biophys. Acta* **261,** 59–62.
Verstraete, W., and Alexander, M. (1973). *Environ. Sci. Technol.* **7,** 39–42.
Voets, J. P., Vanstaen, H., Verstraete, W. (1975). *J. Water Pollut. Contr. Fed.* **47,** 394–398.
Weber, D. F., and Gainey, P. L. (1962). *Soil Sci.* **94,** 138–148.
Weissenberg, H. (1902). *Zentralbl. Bakteriol., Parasitenk, Infektionskr. Hyg., Abt. 2* **8,** 166–170.
Weng, C. N., and Molof, A. H. (1974). *J. Water Pollut. Contr. Fed.* **46,** 1674–1685.
Wesseling, J., and van Wijk, W. R. (1957). *Agronomy* **7,** 461–504.
Wheatland, A. B., Barret, M. J., and Bruce, A. M. (1959). *Inst. Sewage Purif., J. Proc.* pp. 149–159.
White, D. C., and Sinclair, P. R. (1971). *Advan. Microbial. Physiol.* **5,** 173–212.
Wiljer, J., and Delwiche, C. C. (1954). *Plant Soil.* **5,** 155–169.
Wimpenny, J. W. T. (1969). *In* "Microbial Growth" (P. Meadow and S. J. Pirt, eds.), pp. 161–198. Cambridge Univ. Press, London and New York.
Winogradsky, S. (1890). *C. R. Acad. Sci.* **110,** 1013–1016.
Wuhrmann, K. (1963a). *Advan. Biol. Waste Treat., Proc. Conf., 3rd, 1960* Paper No. 3.
Wuhrmann, K. (1963b). *Verh. Int. Ver. Limnol.* **15,** 580–596.
Wuhrmann, K. (1964). *Advan. Appl. Microbiol.* **6,** 119–151.
Wuhrmann, K., and Mechsner, K. (1973). Discussion by K. Wuhrmann on paper by Dawson and Murphy (p. 682, Fig. 1).
Yoshida, T., and Alexander, M. (1970). *Soil Sci. Soc. Amer., Proc.* **34,** 880–882.

The Fermentation Pilot Plant and Its Aims

D. J. D. HOCKENHULL

Glaxo Laboratories Ltd., Ulverston, Cumbria, England

I.	Why a Pilot Plant?	187
	A. Background	187
	B. Simulation, Modeling, and Prediction	189
	C. The Tasks of a Pilot Plant	192
	D. Considerations of Expense	194
	E. Information and Development Strategy	195
II.	The Pilot Plant as an Introduction to Production Management	197
	A. Introduction	197
	B. Activities	200
	C. Conclusion	207
	References	208

I. Why a Pilot Plant?

A. Background

Thirty years ago the technology of fermentation hardly existed in comparison with the highly sophisticated operations of today. Over these years, a new area of activity has been opened up and new means of investigation and realization have been developed.

In the surface cultures of earlier days inoculum was planted into sterile broth and then allowed to grow to maturity with no control beyond that of constant temperature. Everything was preordained, and the laboratory faithfully represented the commercial process.

With the advent of deep culture, things changed. Although the shaken-flask method of laboratory culture gave some indication of plant performance in large stirred vessels, its powers of prediction were limited. Although the method was capable of predicting the performance of new strains with sufficient reliability to facilitate the selection of better performing isolates, it was soon found wanting in other respects. In particular, where ongoing control of nutrient additions became a feature of the industrial process, shake flasks proved to be completely inadequate.

Modern processes entail a much deeper insight into pattern and control. The aim now is to control the culture so that the balance between growth and product formation is maintained at all times throughout the process so as to give the maximum yield of product for a given expenditure of materials, time, wages, overhead, and amortization. In most cases we achieve a steady state of compromise. The culture remains "balanced" until harvest. In certain instances an unstable balance may be more economic. With such, a quick titer is snatched and the culture is terminated in an "exhausted" condition.

Up to now the most sophisticated techniques have advanced very little beyond the controlled addition of sugar and ammonia suggested by Hosler and Johnson (1953) and later elaborated by Hockenhull and Mackenzie (1968). We have hardly touched the hard core of biochemical control. Little has so far been done to manipulate the internal control systems suggested by Demain (1968, 1972). This does not mean we should not try. To meet such aims we need equipment in which such sophisticated work can be carried out. That is to say, we need all the basic sophistication of the best and most flexible production plant plus the extra refinements that are needed for the creation of new techniques altogether.

These new aims indicate a fundamental cleavage between such laboratory tools as the shake flasks and miniature stirred fermentors of 5 liters and upward. Only the latter have adequate simulation value. In them we can act out on a small scale the projects of the future.

The aim of industrial development is to improve the profitability of existing operations as well as to bring new projects to fruition. There is more than one paradox in these objectives. The development of well established products requires constant improvement in technology to retain a competitive position. Hence, with penicillin, a product of more than 30 years' standing, processes are highly sophisticated and, in general, make use of the widest range of ideas and equipment. On the other hand, with a new product the strategy must in the first instance be the application of methods developed in more worked-over fields. At the outset this can often lead to spectacular successes with relatively low costs.

Another aspect is that the most highly developed equipment, generally speaking, is the equipment used for well established major products. The company will have a substantial investment in existing plant. Hence, developments that will work on the existing plant will be more favorably looked on than those requiring more radical approaches. This means, basically, that many of the problems of a development unit are concerned with scaling down rather than the scaling up one associates with chemical pilot plants. The newly developed processes and equipment must be generally compatible with existing installations. This does not, of course, preclude the designing of radically different equipment on the small scale, but does provide a yardstick against which its performance must be measured, so that the capital costs of change may be set against the expected improvements.

The job of a pilot plant is therefore to simulate the behavior of the production unit so that improvements, whether in choice of microorganism, in control techniques, or in novel materials can be evaluated and translated into production terms at minimal cost.

B. SIMULATION, MODELING, AND PREDICTION

Here perhaps one should say a word or two about simulation. What one aims to do is to investigate in miniature the consequences of certain actions upon the behavior of the fermentation system that we propose to operate on the large scale. The experimental variables can be mapped onto the pilot system; if the simulation is successful, results obtained on the small scale will accurately predict the behavior of the production unit. Note here that we are concerned only with such a set of variables as is appropriate to the particular equipment. To be able to carry out an effective simulation we must see that under one set of conditions at least the processes are similar in most respects and coincidental in the most important ones. Thus, we can cut down the variables to a manageable level.

At this point, we should give some thought to the general philosophy of modeling techniques. There are four key notions that one should apply when considering the merits of particular simulation techniques. According to Beer (1972), they are as follows:

1. *Scaling down* in both size and complexity—a model of Shakespeare's birthplace, for instance, could stand on a table and would not be expected to incorporate miniature bricks in equivalent numbers to the building at Stratford-upon-Avon.
2. *Transfer across*, whereby actual parts of actual things are represented again in their relative positions.
3. *Workability*, by which I mean that the model can, in principle, anyway, operate like the original. Thus a model train actually runs round a model railway, and it looks so much like the thing modelled that cine films of models can be substituted for film of actual trains and successfully pretend to be real. That this can be so, although the engine may be driven by clockwork introduces the fourth point—
4. *Appropriateness*. The model is a good model if it is appropriate. Someone watching the film just mentioned is not in the least concerned with how the engine is powered; but an engineering student who dismantled a model railway in a technical college and found an enormous coiled spring inside would not be impressed.

The first notion, that of *scale-down*, is basic to any ideas on subplant scale work we may want to carry out. A large number of small experiments are much cheaper than one large production-scale run, and less output is put at risk. Moreover, the greater the disparity in size, the less reliable is the information, by and large. On the other hand, when the experimental unit is smaller than 1 gallon, it becomes impossibly difficult to operate and one is forced to use the much less appropriate shake flasks, which will simulate effectively only certain aspects of the fermentation. The aim of the next few paragraphs will be to set the advantages of scale against the three remaining criteria: *transfer across, workability,* and *appropriateness.*

Transfer across (also called scale-up) presents few problems. Most of the fermentation information has one-to-one correspodence between two particular scales so that modes of operation can be chosen that are appropriate to the sets under scrutiny. Down to the 1-gallon stirred fermentor, the correspondences are good in most cases. At the lower end of the scale, however, the correspondence is starting to break down. Difficulties arise at this level especially because nutrient feed techniques are extremely difficult to operate truly continuously on a small scale. There are also difficulties in sample abstraction that badly upset the volume and concentration of the culture. In addition, the properties of the aeration-agitation system change markedly with scale and may render highly suspect any deductions made. A particularly upsetting aspect is that, in the smallest stirred vessels, the solution of oxygen at the top surface of the broth contributes almost half the total oxygen solution rate, whereas in large tanks this contribution is negligible. In addition, should scale-up be made on a linear basis, other criteria, such as power consumption, shear rate, tank circulation time, have different dimensions and vary according to different rules. A particular effect of this is that the limiting factors on one scale may well differ from those on another, and an apparently similar investigation may really be about two different things. However, in spite of this, even the superficially very different shake-flask culture may supply correspondence adequate for the solution of certain problems.

Whereas *transfer-across* is concerned with the legitimacy of the mapping and with the correspondence established, *workability* is concerned with the usefulness of work, with predicting the outcome of an experiment on the large scale by what happens on the smaller. In practice, discussion of *transfer-across* in isolation from *workability* is difficult, for as soon as we quote concrete examples of the one, the other is implied. In general, however, *transfer-across* is a criterion we apply to the two experimental units as a whole—i.e., we compare the plant fermentation equipment with the 5-liter laboratory fermentor in the light of all possible problems. *Workability* is a more particularized criterion and refers to the usefulness of the setup in solving a particular problem or exploring a more limited set of relationships.

Let us look once more at the example of shake-flask culture and its usefulness in the production of penicillin. Although the flask system is a closed one in that no further nutrient additions are made after inoculation, the behavior of the culture can be made to imitate that in a production stirred tank over a considerable period. The use of addition kits to supply the large fermentors with continuous carbohydrate nutrient is replaced, for the flasks, by the initial incorporation of lactose which is then broken down slowly in a controlled way by the organism to supply itself with hexose sugar at just the right slow, steady state. This technique is still useful when the purpose of the experiment is to distinguish between the yield potential of mutated or freshly isolated strains of *Penicillium*. Without the facility for scaling down both size and the amount of individual attention given to each culture, it would have been impossible to have selected so many improved organisms as we have. In fact, it is difficult to dissociate the notion of *workability* from that of *predictive value*, for what we have just discussed is the ability of a low-cost experiment in a small vessel to predict the behavior of organisms on the large scale.

Before we digress, however, we should discuss our last criterion, *appropriateness*. There have been many criticisms of shake-flask techniques on this particular score. Objection to the use of shake-flasks in the selection of *Penicillium* strains has been made principally on the ground that the lactose technique may impose a different kind of limitation from continuously fed glucose. This might lead to nonrecognition of mutants which could do better under conditions of abundant carbohydrate feeding. Furthermore, the mechanism whereby oxygen is transferred to shake-flask culture is very different to that obtaining in a stirred tank. In practice, however, the correct choice of medium in the one instance and of scale in the other have so far given good correspondences. This is all one can say; there is always the finite chance that, at some time in the future, the technique will impose an absolute limitation not inherent in the large installations. Happily, up to the present time, what we know about the internal workings of the black box which is the organism, does not support these fears. However, to be critical of the appropriateness of one's simulation techniques is no bad thing.

In the end, the construction of any model system depends on its *predictive* value as compared with its cost, that is to say on its *predictive efficiency*. (How good or bad are the ideas fed into the system does not matter if the small-scale equipment does not effectively reflect the possibilities open to us on the plant of the future. Further, unless the ideas are suitably programmed to take account of the logic of the fermentation process, i.e., unless the mental model is sufficient in cross references and correspondences, transfer across to the realities of the physical process will not be possible.) In fact, Beer (1966) says, "The Scientific model is a homomorphism on to which two different situations are mapped, and which actually defines the extent to which they are structurally identical. What is dissimilar about the original situations is not reflected in the mapping because the transformation rules have not specified an image in the set the model constitutes for irrelevant elements in the conceptual sets. If the transformation has ignored, as irrelevant, elements that are in fact relevant, then the model will lose in utility but it cannot lose in validity. This is its particular strength. Now the measure of utility is vitally important, but it may be accomplished by straightforward testing of the *predictive value* of the scientific model. If this predictive value is low, we had better start again. But, if it is high, we can use it with confidence and without further worry as to the appropriateness or otherwise of the analogy drawn at the conceptual level."[1]

We have discussed in very general terms the idea that the pilot plant is a model of the production plant of the future and have discussed various notions about how these entities should relate one to the other. Although examples have been drawn from our experience to expand and make more concrete those ideas, we still have to categorize the various kinds of work a pilot plant must take on.

C. The Tasks of a Pilot Plant

First of all, laboratory work will need translation to the kind of equipment used on a plant. Thus, in the case of a new antibiotic, say, a fermentation carried out in a shake flask will need translation to stirred vessels. Because of the lack of formal similarity in the techniques, there is a great deal of arbitrariness in this scaling-up process. Hopefully, however, it should be possible to develop a batch type of process which works more or less satisfactorily on all scales. At this point it is important to achieve formal similarity between inputs and outputs on all scales, so that work is based on a common pattern. This is especially important when we are considering the total gestalt of the process. If we can describe the activity of the larger vessels in the language of the smaller, we shall have established a launching pad for further excursions. The ensuing stage of optimization does not differ radically from that already made at the shaken-flask stage. In order to fit into the limitations of the stirred tank fermentors, and to take advantage of their special qualities and control equipment, a great deal of highly pedestrian and empirical optimum seeking is the order of the day. Composition of the medium is altered, and alternative sources of raw materials or even substitute materials are tested at many levels and in many combinations. The age, amount, and other qualities of the inoculum are changed until the best

[1] From Beer: Decision and Control. Copyright © 1966 John Wiley & Son Ltd.

results are obtained. Different aeration and agitation regimes are tested, the best being selected as well as the pattern of response being fully documented. Important at this level is that the stirred vessels and the shaken flasks shall be good models of each other, for the more work can be carried out in the former, the less it costs.

Sooner or later, however, the closed system must be opened. By controlling such factors as pH, sugar input, and nitrogen usage a culture can be kept in the productive state for a longer period than in the batch process. In fact, in many instances (cf. Hockenhull and Mackenzie, 1968) a quasi-steady state may be achieved during which the culture is held so as to make antibiotics or other products at a very high rate almost indefinitely. This is usually effected by controlling one or more nutrient feeds so as to keep the growth rate at the desired value. This type of work can be carried out effectively only in stirred fermentors of such a size that continuously measured additions of nutrients can be made.

Such development inevitably leads to further sophistication. More and more necessary becomes the need to develop control systems whose basis is the observation that certain characteristics of the fermentation change as conditions become more or less suitable for product formation. For example, the griseofulvin fermentation is kept in maximum production by raising or lowering the sugar feed-rate as the pH value rises or falls (Hockenhull, 1962). Other criteria are being sought continually, and similar control systems evolved. While most of these controls are at present manually operated, some progress has been made toward automation. In the future it may be necessary to satisfy several criteria at once, or, perhaps, to use a parameter obtained by combining a number of different observations. Further one may need to take account of the history of the culture in deciding how to program its control. In these circumstances we can envisage electronic computation playing a much larger part in our activities.

All this is the larger context of the pilot plant. Into it we have to fit many mundane activities. Strains selected on shake flasks may not always be risked straight into the production lines. A change in process may not be acceptable to the various Food and Drug Administrations until they are satisfied that the product is unchanged in quality: satisfying them would entail the production of large-scale samples on which to perform the necessary tests. Different samples of materials or alternative ingredients have to be tested for productivity under conditions as near as possible to those on the plant.

Introduction of a new idea directly on to the production plant is risky and may prove expensive. As adequately equipped and staffed pilot plant can save money and make more. First of all, there is available a large number of small, relatively inexpensive units so that one may risk "long shot" experiments without too many qualms. Second, the staffing of a pilot plant is generally more lavish (directly or indirectly) than a Works unit could tolerate. This means that comparison and optimization can be more quickly carried out. Third, work on the more sophisticated areas of control can be carried out with much cheaper and cruder equipment, partly because one can take risks as each unit is relatively cheap to run. Rarely does one find commercially available new equipment that is entirely satisfactory. Some must be adapted. Some equipment must even be redesigned in the field. The prototypes of such inventions are frequently crude and their fitting a "lash-up." This kind of engineering, often ample for preliminary studies on feasibility, would not be tolerated on a production-scale plant, particularly as the special safety precautions needed might interfere with the smooth running of a fermentor hall. On small vessels one can take risks. The high degree of supervision and the minute-to-minute attention make up for the defects and enable us to "debug" the equipment before we install it on the factory floor.

D. Considerations of Expense

However, an adequate pilot plant is plainly an expensive installation, especially in regard to staff. In the last analysis, the money that we hope to make or save by inventing, altering, and improving processes will determine our expenditure. It will also, taken into account with past experience, determine the way in which our budget is spent. We may argue as to whether floor space should be chosen at the expense of equipment, how many of each size of vessel we should have, whether instruments can replace process workers or whether they will actually increase staffing levels, and so on. This is not all. If the kind of exploratory work is suitable and the risk of production loss small as compared with the cost of pilot experimentation, it may be economic to cut out the pilot plant stage and do the work directly upon the main plant. This situation has, one must admit, a political aspect. For example, it can arise where there is a lack of mutual credibility, holding up the transfer of laboratory results to the main plant, or where the development unit drags its feet in doing the appropriate tests before the plant is ready for the process change. The kind of decision, of which this is an extreme example, may occur in the pilot plant itself. We may have to decide whether we can scale directly from 5 to 100,000 liters or alternatively whether we will need to use vessels of intermediate size, up to, say, 5000 liters. Briefly, my own company has an array of such intermediate sizes. Their distribution reflects the experience and needs of the past 20 years. Such needs, however, change with time and are constantly reviewed in relation to the current situation.

There are other functions that a pilot plant carries out. It acts as a training ground for potential production-plant managers, for here they can learn all the aspects of technical management while putting less at risk than on the main plant. This aspect is discussed later in this chapter.

I would say, however, that the pilot plant is the place where many skills and possibilities meet. It is a frontier at which many technical languages are used with rather less than perfect accuracy. It is a place where the graduate learns "process sense" (if that is not handed down in the genes). One learns new skills and new languages, particularly those of systems theory and cybernetics. And, not least, it is where we sort out the men from the boys.

E. Information and Development Strategy

The pilot plant is, particularly, the chief clearing house for technical information. It permits a continual traffic between the production unit and the laboratory scientists. Figures 1–3 show just how much informa-

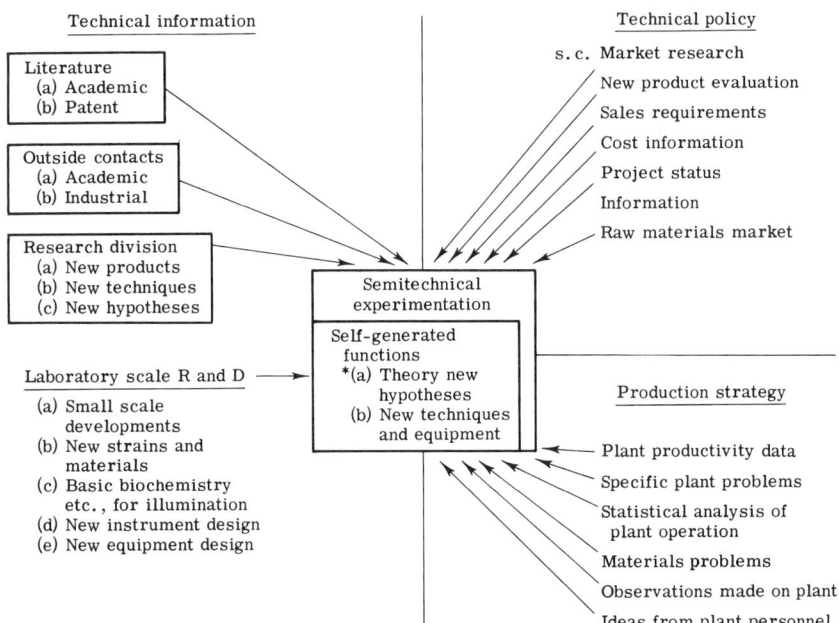

Fig. 1. Incoming pilot plant information. R and D, research and development.
* Intuition, hypothesis, experiment, deduction, design.

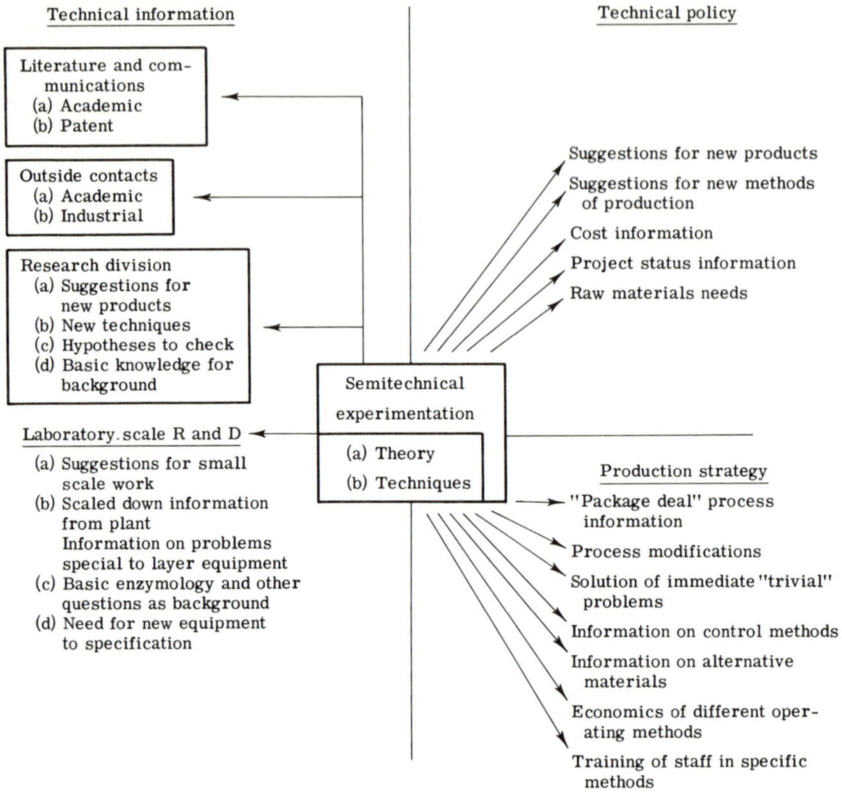

Technical information

Literature and communications
(a) Academic
(b) Patent

Outside contacts
(a) Academic
(b) Industrial

Research division
(a) Suggestions for new products
(b) New techniques
(c) Hypotheses to check
(d) Basic knowledge for background

Laboratory scale R and D
(a) Suggestions for small scale work
(b) Scaled down information from plant
Information on problems special to layer equipment
(c) Basic enzymology and other questions as background
(d) Need for new equipment to specification

Semitechnical experimentation
(a) Theory
(b) Techniques

Technical policy

Suggestions for new products
Suggestions for new methods of production
Cost information
Project status information
Raw materials needs

Production strategy
"Package deal" process information
Process modifications
Solution of immediate "trivial" problems
Information on control methods
Information on alternative materials
Economics of different operating methods
Training of staff in specific methods

FIG. 2. Outgoing pilot plant information.

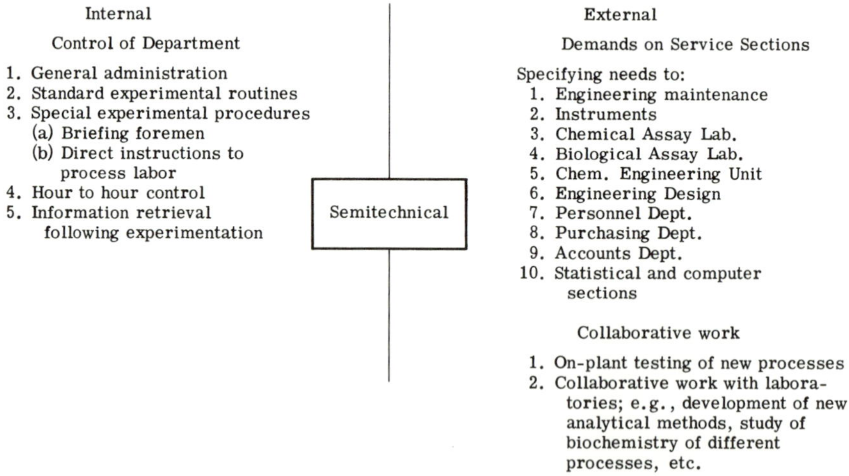

Internal

Control of Department

1. General administration
2. Standard experimental routines
3. Special experimental procedures
   (a) Briefing foremen
   (b) Direct instructions to process labor
4. Hour to hour control
5. Information retrieval following experimentation

Semitechnical

External

Demands on Service Sections

Specifying needs to:
1. Engineering maintenance
2. Instruments
3. Chemical Assay Lab.
4. Biological Assay Lab.
5. Chem. Engineering Unit
6. Engineering Design
7. Personnel Dept.
8. Purchasing Dept.
9. Accounts Dept.
10. Statistical and computer sections

Collaborative work

1. On-plant testing of new processes
2. Collaborative work with laboratories; e.g., development of new analytical methods, study of biochemistry of different processes, etc.

FIG. 3. Executive and administrative communication areas.

tion is involved and the high degree of complexity embodied in it. The pilot plant staff should be able to reduce this welter of information to a manageable level. They must filter out information, or process it, until the total variety is sufficiently small to be used constructively. Often enough, production units are not interested in the details of experiments so long as the results give them an adequate basis for their own operations. They must therefore receive clear, adequate, and usable information on the processes they carry out themselves.

We have said earlier that the pilot plant simulates the production plant *in posse* rather than *esse*. It prefigures the production processes of the future and gives standards against which to measure progress. As a model it should have critical predictive value. At the same time the operating pilot plant must set realistic targets for the production plant. In fact, the pilot plant is a metasystem with respect to the main plant, and as such its communications give a metalinguistic commentary on the current operational outputs.

Because this is so, the pilot plant should be controlled in the same place as the forward production policy of the firm. In regard to new products and processes it is in a decision-making position, for it can grasp the pattern of existing production and envisage what is possible for the future. This is a strong reason for having experimental plant in the factories. Again, their strategic role is a reason why the style of a fermentation pilot plant is so different from that of, say, chemical pilot plants. For, unlike the latter, for the most part the fermentation pilot plant plays a continuously generative and creative part.

II. The Pilot Plant as an Introduction to Production Management

A. Introduction

The fermentation production department relies on its pilot plant counterpart for information and new processes. Besides this customer–supplier relationship, the structures, functions, and organizations of the two units have a great deal in common. One may indeed map many of the functions of the development unit onto those of production. Bearing in mind their differing tactical aims, many activities of the one are simulated in the other. Tables I and II, which summarize the duties of a manager in each of the two areas, show this. For reasons of workable brevity, the correspondences between these tables are not wholly explicit: in the following pages I aim to bring them out in more detail. At the same time I shall develop the theme that, as a training ground for production personnel, the development unit can give something extra over and above what might be learned as a junior in the production unit itself. The latter introduction to training is too often deficient in deep explanation and approximates, sadly, to the process known in management circles as "sitting by Nellie."

TABLE I
JOB DESCRIPTION: UNIT HEAD: FERMENTATION PILOT PLANT

Main objectives	Areas of responsibility
1. Production of improved processes for products assigned to the unit, communication of all progress to departmental head	a. Full employment of space and equipment b. Correct balance between study of "how to" and "how did" of fermentation c. Correct allocation of graduate attention to ensure 24-hour cover
2. Arranging nonroutine maintenance of equipment, ordering and commissioning new equipment, progress chasing for these activities, delegation of routine maintenance	Up to expenditure as laid down for the position
3. Liaison with main production unit engaged on the same products	a. Guiding the production section head on the establishment of new processes, on plant experimentation b. Maintaining contact with current plant operation
4. Liaison with laboratory units engaged on process development, testing of new strains, maintaining two-way contact, feeding in problems to the laboratory, and applying their results to fermentation development	Advisory only apart from strain testing
5. Ensuring that the foreman of the pilot plant area under his supervision is carrying out his management of men and equipment to the best of his ability	Complete responsibility
6. Keeping up to date with current trends of thinking in biochemistry, microbiology, and fermentation technology	Complete discretion so long as it does not consume too much time as against other responsibilities
7. Training of assistant unit head	Delegation of enough work to keep him stretched but not overworked
8. Ensuring that other graduates in the department are fairly granted access to equipment and staff under his control	Advisory except when cooperation breaks down
9. Organization of information to staff on administrative and technical matters; dealing with personnel problems	Complete responsibility

TABLE II
Job Description: Unit Head: Fermentation Department

Main objectives	Areas of responsibility
To produce a set output of certain fermentation products 1. Correct program scheduling; correct implementation.	Targets are set for him which he must meet a. Setting correct number of seed and production vessels; dispatching harvests to recovery unit at right rate b. Routine control of fermentation by directing technicians' work and regular inspection of process c. Modification of vessel scheduling to ensure maximum output of product
2. Correct working practices and conditions	a. Safe working; regular inspections and reporting appropriately to outside bodies b. Understanding of plant processes and modifications throughout unit
3. Equipment for the appropriate products is in top condition	a. Plant inspection of vessels, valves, pipework, etc., for faults b. Having plant available for the engineers to make "scheduled maintenance" checks c. Projects for repair and replacement are put forward as needed
4. Existence of authorized documents covering fermentation processes under his control	a. Manufacturing guide to official format b. Regular updating of manufacturing guides c. Manufacturing guide amendments to cover temporary changes
5. Correct actions in emergency, e.g., power failure, fire, contamination, loss of temperature control, unsatisfactory raw materials, and failure of equipment	a. Deciding on priorities and applying correct action b. Action to prevent recurrence
6. Cost control	a. To make what decisions still remain within the limits of the constraints already set b. Testing of cheaper and more effective starting materials
7. Initiation of limited experimental work	a. Initiation of paper work for modifications b. Slight modifications on large production vessels
8. Liaison with recovery units	a. Close liaison to give notice of possible difficulties and to check performance afterward b. General discussion on quality control
9. Liaison with research and development (R and D) units	a. Information on current problems and difficulties with current process b. Use of R. and D. information in improving performance
10. Knowing present state of the art	Reading and otherwise acquiring background of contemporary knowledge from all sources and in general; being especially up-to-date in own area
11. Reporting work	a. Weekly, monthly, and quarterly reports (routine) b. Project reports when called for c. Adequate paper work to back oral communication d. Adequate briefing of superiors verbally and in writing

B. ACTIVITIES

Although the various activities are difficult to classify unambiguously, we may group most of them under the following headings—not necessarily in order of importance: (1) Techniques, (2) Control, (3) Social Organization, (4) Equipment Management, (5) Decision Making, (6) Innovation and Creation, (7) External Relations, (8) Planning and Budgeting.

1. Techniques

Any process examined on a pilot plant must have a reference standard. This is most often a simulation of the current production plant processes. If improvements are to be useful on the main plant, the experimenter should have a clear understanding of the degree of homology between the two types of equipment. This in turn gives a fundamental appreciation of the variables of the plant process and their control. This information is most specific where it concerns the developer's special interest currently, but he can acquire an understanding of other processes at the same time because of the common principles involved. The developer should have had close contact with—and occasionally have stood in for—his colleagues with responsibilities for other products. One would hope also that he was fortunate enough to have worked on a number of important projects himself. Even in pilot plants engaged entirely in developing new products not yet produced on a factory scale, there will be common areas of information, and the technology may well have to conform substantially to such plant as is in current use.

So much for the background. The developer should see more than this. In the course of experimenting he should see the effect of variations on the normal parameters. Not only does this alter the process efficiency, but it distorts the relationship between the variables. He therefore sees how a plant would behave under abnormal conditions, such as occur from operator error or instrument failure. He can thus build up a repertoire of restorative and remedial actions appropriate to the transfer functions of the perturbations and response rates of the systems involved (i.e., trouble shooting). Further, he gains an idea of the areas in which the plant is relatively stable and those in which drastic action must be taken quickly. In fact, he has more opportunities to get the "feel" of the process than he would when working under conditions in which the value of the end product might make him unwilling to take action without further information, by which time it might well be too late. Furthermore, as his actions are freed from the restraints of a production target, and as he is venturing into uncharted areas, catastrophes may occur which he must deal with on the spot because no one else can.

Successful action in such emergencies brings confidence, and the ability to deal with similar alarms and excursions in the main plant, where the freedom of action allowed may well be less and the cost of error correspondingly more.

2. Control

Control of the actual production processes has several distinct aspects. These are conveniently grouped under the following headings: (a) overriding control of the standard process operations as laid down in the manufacturing guides; (b) programming, scheduling, and dealing with emergent contingencies; (c) bringing new developments to fruition.

The physical operations themselves are carried out by a work force who should be disciplined, controlled, understanding, adequately instructed, and well motivated. We shall examine the production situation in the light of these considerations, and hope to show how experience gained on an experimental plant will prove useful in the more disciplined atmosphere of production.

On a fermentation plant the batching of ingredients, their sterilization, and the running of the fermentation are all carried out by process operators, normally from standard instructions. Supervisory grades will see that this is done properly and at the right time. They will also draw attention to such special variations as when experiments are to be done or emergencies are to be dealt with. Carrying prime responsibility and controlling all these functions is the manager, who has other requirements to fulfill in addition to an overall appraisal of plant control. He must know his process and where it is likely to give difficulties and at what time these should be especially guarded against. He must know where the work load is heaviest and organize ahead for this. In this way, he will know what to ask for and when.

This is indeed a most important role. The giving of clean, unambiguous, and ample instructions, both verbal and written, is an obligate skill. It should have been learned on the pilot unit, where many more procedural changes have to be made and where accurate translation of requests into operations is not reinforced by the repetitive pattern of the standard fermentation procedure in the manufacturing guide.

In addition, the production plant manager has to meet special requirements. He must plan for maximal or cheapest output to fit in with the recovery plant and its schedule. He must deal with breakdowns and think ahead to future changes of plan and their effect on the work force. These must be communicated and explained in such a way that the changes and the reasons for them are understood by the staff. The ability to explain to the workers in the development unit the necessity for changing from one problem or product to another should come in useful here.

The hierarchical character of a manager's job gives him a full picture of the way his plant fits into the working of the factory as a whole. He is judged not only on how his people work, but on the targets he sets and achieves in the higher level of planning, that of the factory as a whole. Managing a development unit provides the right foretaste for this because he must foresee the future needs of the plant and explain them to his colleagues and subordinates in appropriate terms.

The plant manager, as we have discussed in an earlier section, must be technically "on top" of his process. Not only this, but he must be able to explain to, or rationalize for, his workers the operations required of them. He must also brief his workers so that the systems with which they are working "make sense." For, although what actually happens in a fermentor is seen "through a glass, darkly," there are nevertheless patterns of behavior that link input and output. The "don't ask why, just do it" system of management is not efficient. This applies especially to operations complicated by the unpredictability of biological systems. In an experimental unit, explanatory briefing has to be carried out more often than on the main plant. Moreover, because there must be some kind of logic in explaining the frequent changes in plans and ideas, the explanations must often be more fundamental and complex than would be needed on the main plant. Thus the training obtained in this area should be adequate to deal with the technical problems of the production plant.

3. Social Organization

In the foregoing section we have touched briefly upon one aspect of man management. There are many others, most of them more important. As we have already stated, the sectional managers in a pilot plant have under them staff of many grades from process operators through to foremen or laboratory technicians. The manager must manage this staff so that they perform his wishes efficiently, promptly, and with good cheer. He must first of all break the job down and allocate it appropriately to the staff he has available; he must see that his supervisors adequately understand what is delegated to them and in turn understand how much individual responsibility each man in their teams is capable of. This means that the unit head must have some idea of what work loads are possible to his team and the discipline and persuasiveness to get the men to achieve them. He must, also, deal with absences through sickness, holidays, and other causes planned and unplanned. He must be capable of using substitutes—often inadequately trained—which may mean altering the tempo of the work or reallocating jobs so that the more experienced men can cover the deficiencies of a temporary workmate. He must be a father to his men and support their case when personal circumstances indicate the granting of compassionate leave. He must look after their career progression, sending them on the appropriate courses or arranging instruction himself. He often has to write job specifications with them and relate these to general factory policy. He must make annual assessments and deal with recruitment, all within the factory context. He must also see that his department runs substantially according to the conventions of the factory. In addition, he is responsible for safe working practice both in accident prevention and good housekeeping, for the avoidance of accidents and their occasion, and for seeing that the lessons taught are practiced. All these aspects arise on the experimental plant, although they do not give as much difficulty nor are they as complicated as on the main plant. With pilot plants, the ratio of supervisory staff is higher, because there is a higher ratio of men to equipment, and because the men have applied for and been selected for their jobs because they like change and do not mind their routines being altered in certain areas. This general setup does mean that, although the pilot and main plants have similar organizational and management problems, the former are more easily dealt with.

4. Equipment Management

In a pilot plant one has, in theory at least, an effective imitation of the main production unit. Most of the properties of the production equipment should be effectively mapped onto the pilot apparatus. Not all products scale up or down in the same way, and therefore a considerable degree of flexibility must be available. In addition, the pilot plant must mirror the production plant of the future and should be capable of far more sophistication.

New equipment must be specified with care and well maintained, in spite of being often used in abnormal conditions. This trains the manager in the skills of specification, ordering, progress chasing, and job inspection. Maintenance must be scheduled and the equipment kept in working order. If one can do this in a pilot unit, often in the face of lower priorities, one can certainly manage it on the larger plant, where departures from routine should be much less frequent.

What skills are particularly needed in this context? Certainly an important one is the ability to deal with engineers and instrument specialists. As a rule these are organized as members of service departments whose functions are shared among many groups of the factory. Thus, the work required has to be clearly defined and a case made for the appropriate expenditure and priorities. The job has to be specified accurately so that it is not only right with respect to engineering, but does the job it is intended to do. This area is always a difficult one, as there are semantic differences between scientists and engineers. However, if a manager has been able to keep a pilot plant up to date and in working order, he should cope well with similar problems on the main plant. He must, of course, look for special difficulties connected with size and the interrelatedness of his installations, particularly when his masters pressure him to meet output targets "now," and the temptation is to put off routine maintenance for yet another cycle.

5. Decision Making

In the course of experimental work there is continual readjustment of aims. This arises from the success of the program or the reverse. In addition, there may be changes in the relative importance of the various investigations; there may occur a need for special modifications requiring alteration by the engineers; in addition, pressures from the production unit may require an interruption of program. Again, when an experiment goes badly wrong, one must decide whether to end it or to run it on in hopes of salvaging information.

Similar decisions must be made on the production plant, although in the main they have a slightly different emphasis. How does one schedule in a time of high contamination risk? How does one deal with gaps in the program? What happens when a tanker of raw material fails to arrive? What batches does one set at most risk when there is a shortage of cooling water? How long does one risk a mechanically unsound fermentor?

In addition, there are problems of control and steady operation. Often in experimental batches there may be unpredictable and rapid changes, such as a sudden fall or rise in pH value. Sometimes one has to recognize what is happening to the process solely on the strength of a telephone message from a process operator. Instinctively one learns to deal with the emergency effectively even if it occurs at 3 A.M. Similar decisions occur on the plant; and with experience, one learns to anticipate what may go wrong and correct it before it happens. Further, in talking about decisions one must remember that a decision entails a course of action taken on inadequate information. One can develop this facility for prompt action more effectively on experimental plant because the penalties of a wrong decision are less (and there might even be the premium of extra information). Experience of this kind makes one aware of what risks one can take and what penalties might accrue from tardy action. It also makes one far readier to accept experimental procedures when one has the "feel" and can make the decision to abort an experiment in mainstream or modify it in some way should it show signs of incipient disaster.

6. Innovation and Creation

Work on the pilot scale under a variety of experimental conditions exposes the practitioner to a wide variation of behavior. In many instances unforeseen outcomes can be related back to the process and become the basis for future experiments. This training enables one to spot, under the much more uniform conditions of the production plant, similar effects that would otherwise have passed unnoticed. These can be the basis of on-plant experiment or may lead to requesting the research and development (R and D) units to do something about it.

This kind of training helps a production man too in deciding whether the R and D effort is being as helpful to him as might be. It helps also in developing a tactful approach to the plant supervisor when suggesting new lines of investigation that might prove useful to him. Such training also helps the future production plant manager to see experimental work from the viewpoint of the Development Unit. This may prevent the Production Department from asking for a line of research that cannot possibly give conclusive results, but only a good deal of tooth gnashing. An example of this is the relative assessment of two rather similar grades of medium ingredients for which a 5% difference may take 50 pairs of comparisons. (In this instance there may well be a case for on-plant experimentation using an evolutionary operational procedure.)

Further, the production manager can figure out possible improvements which, after being checked perhaps on the pilot plant, may be readily adopted on the main plant. Where these involve equipment or working methods, he has often to do his own innovation without outside assistance. The experimental disciplines he has learned should then serve him in good stead.

Finally, the production man should be more sympathetic to suggested changes and can therefore help make the small adjustments that will fit the process to the large-scale equipment, if that is possible. In fact, having pilot plant training, a production manager should work better with his opposite number in R and D on any commissioning of new processes and plant they have to do. His experience should make him welcome the joint attack on the problem in hand by Production and Development units, rather than being left to sort out the snags himself or not being consulted.

7. External Relations

Working on a pilot plant entails a considerable amount of work on equipment. Basic equipment has to be designed; alterations have to be made (e.g., provision of higher power agitators); ancillary measuring and metering apparatus must be installed and often designed. All this has to be kept in working order. For the great part of these activities one relies on service engineers. One learns how to specify one's needs unambiguously, how to convey this to the design side, how to check their views and make sure that any "improvements" they suggest can be incorporated without losing sight of the purpose of the job, and, finally, how to chase the job's progress to ensure that it is available when wanted. In addition, one learns the principles of scheduled maintenance and learns to be responsible for the condition of one's plant by one's own observations and by stimulating the observation of one's staff so that breakdowns occur minimally. All these skills are needed by the plant manager and all can be learned on a pilot plant.

I cannot stress too highly the value of good personal relations. This is especially so when engineers are concerned. It is most necessary to get the engineers, both design and maintenance, interested in the process and actively contributing their ideas on process and equipment improvements. Time, a commodity more freely available in a development unit is needed to achieve this, but, once mutual respect has been built up, such rapport can be useful in the area of production. The future production man should therefore secure the cooperation of his future engineers while he is learning the job in the development unit.

Similar relationships should be formed with other services—with the analytical chemists, buyers in the Purchasing Department, the man who runs the computer, and the all-powerful accountant. In engaging with them effectively, we affirm that we all work for the same company and we encourage team spirit throughout the organization.

8. Planning and Budgeting

Development and production have a lot in common in this area. The basic elements of both are the current costs of the processes and of the finished products, and a realistic target for what each should be in a certain time hence.

On the basis of these figures the development manager decides how great an improvement he must make to cover his expenditure to the required ratio. He must do this with all his products and with each idea in order to grade them in order of priority and to allocate effort for them.

The production man uses the figures in a different way. He alters his process to make the best use of his materials if their current cost should change, if the volume of business should expand or contract, and so on. He must also cost out each new method suggested by R and D (as indeed they should have done) to see what new plant is needed, how much it costs, and how soon the improvements in the process will bring financial returns to pay for it.

As has been implied earlier, the planning and scheduling activities on the pilot plant and the production unit pose similar problems, especially when a large number of products are being made at the same time.

Finally, records have to be kept. All the information on each piece of development must be systematically and retrievably filed both in the individual's notebook and centrally. The production plant man must do the same, not only because of the need to guide the ongoing production process, to establish production norms, and to carry out costs analyses, but also because such records are a legal requirement to satisfy the government inspectorate and licensing authorities both at home and in other countries of sale.

C. Conclusion

We have seen in the above sections how useful is a spell on the pilot plant to an intending production manager. Although the simulation of the latter job by the former is by no means complete, I venture to suggest that in every major activity of each there is a fair degree of correspondence. There is enough similarity, too, between the two jobs as a whole as to provide a logical structure on which one can build the special expertises of the plant. One can therefore set about learning the special and more important duties of running a production unit without at first worrying unduly about mechanisms and details. And the man who starts the production job with additional information and insights can give a more creative performance and is enabled to improve himself and his process continuously.

Last of all, when the young aspirant transfers from a development to a production unit, a long and arduous process of learning and retaining is before him. One hopes that he has learned self-reliance and self-criticism, so that he realizes his need to teach himself. Not only this, one hopes that he has had firmly drilled into him the responsibility for doing so. From the very beginning of his training with the pilot plant, he should never be allowed to seek refuge in: "nobody told me, so I can't be expected to know that." On the contrary, the curiosity and the will to understand should be encouraged and rewarded, especially when it leads to successful project work. This attitude is not something that should be left to chance—the responsibility of the boss is to see that his men are adequately trained in this as in other areas.

Acknowledgments

I would like to thank my colleagues Messrs. Foster, Hargraves, Coates, and Hunt, for their helpful discussions and advice.

References

Beer, S. (1966). "Decision and Control," p. 113. Wiley, New York.
Beer, S. (1972). "The Brain of the Firm," p. 111. Penguin Books, London.
Demain, A. L. (1968). *Lloydia* **31**, 395.
Demain, A. L. (1972). *J. Appl. Chem. Biotechnol.* **22**, 345.
Hockenhull, D. J. D. (1962). *U.S. Patent* 3,069,328.
Hockenhull, D. J. D., and Mackenzie, R. M. (1968). *Chem. Ind. (London)* p. 607.
Hosler, P., and Johnson, M. J. (1953). *Ind. Eng. Chem.* **45**, 871.

The Microbial Production of Nucleic Acid-Related Compounds

KOICHI OGATA

*Department of Agricultural Chemistry,
Kyoto University, Kyoto, Japan*

I.	Introduction	209
II.	Production of 5'-IMP and 5'-GMP by the Enzymic Hydrolysis of RNA	210
	A. Enzymes Hydrolyzing RNA to 5'-Nucleotides	211
	B. RNA as an Industrial Raw Material	211
	C. Hydrolysis of RNA	212
III.	Production of 3'-, 2'-Nucleotides and 5'-Deoxynucleotides by the Enzymic Hydrolysis of RNA and DNA	213
	A. 3'-Nucleotides	213
	B. 2'-Nucleotides	214
	C. 5'-Deoxynucleotides	214
IV.	Excretion of RNA Derivatives	214
	A. Nucleotides and Nucleosides	214
	B. Cyclic AMP	215
V.	Fermentative Production of Nucleosides, Nucleotides, Ribose, Orotic Acid, and DNA	216
	A. Inosine Fermentation	217
	B. 5'-IMP Fermentation	217
	C. AICAR Fermentation	223
	D. 5'-XMP Fermentation and Its Conversion to Guanine Nucleotides	223
	E. Ribose Fermentation	224
	F. Orotic Acid Fermentation	224
	G. Formation of DNA	225
VI.	Salvage Synthesis of Nucleosides and Nucleotides	225
VII.	Conversion of Nucleosides	226
	A. Phosphorylation of Nucleosides	226
	B. Production of 5'-GMP from AICAR	227
	C. Transribosylation of Nucleosides	227
VIII.	Formation of Nucleoside Derivatives	230
	A. Nucleoside Polyphosphates	230
	B. Sugar Nucleotides and Cytidine Derivatives	231
	C. FAD and NAD	234
IX.	Production of Coenzyme A	236
	A. The Biosynthesis of CoA and Its Isolation	236
	B. The Biosynthesis of Intermediates of CoA	240
X.	Conclusion	240
	References	241

I. Introduction

In Japan, extracts of dried seaweed (*Kombu*) and the muscle of fish (*Katsuobushi*) are widely used as food seasonings. Ikeda and Kodama have demonstrated that the good flavor of the former comes from monosodium glutamate and that of the latter from a compound of inosinic acid.

The microbial production of monosodium glutamate from glucose was demonstrated in 1957 by Kinoshita and co-workers and by Asai and co-workers. Small amounts of inosine 5'-monophosphate (5'-IMP), originating from ATP, have also been produced from the juice of boiled fish and squid muscle. Since 1962, several factories have been carrying out large-scale production of 5'-IMP and guanosine 5'-monophosphate (5'-GMP) using the enzymic hydrolysis of RNA (hydrolyzing method). A wide range of research on the production of 5'-IMP, 5'-GMP, and their related compounds has since been carried out, and many reports show that the biochemical mutants (auxotrophs) of microorganisms are capable of accumulating appreciable quantities of these substances in their culture media (fermentation method).

At present, the industrial production of nucleotides, especially 5'-IMP and 5'-GMP, by the hydrolyzing and fermentation methods comes to more than 3000 tons per year. The microbial production of derivatives of nucleosides and nucleotides, e.g., the sugar nucleotides, adenosine 3',5'-monophosphate (3',5'-cAMP) and coenzymes has also been mainly investigated by Japanese researchers.

This review will be limited to the various new techniques established for the industrial production or laboratory preparation of nucleotides, nucleosides, and their related substances.

II. Production of 5'-IMP and 5'-GMP by the Enzymic Hydrolysis of RNA

Kuninaka (1) and Nakao and Ogata (2) found that the flavor of nucleotides is attributable to their purine bases, the OH^- and PO_4^{3-} groups at the 6 and 5' positions of purine, respectively (Fig. 1), and

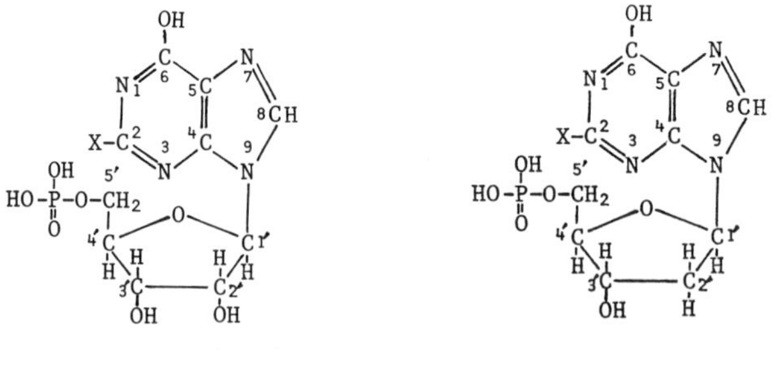

(A) 5'-Nucleotide   (B) 5'-Deoxynucleotide

Fig. 1. The chemical structure of flavor nucleotides. (A) X: H=5'-IMP, NH_2=5'-GMP, OH=5'-XMP; (B) X: H=5'-dIMP, NH_2=5'-dGMP, OH=5'-dXMP.

is unrelated to H⁺ or OH⁻ at the 2′ position. 5′-GMP has about 3 times more flavor than 5′-IMP, and 5′-AMP and xanthosine 5′-monophosphate (5′-XMP) have scarcely any flavor.

The historical development of nucleotides as flavoring agents has been reviewed by Kuninaka (*3,4*) and Ogata (*5*) in Japanese, and recently by Demain (*6*) and Ogata (*7*) in English.

A. Enzymes Hydrolyzing RNA to 5′-Nucleotides

Snake venom and cow intestinal mucosa are known to contain a phosphodiesterase which hydrolyzes RNA to form 5′-nucleotides. Some 5′-nucleotides have been obtained from RNA as chemical reagents using these enzymes, but little consideration has been given to their use in the industrial production of the food additives 5′-IMP and 5′-GMP.

A microbial enzyme was first obtained from a water extract of a solid culture of *Penicillium citrinum* on wheat bran by Kuninaka *et al.* (*8,9*). Ogata *et al.* (*10*), after extensive studies, found that similar enzymes are distributed in the culture broth of a variety of microorganisms belonging to the ascomycetes, fungi imperfecti, actinomycetes, and bacteria. These enzymes which hydrolyzed RNA to 5′-nucleotides, also degraded DNA to 5′-deoxynucleotides, except for the enzyme from *Bacillus subtilis* (*11*). Presently, the enzymes used industrially are obtained from cultures of *P. citrinum* and *Streptomyces aureus*.

Saruno *et al.* (*12*), Tone *et al.* (*13*), Mouri *et al.* (*14*), and Fujimura *et al.* (*15*), respectively, have reported that 5′-phosphodiesterase is produced by *Monascus purpureus, Phoma cucurbitacearum,* mushrooms, and *Pellicularia* sp. In addition, Nakagiri *et al.* (*16*) have reported the enzyme in the barley rootlets used in beer brewing.

The model structure of RNA and the degradation of 5′- and 3′-phosphodiesterase are illustrated in Fig. 2.

B. RNA as an Industrial Raw Material

The source of RNA for industrial use is yeasts, which are harvested from the waste of the sulfite pulp process or are cultivated specifically for their RNA. The RNA content of yeasts varies by 10–15% of the dried cell weight.

RNA is extracted with a diluted alkaline NaCl solution and is precipitated by acidification or the addition of an organic solvent to the extracted solution. The precipitated RNA is dried and stored. With heated yeast cells, the process of the RNA extraction is omitted, and the cells are directly added to the enzyme solution.

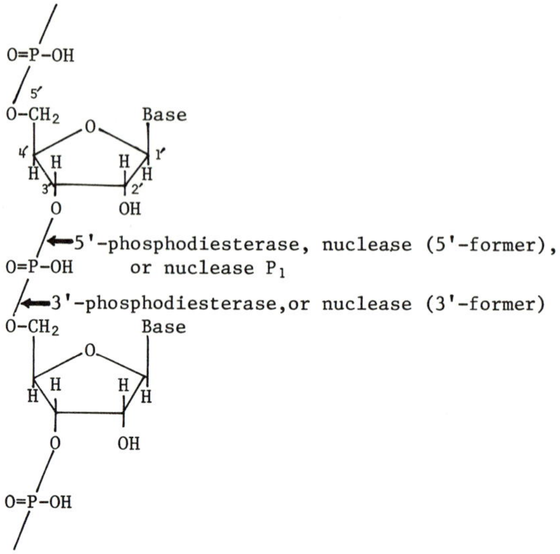

FIG. 2. The structure and degradative mode of RNA.

C. Hydrolysis of RNA

Enzyme solutions extracted from solid cultures or from the culture filtrate contain many enzymes that hydrolyze RNA, nucleotides, and nucleosides, as illustrated schematically in Fig. 3.

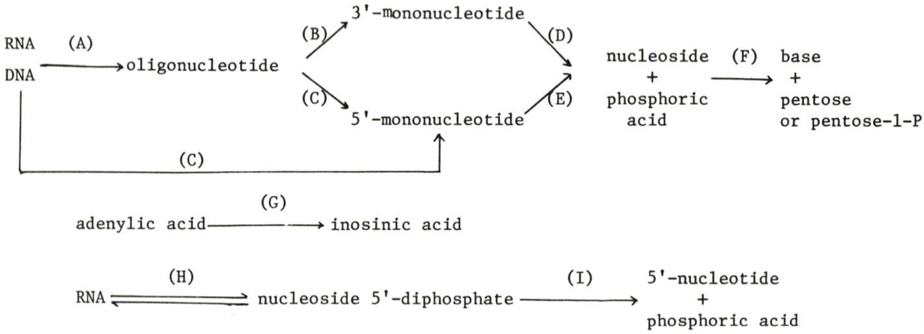

FIG. 3. Enzymes participating in the degradation of nucleic acids: (A) RNase, DNase, nuclease; (B) 3'-phosphodiesterase or nuclease (3'-former); (C) 5'-phosphodiesterase, nuclease (5'-former) or nuclease P_1; (D) 3'-nucleotidase or nonspecific phosphomonoesterase; (E) 5'-nucleotidase or nonspecific phosphomonoesterase; (F) nucleosidase (nucleoside hydrolase, nucleoside phosphorylase); (G) adenylic acid deaminase; (H) polynucleotide phosphorylase; (I) nucleotide pyrophosphatase.

Kuninaka (17) increased the yield of 5′-nucleotides to about 90% of the original RNA by hydrolyzing the RNA at 65°C and pH 5.0 which inactivates the heat-labile phosphomonoesterase. With *S. aureus*, the 5′-AMP in the hydrolyzate is deaminated to 5′-IMP by a 5′-adenylic acid deaminase simultaneously produced in the culture broth. When the enzyme solution lacks 5′-adenylic acid deaminase, the enzyme should be added to the RNA-hydrolyzed solution.

Aida et al. (18) and Fujishima et al. (19) have obtained, respectively, 5′-adenylic acid deaminase from *Microsporium adedovinai* and *Aspergillus melleius*. Yoneda (20) purified the RNA-hydrolyzing enzymes of *S. aureus* and determined that RNA was hydrolyzed to 5′-nucleotides using the cooperative activity of nuclease and 5′-phosphodiesterase. Kuninaka et al. (21) have declared that *P. citrinum* produces two kinds of phosphodiesterase, which they call nuclease (5′-former) and nuclease (3′-former). The 3′-former is a thermolabile enzyme and the 5′-former is a thermostable enzyme, which markedly increases its activity on being heated in the presence of Zn^{2+} or Veronal acetate.

Fujimoto et al. (22,22a) purified the 5′-former enzyme of *P. citrinum* to the homogeneous state using ultracentrifugation and found that it acts on RNA and on native and heat-denatured DNA, respectively, to produce nucleoside and deoxynucleoside 5′-monophosphates, and on nucleoside 3′-monophosphates and nucleoside 2′-monophosphates to produce, respectively, nucleosides and inorganic phosphorus (P_i). The *Penicillium* 5′-former nuclease is hereafter called nuclease P_1 for convenience.

Nucleotides from the hydrolyzate of RNA are extracted by the usual method involving active charcoal and an ion-exchange resin column chromatography. Solutions of IMP and GMP are dried as mixtures or separately. Although almost all the cytidine 5′-monophosphate (CMP) and uridine 5′-monophosphate (UMP) are waste products at present, the fermentation method is not always economically superior to the hydrolyzing method because the residual protein of the yeasts is fully utilized as fodder in the latter method.

III. Production of 3′-, 2′-Nucleotides and 5′-Deoxynucleotides by the Enzymic Hydrolysis of RNA and DNA

A. 3′-Nucleotides

Kuninaka (23) has reported details of the hydrolysis of RNA by the enzyme of *Aspergillus oryzae* via the 3′-nucleotides. Nakao and Ogata (24,25) have demonstrated that many strains of molds, bacteria, and yeasts produce 3′-phosphodiesterase and that the enzyme from the culture filtrate of *Rhodotorula glutinis* gives 3'-nucleotides in yields above 90% of the RNA when the reaction is run at 60°C.

B. 2'-NUCLEOTIDES

Igarashi and Kakinuma (26,27) have found a 3'-mononucleotidase with a high substrate specificity in the culture filtrate of *B. subtilis*. Using this enzyme, Kakinuma and Igarashi (28) have proposed a preparation method for 2'-nucleotides. The nonenzymic alkaline hydrolyzate of RNA contains 2'- and 3'-nucleotides in nearly the same ratio. The 3'-nucleotides are converted to their respective nucleosides by the addition of a culture filtrate of *B. subtilis* to the alkaline hydrolyzate of RNA, then pure 2'-nucleotides are separated by extraction with a resin.

C. 5'-DEOXYNUCLEOTIDES

The 5'-phosphodiesterases of *P. citrinum* and *S. aureus* show hydrolytic activity on denatured DNA forming 5'-deoxynucleotides. A detailed study to obtain 5'-deoxynucleotides has been carried out by Nakao and Ogata (29) with a culture filtrate system using *Aspergillus quercinus* and DNA from salmon sperm. The optimum pH of the hydrolysis is 8.5–9.0, and the yield of 5'-deoxynucleotide from DNA about 40%.

IV. Excretion of RNA Derivatives

A. NUCLEOTIDES AND NUCLEOSIDES

Numerous reports have shown the excretion of UV-absorbing materials or nucleotides from microbial cells. Demain (6) summarized the widespread occurrence of this excretion under a variety of conditions as follows: (a) temperature—chilling and heating cells; (b) irradiation—UV-ray and X-ray; (c) effect of an antibiotic or detergent; (d) starvation; (e) excretion in the growth media.

These studies are interesting because of their physiological significance. In many cases, the enzyme concerned in the degradation of intracellular RNA was phosphodiesterase or polynucleotide phosphorylase. When preparing nucleotides, Nakao et al. (30) reported that numerous species of yeasts excrete degradation products during incubation in buffers. The types of excreted nucleotides differ according to the strain. For example, *Schizosaccharomyces pombe* and *Rhodotorula pallida* excrete bases, nucleosides, 5'-nucleotides, and oligonucleotides. *Endomyces decipiens* excretes 3'-nucleotides, bases, and nucleosides. *Saccharomyces miso* excretes 3′-nucleotides at pH 4.0 and 5′-nucleotides at pH 10.0, in addition to bases, nucleosides, and oligonucleotides.

Tsukada and Sugimori (31) have reported the excretion of 5′-nucleotides from the cells of a mutant of *Saccharomyces rouxii*. Further studies with the cell-free extract showed that the production of 5′-nucleotides is the result of polynucleotide phosphorylase action.

Watanabe et al. (32) have found that *Torulopsis xylinus* excretes 5′-nucleotides when its cells are suspended in the acetate of Tris buffer (pH 5.0 or 7.0) with surfactants such as sodium lauryl sulfate or sodium dodecyl benzene sulfonate.

Recently, Kitano et al. (33) reported that borate buffer is very useful for the excretion of 5′-nucleotides. Moreover, the addition of a surfactant, cetyl trimethyl ammonium bromide, in the presence of borate buffer produces the complete degradation of endogenous RNA into 5′-nucleotides. All the 5′-nucleotides formed were excreted outside the intact cells.

Extensive studies have been made on the accumulation of nucleotides and nucleosides in the culture filtrates of numerous wild strains of bacteria, yeasts, molds, and *Streptomyces* (34–40). Although many strains accumulated 5′-nucleotides or nucleosides, results were not successful when compared with the quantity of 5′-nucleotides produced by auxotrophs.

B. Cyclic AMP

Cyclic AMP (cAMP, adenosine 3′,5′-monophosphate) is a physiologically important nucleotide. It is active as the second messenger of animal hormones and is the activator of protein kinase, etc. Extensive studies have been made of the distribution and function of cAMP in animals, plants, and microorganisms. In microorganisms, Okabayashi et al. (41) and Makman and Sutherland (42), respectively, have found cAMP in the culture filtrate of *Brevibacterium liquefaciens* and in the cells of *Escherichia coli* (Crooks strain).

Okabayashi et al. (43,44) have reported a larger accumulation of cAMP on the addition of DL-alanine or asparagine as the N source with glucose or other carbon sources. The quantity of cAMP in the culture filtrate of *B. liquefaciens* was about 1 gm/l. Almost the same result was obtained with the nongrowing system of this organism. The content of cAMP in cells of *E. coli* increased when the glucose in the medium was consumed. cAMP was excreted from the cells when fresh glucose was supplied (45). The *E. coli* B group, at a low concentration of glucose, accumulated 2–4 μmoles of cAMP in its broth with the complete exhaustion of glucose (45).

Recently, Ishiyama et al. (46) have found that *Corynebacterium muri-*

septicum No. 7 and *Microbacterium* sp. No. 205 isolated from soil, accumulate more than three to four times as much cAMP in their culture filtrates as is accumulated by *E. coli* or *B. liquefaciens*, when they were cultured in a medium containing adenine, hypoxanthine, or other purine derivatives. Although these bacteria did not require purines or their derivatives, they scarcely excreted cAMP when cultured in a medium without these compounds. The addition of DL-alanine did not cause a significant increment in cAMP production. The effective carbohydrates were glucose, fructose, maltose, sorbitol, and mannitol, and the amount of accumulated cAMP was 3.1–3.7 mg/ml. Two hypotheses for the role of adenine, hypoxanthine, or their derivatives in cAMP production were proposed:

1. The end product (cAMP) of the *de novo* synthesis system in cAMP biosynthesis, or the intermediate (5'-IMP or 5'-AMP), strongly represses the enzyme system at an early stage in the *de novo* synthesis system of cAMP. But the salvage synthesis system in the cAMP biosynthesis from adenine, hypoxanthine, or their derivatives is not subjected to feedback inhibition by the end product or the intermediate owing to natural genetic lesions in the isolated microbes. A large amount of the extracellular cAMP is thought to be a waste product of the purine metabolism of these microbes.

2. If any one of the affected function of these isolated microbes is not restored, these microbes cannot maintain their normal lives, even though some of the functions are restored. cAMP is required for the restoration of some of the affected functions, but not for the restoration of all of them. Furthermore, the pathway of the *de novo* synthesis of cAMP is limited in its early stages due to natural genetic lesions or to culture conditions. Consequently, the pathway for the salvage synthesis of cAMP from adenine, hypoxanthine, or their derivatives is extended.

Tomita and Suzuki (47) have found that cAMP was extracellularly accumulated by *Arthrobacter roseoparaffineus* KY 4301 and *Micrococcus paraffinolyticus* KY 4306 when grown on *n*-alkane mainly containing C_{12}–C_{14} as the sole carbon source. The addition of purine or its derivatives to the *n*-alkane medium did not stimulate the production of cAMP in contrast to the previous finding. The addition of glucose at a concentration of 2%, from the beginning of fermentation did not change the production of cAMP. The amount of cAMP accumulated was 1.0–1.4 mg/ml.

V. Fermentative Production of Nucleosides, Nucleotides, Ribose, Orotic Acid, and DNA

Many excellent reviews have been published on nucleotide biosynthesis (48–50). Magasanik (51) has classified purine nucleotide auxotrophs according to the type of genetic block in the biosynthetic pathway as shown in Fig. 4. It is possible to obtain small amounts of nucleotide or nucleoside which accumulate in relation to the blocked points using these mutants.

Techniques for obtaining auxotrophs and their industrial utilization have been advanced by amino acid fermentation, e.g., lysine fermentation by the homoserine auxotroph. Numerous auxotrophs have also been obtained from the various gram-positive bacteria, as seen in Table I

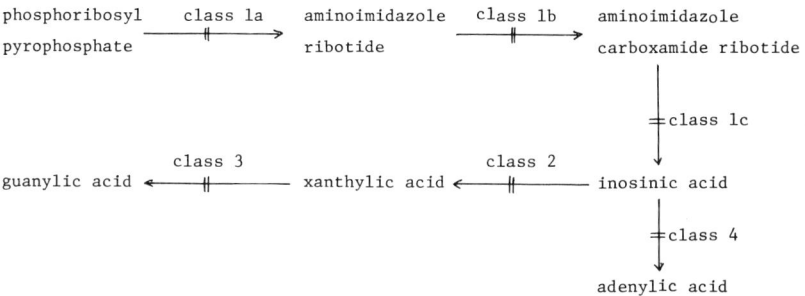

Fig. 4. Classification of purine nucleotide-synthesizing mutants.

(52–116). Presently inosine, 5'-IMP, 5-amino 4-imidazole carboxamide riboside (AICAR), and ribose are produced by these auxotrophs on an industrial scale.

A. Inosine Fermentation

Aoki et al. (69–74) have induced a strain of adenine-requiring B. subtilis which requires adenine, histidine, and tyrosine. They demonstrated that the inosine-producing B-4 strain is distinct from its parent, which produces hypoxanthine. The strain belongs to class 4 in Fig. 4. Its product, 5'-IMP, is dephosphorylated after it has passed through the cell membrane. The amount of accumulated inosine was 7 gm per liter of the culture filtrate.

Nara et al. (76) simultaneously found that the adenine and lysine auxotroph of B. subtilis produces inosine.

B. 5'-IMP Fermentation

Furuya et al. (54) found that 5'-IMP was accumulated in cultures of Brevibacterium ammoniagenes KY 13102 which require adenine and have a very weak phosphomonoesterase activity. A concentration of Mn^{2+} in the medium is essential for the permeability of 5'-IMP through the

TABLE I
ACCUMULATION OF NUCLEIC ACID-RELATED COMPOUNDS BY MICROBIAL MUTANTS[a]

Strain	Mutant properties	Product	Accumulation (mg/ml)	Added material	Workers	References
Brevibacterium ammoniagenes	Adenine requiring	5'-IMP Hypoxanthine	Over 5		Nara et al. (1967, 1968)	52,53
	Adenine requiring and permeability mutant (manganese insensitive)	5'-IMP 5'-IMP	12.8	L-Proline, 1.0 gm/l Excessive manganese	Furuya et al. (1968) Furuya et al. (1969)	54 55
	Adenine-guanine requiring	5'-IMP 5'-XMP	8–9 6	—	Misawa et al. (1969)	56
	Guanine requiring	5'-XMP	4	—	Misawa et al. (1969)	57
	Uracil requiring	Orotic acid	6.5	—	Skodova and Skoda (1969)	58
	Adenine requiring and 6-mercaptoguanine resistant	Inosine 5'-IMP	9.3	Hypoxanthine, surface active agent (salvage)	Furuya et al. (1970)	59
	Adenine requiring and permeability mutant (manganese insensitive)	5'-IMP Hypoxanthine	7.5 2	Excessive manganese	Kato et al. (1971)	60
	Decoyinine resistant	5'-Guanine nucleotides	—	XMP (salvage)	Furuya et al. (1971)	61
			9.7	Glucose[b]	Furuya et al. (1973)	62
Bacillus subtilis	Adenine requiring	5'-IMP Inosine Hypoxanthine	1.1 5.0 0.5	—	Uchida et al. (1961)	63

Mutation	Product	Amount	Precursor	Reference	Ref.
Purine requiring	Adenosine	1	Adenine (salvage)	Hara et al. (1962)	64–67
	AICAR	2.6		Shiro et al. (1962)	68
Adenine-histidine requiring	Inosine	7		Aoki et al. (1963, 1967, 1968)	69–74
Adenine requiring	S-AICAR derivatives, hypoxanthine	2	—	Fujiwara et al. (1963)	75
Adenine and lysine requiring	Inosine	2.7		Nara et al. (1963)	76
Adenine requiring	Hypoxanthine	0.8		Fujimoto et al. (1965)	77
Adenine requiring	5'-IMP	—		Fujimoto and Uchida (1965)	78
Adenine requiring	Inosine	1.0	Hypoxanthine (salvage) or glucose[b]	Yamanoi et al. (1965, 1966)	79–81
	Hypoxanthine	34.0			
	Inosine	5.3[b]	Various bases (salvage)		
	Various purine ribonucleosides		Hypoxanthine (salvage)		
Guanine requiring	Xanthosine	8.9	—	Fujimoto and Uchida (1965)	82
Guanine-adenine requiring	Xanthosine		—	Fujimoto et al. (1966)	83
Adenine requiring	5'-IMP	—	—	Fujiwara et al. (1967)	84
Adenine requiring and 8-azaguanine resistant	Guanosine	4	—	Konishi and Shiro (1968)	85
	Inosine	3			
	Inosine	2			
	Xanthosine	5			
L-Isoleucine requiring	Adenosine	1.3	—	Konishi et al. (1968)	86
Adenine requiring	Inosine	12	—	Momose and Shiio (1969)	87
Adenine requiring and 8-azaxanthine resistant	Inosine	7–8			
	Guanosine (Xanthosine)	3–4 (7–8)			

(Continued)

TABLE I (Continued)

Strain	Mutant properties	Product	Accumulation (mg/ml)	Added material	Workers	References
Bacillus subtilis						
	Adenine-hypoxanthine requiring and 8-azaguanine or 8-azaxanthine resistant	Guanosine Xanthosine Inosine	—	—	Komatsu et al. (1970)	88
	Amino acids requiring and 8-azaguanine resistant	Inosine Guanosine Guanosine	—	Hypoxanthine (salvage) Inosine, adenosine or guanine (salvage)	Kanamitsu (1970)	89
	Adenine-guanine requiring	5'-IMP 5'-XMP	—	—	Akiya et al. (1972)	90
	Adenine requiring and 8-azaguanine resistant	Inosine	16–18		Ishii and Shiio (1972)	91
	Adenine-leaky	Guanine and guanosine	12.5	Hypoxanthine or inosine (salvage)	Midorikawa et al. (1972)	92
	Adenine requiring	5'-IMP	—	Purine derivatives (salvage)	Midorikawa et al. (1973)	93
Bacillus megaterium						
	Purine requiring	AICAR	12	—	Kinoshita et al. (1968)	94
Bacillus pumilus						
	Adenine and phenylalanine requiring	AICAR	21	—	Shirafuji et al. (1968)	95

Bacillus sp.	Adenine requiring	Inosine	—	Hirano et al. (1967)	96
	Adenine requiring and 8-azaguanine resistant	Guanosine Inosine	5 13	Nogami et al. (1968)	97
	Shikimic acid requiring or D-gluconate-nonutilizing	Ribose	35	Sasajima et al. (1970, 1971, 1974)	98–99a
	Xanthine requiring	Adenosine	Over 16	Haneda et al. (1971, 1972)	100,101
	Adenine requiring	Guanosine	—	Komatsu et al. (1972, 1973)	102,103
Corynebacterium glutamicum	Guanine requiring	5'-XMP	2.8	Misawa et al. (1964)	104,105
	Adenine requiring	5'-IMP	—	Nakayama et al. (1964)	106
	Adenine requiring and 8-azaguanine resistant	5'-IMP 5'-IMP Hypoxanthine	1.8 4.0 4.7	Hypoxanthine (4 mg/ml)	
	Pyrimidine requiring	Orotic acid Orotidine	6 3	Sato et al. (1965)	107
	Guanine-adenine requiring	5'-XMP	3–4	Nakayama et al. (1965)	108
				Demain et al. (1965)	109
	Adenine and amino acids requiring	5'-GMP 5'-IMP	Each 1	—	Demain et al. (1966)	110
Corynebacterium petrophilum	Adenine requiring	Inosine	1.6	Hydrocarbons	Iguchi et al. (1966, 1967)	111,112
Arthrobacter paraffineus	Uracil requiring	Orotic acid Orotidine	8–9(6) 7–8(3)	Sorbitol (n-paraffin)	Kawamoto et al. (1970)	113

(Continued)

TABLE I (Continued)

Strain	Mutant properties	Product	Accumulation (mg/ml)	Added material	Workers	References
Aerobacter aerogenes	Guanine requiring	Xanthosine	6	—	Nakayama et al. (1963)	114
Streptomyces showdoensis	Uracil requiring	Orotic acid Orotidine	1.2 0.8	—	Ozaki et al. (1972)	115
Candida tropicalis	Adenine or hypoxanthine requiring	Orotic acid	—	—	Watanabe et al. (1968)	116

[a] AICAR: 5-amino 4-imidazole carboxamide riboside; 5'-GMP: guanosine 5'-monophosphate; 5'-IMP: inosine 5'-monophosphate; 5'-XMP, xanthosine 5'-monophosphate.
[b] 5'-Guanine nucleotides from glucose were directly accumulated by mixing cultivation of a 5'-XMP-accumulating strain (adenine–guanine auxotroph) (57) and a 5'-XMP-converting mutant (decoyinine resistant).

cell walls and membranes. The amount of accumulated 5′-IMP was 12.8 gm per liter of the culture filtrate.

C. AICAR Fermentation

Shiro et al. (68) have reported the accumulation of AICAR in the culture filtrate of a purine-requiring mutant of B. subtilis. Kinoshita et al. (94) also obtained a yield of about 12 gm of AICAR per liter in the cultivation of a purine-requiring mutant of Bacillus megaterium.

Shirafuji et al. (95) have isolated nonexacting purineless mutants from an inosine-forming adenine auxotroph of Bacillus pumilus, some of which accumulate AICAR. They have also isolated a number of mutants from the AICAR-forming purineless mutant which require adenine specifically. Half of them accumulated a large amount of AICAR, as compared with their parents, the nonexacting, purineless auxotrophs. The amount of AICAR produced was 20 mg/ml.

D. 5′-XMP Fermentation and Its Conversion to Guanine Nucleotides

Demain et al. (110) have reported that a revertant derived from an adenine and xanthine auxotroph of Corynebacterium glutamicum accumulated 1 gm of 5′-GMP per liter, but this productivity is not enough for industrial production.

Demain et al. (109) have obtained a guanine-adenine auxotroph of C. glutamicum, which accumulated 3–4 gm of 5′-XMP. Misawa et al. (56,57) reported that the accumulations of 5′-XMP by a guanine auxotroph and by an adenine-guanine auxotroph of Brevibacterium ammoniagenes were 4 gm/l and 6 gm/l, respectively. Microbial conversion of 5′-XMP to guanine nucleotides provides a new industrial procedure for the fermentative production of guanine nucleotides. Therefore, as an industrial procedure, the microbial conversion of 5′-XMP to 5′-GMP should be an efficient, new procedure for the fermentative production of guanine nucleotides.

Furuya et al. (61) have obtained a decoyinine-resistant mutant of B. ammoniagenes. Decoyinine is an antibiotic with an adenosine-like structure, which is reported to specifically inhibit XMP aminase (116a). This mutant was found to efficiently convert 5′-XMP to guanine and guanine nucleotides. Furthermore, Furuya et al., (62) devised an improved process: the direct production of 5′-guanine nucleotides from a carbohydrate by mixed cultivation of the 5′-XMP-accumulating strain and the 5′-XMP-converting mutant. An examination of the various conditions for mixed cultivation showed that the seed ratio of the two strains and the feeding of glucose and urea during cultivation had profound effects on the 5′-guanine nucleotides. A maximum concentration of 9.67 mg of 5′-guanine nucleotides (GMP, 2.47; GDP, 1.51; and GTP, 5.69 mg) per milliliter was obtained directly from glucose in the mixed culture.

E. Ribose Fermentation

Sasajima et al. (98–99a) isolated three mutants of a *Bacillus* sp. while studying the effects of mutation in carbohydrate metabolism during the synthesis of purine derivatives. The parent strain of *Bacillus* sp. required adenine and xanthine for its growth, was resistant to 8-azaguanine and was excellent for inosine formation. Two of its mutants required shikimic acid for their growth and one mutant could not utilize D-gluconate. These mutants accumulated a large amount of D-ribose in a culture medium consisting of 12.5% glucose, 2.5% dried yeast, 1.5% $(NH_4)_2SO_4$; 0.5% $CaHPO_4$; 0.5% $Ca_3(PO_4)_2$, and 1.0% $CaCO_3$ (pH 7.0). The amount of D-ribose was about 35 mg/ml after 6 days of cultivation.

The shikimic acid-requiring mutants lacked D-sedoheptulose 7-phosphate; D-glyceraldehyde glycolaldehydetransferase and the D-gluconate-nonutilizing mutant lacked D-ribulose-5-phosphate 3-epimerase.

D-Ribose is now produced industrially by a fermentation method using these mutants.

F. Orotic Acid Fermentation

Since Michelson et al. (117) found that orotic acid was produced in a culture filtrate of the pyrimidine-requiring mutant of *Neurospora crassa*, many of these mutants have been shown to produce orotic acid. In Japan, orotic acid is industrially produced for medical use.

Nakayama et al. (108) have reported that the pyrimidine-requiring mutants of *B. subtilis*, *Corynebacterium glutamicum*, *C. rathayi*, *Brevibacterium ammoniagenes*, and *Escherichia coli* produce appreciable amounts of orotic acid in their culture filtrates. *C. glutamicum* produced 3.0 mg/ml of orotic acid when cultivated in a medium containing 50 μg of uracil per milliliter.

Kawamoto et al. (113) studied the fermentative production of orotic acid and orotidine by a uracil-requiring mutant of *Arthrobacter paraffineus*. They cultivated the strain of 5 days in a medium containing a 10% n-paraffin mixture (C_{10}–C_{15}), inorganic matter, corn steep liquor, thiamine, and 200 μg of uracil per milliliter. The production of orotic acid and orotidine was about 5.5 mg and 2.2 mg per milliliter, respectively. When the n-paraffin was replaced by 10% sorbitol, about 9 mg of orotic acid and 8 mg of orotidine were produced per milliliter in a 5-day culture.

Machida et al. (*118,119*) have also reported the accumulation of orotic acid and its mechanism with the wild strain of *E. coli* K12.

G. Formation of DNA

Recently Tomita and Suzuki (*120*) found that some hydrocarbon-utilizing bacteria grown on *n*-paraffin as the sole source of carbon accumulated a large amount of DNA extracellularly, which was free from intact cells and slime material. Various slime layers of halophilic bacteria or other bacteria are known to contain DNA (*121–123*).

Pseudomonas fluorescens was cultivated in a medium consisting of 10% *n*-paraffin (C_{12}–C_{14}) as the carbon source, inorganic material, corn steep liquor, and yeast extract. The amount of DNA produced in the supernatant of the broth was 0.1 gm to 0.6 gm per liter. DNA extracted from the culture filtrate contained less than 5% RNA. The base composition of the DNA accumulated in the medium was the same as that of intracellular DNA. A strain of *Arthrobacter* accumulated DNA in a glucose medium while very little DNA was accumulated in a *n*-paraffin medium.

The mechanism for the accumulation of DNA is thought to be the release of DNA with or without cell lysis. The accumulation of DNA by *Pseudomonas* was initiated when cell growth (estimated by the absorbancy at 660 nm) began to decline. The purification method of Rudin and Albertsson (*123*) was most suitable for the large-scale isolation of DNA since contamination by RNA was negligible, and less fragmentation was found during purification.

VI. Salvage Synthesis of Nucleosides and Nucleotides

Various conversion routes for purine derivatives have been reviewed by Moat and Friedman (*49*). Kojima et al. (*64–67*) devised the production of 5′-IMP from chemically synthesized adenine. They obtained a purine-requiring mutant (st^r, try$^-$, pur$^-$) which was induced secondarily from the *B. subtilis* Marburg strain (st^r, try$^-$). About 1 gm of adenosine was formed by the addition of 1 gm of adenine per liter to the fermentation of the mutants. This value was equivalent to a 58% yield and was later increased to 80%. The adenosine formed was phosphorylated by the adenosine phosphokinase of *Saccharomyces carlsbergensis*. The resulting adenylic acid was deaminated by nitrite or the adenylic acid deaminase of the mold. The yield of 5′-IMP-Na$_2$ from 100 gm of malononitrile was 59 gm. Two different pathways, (A) and (B), were assumed to operate in this type of adenosine formation (Fig. 5).

Nara et al. (*124,125*) found that *Micrococcus sodonensis, Arthrobacter citreus,* and *Brevibacterium insectiphilum* convert hypoxanthine to 5'-IMP. Nara et al. (*53*) also reported that *B. ammoniagenes* accumu-

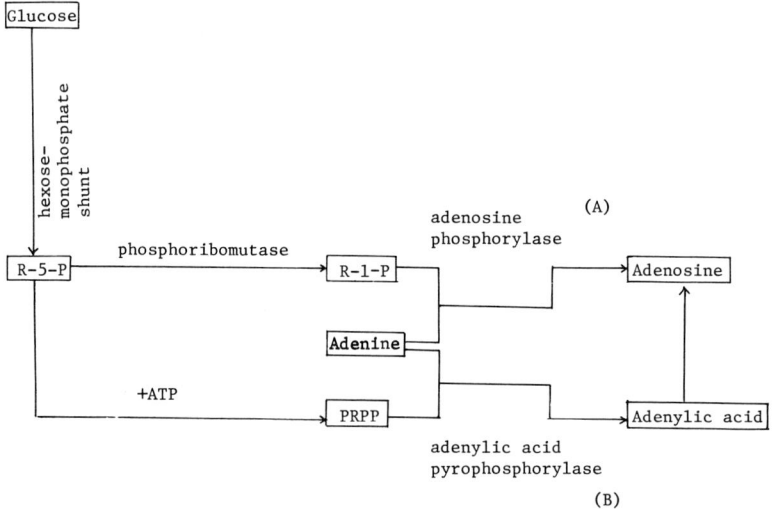

Fig. 5. Outlines of tentative pathways for adenosine fermentation by the purineless auxtotroph of *Bacillus subtilis* Marburg. R-1-P: ribose 1-phosphate; R-5-P: ribose 5-phosphate; PRPP, phosphoribosyl pyrophosphate.

lated over 10 mg of 5'-IMP per milliliter on the addition of 4 mg of hypoxanthine per milliliter and the GMP and AMP from guanine and adenine, respectively.

Yamanoi et al. (*79–81*) have found that xanthosine, guanosine, and inosine are produced from their respective bases, by the mutant of *Bacillus subtilis* which produces inosine. This process is suitable for preparing the nucleosides of chemically synthesized nonnatural purines, e.g., 6-methoxypurine, N^2,N^2-dimethylguanine, etc.

VII. Conversion of Nucleosides

A. Phosphorylation of Nucleosides

Two enzymes related to the phosphorylation of nucleosides are known: nucleoside kinase (*126,127*) and nucleoside phosphotransferase (*128*). Nucleoside kinase has been used in the phosphorylation of adenosine as reported in Section VI.

Katagiri et al. (129,130) and Mitsugi et al. (131–138) have made extensive reports on the nucleoside phosphotransferase of bacteria. They divided the bacteria into two groups. Group A forms 5′-nucleotides by transphosphorylation between a nucleoside and another 5′-nucleotide. Group B mainly produces 3′(2′)-nucleotides from a nucleoside and a 3′(2′)-nucleotide. Both the A and B groups can use p-nitrophenyl phosphate as a phosphoryl donor in forming nucleotides.

In *Pseudomonas, Flavobacterium,* and *Serratia*, inosine and guanosine are phosphorylated, using p-nitrophenyl phosphate as the phosphoryl donor, into 5′-IMP and 5′-GMP, respectively, with a yield of about 80%.

With respect to the chemical phosphorylation of inosine or guanosine, Hikino et al. (139) have devised a selective phosphorylation of the 5′-position which gives good yields using a simple process.

B. Production of 5′-GMP from AICAR

Kinoshita et al. (94) and Yamazaki et al. (140) have reported a method for the industrial production of 5′-GMP-Na$_2$. As shown in Fig. 6, this method is the combination of two processes, fermentation and chemical syntheses.

The fermentation process is that which produces AICAR (II) from glucose (see Section V,C) with a nonexacting purineless mutant of *B. megaterium*. (II) is separated from the culture filtrate (12 gm/l) by an ion-exchange resin and is treated with acetone in the presence of acid to give its 2′,3′-O-isopropylidine derivative (III). (III) is then allowed to react with sodium methylxanthate at 180°C to give 2′,3′-O-isopropylidene-2-mercaptoinosine (IV). (IV) is oxidized with hydrogen peroxide to give the sulfoinosine (V). (V) is heated with aqueous ammonia to afford 2′,3′-O-isopropylideneguanosine (VI). (VI) is phosphorylated with phosphoryl chloride, followed by the treatment with aqueous acid to give 5′-GMP, which is isolated as 5′-GMP-Na$_2$ in a yield of over 40%, based on (II).

C. Transribosylation of Nucleosides

Bacterial enzymes are known to catalyze the nucleoside-N-ribosyl or deoxyribosyl group transfer reaction.

There have been many investigations of these reaction mechanisms.

$$\text{X-R} + \text{P}_i \rightleftharpoons \text{R-1-P} + \text{X} \tag{1}$$
$$\text{R-1-P} + \text{X}' \rightleftharpoons \text{X}'\text{-R} + \text{P}_i$$
$$\text{X-R} + \text{X}' \rightleftharpoons \text{X}'\text{-R} + \text{X} \tag{2}$$

where X and X′ represent the purine and pyrimidine bases, respectively, and R represents ribose or deoxyribose.

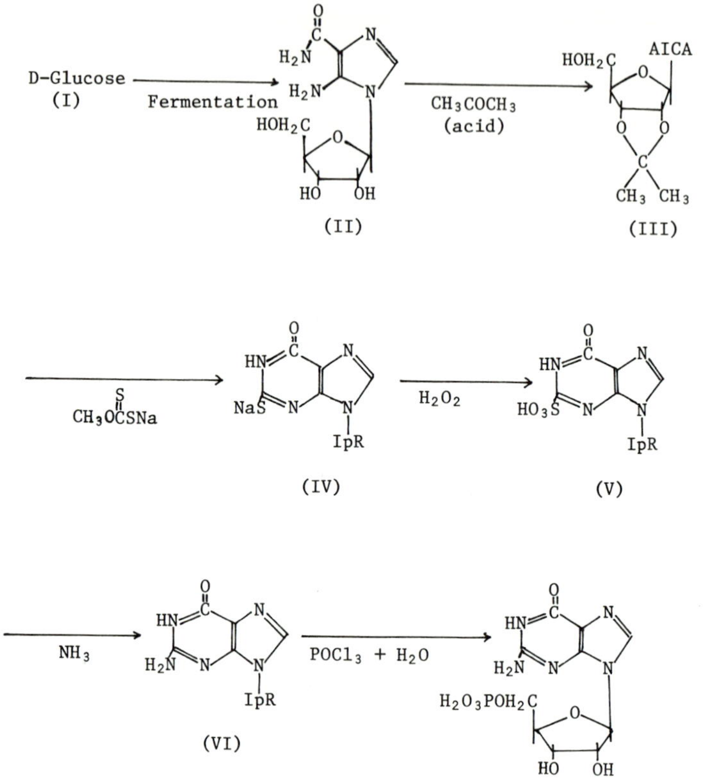

FIG. 6. The production of 5'-guanylic acid by biological and chemical syntheses.

Reaction (1) is catalyzed by nucleoside phosphorylase. Some investigations have been made on the deoxyribosyl group transfer reaction of one purine or pyrimidine nucleoside to another, and on the ribosyl group transfer between purine nucleosides. The enzyme (2) which catalyzes the interconversion of deoxyribosylnucleosides, has been named *trans*-N-deoxyribosylase. Little is known of either reaction (1) or reaction (2) on the enzymes that catalyze the ribosyl group transfer between the pyrimidine and purine nucleosides.

Sakai et al. (141–143) have found that there is a nucleoside-ribosyl group transfer reaction between pyrimidine and purine nucleosides in cells of a large variety of bacteria. This reaction is interesting not only for its physiological significance, but because it is the means of production of inosine or guanosine from hypoxanthine or guanine and uridine or cytidine which are the wastes of 5'-IMP and 5'-GMP production in the hydrolyzing method.

Sakai et al. (143) have also purified the enzymes which catalyze the ribosyl group transfer from cells of *Corynebacterium sepedonicum* and

Erwinia carotovora and have demonstrated that in most bacteria, the trans-*N*-ribosylation between the pyrimidine and purine nucleosides is catalyzed by two kinds of nucleoside phosphorylases (pyrimidine nucleoside phosphorylase and purine nucleoside phosphorylase) and that ribose 1-phosphate serves as the intermediate, as shown in Fig. 7.

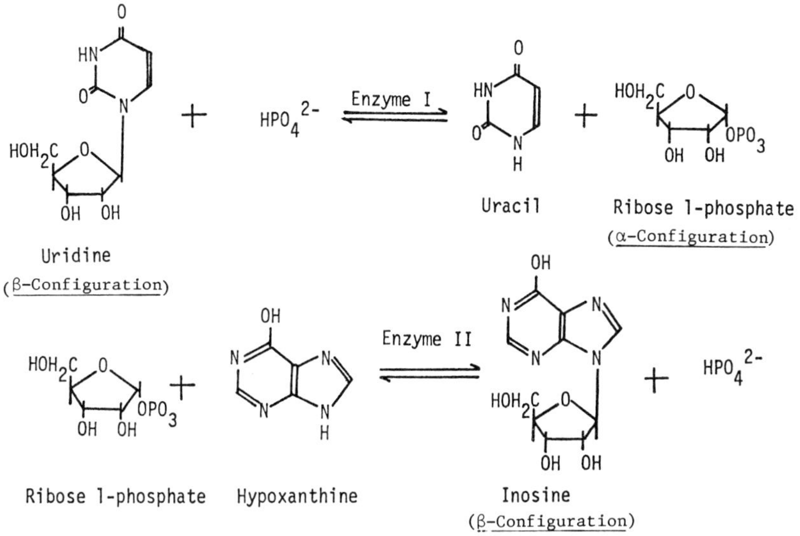

FIG. 7. The nucleoside-*N*-ribosyl group transfer reaction between uridine and inosine. Enzyme I: pyrimidine nucleoside phosphorylase; Enzyme II: purine nucleoside phosphorylase.

In this reaction, α-ribose 1-phosphate is formed by the action of pyrimidine nucleoside phosphorylase through the phosphorolysis of pyrimidine-β-riboside. Purine-β-riboside is then synthesized by the action of purine nucleoside phosphorylase. The reaction proceeds only in the presence of inorganic phosphorus. The purine nucleoside phosphorylases obtained from these organisms catalyze the formation of inosine, guanosine, adenosine, and xanthosine.

The question of whether each riboside formation is catalyzed by different enzymes has not been answered, since no studies on the homogeneity of these enzyme preparations have been performed. This reaction is not presently utilized for the industrial production of inosine or guanosine, since a sufficient amount of hypoxanthine or guanine cannot be supplied.

Tanaka and Nakazawa (*144*) have reported that *Brevibacterium ammoniagenes* produces allopurinol ribotide in a medium containing allopurinol (4-hydroxy[3,4-d]pyrazolopyrimidine), a synthetic analog of hypoxanthine. This compound is used as a medicine for gout because of its inhibitory activity against the xanthine oxidase of allopurinol.

Sakai et al. (145) applied this reaction to the ribosylation of allopurinol. The allopurinol riboside was produced through the N-ribosyl transfer reaction from uridine to allopurinol, using a cell-free extract of *Erwinia carotovora* as the enzyme source. The yield of allopurinol riboside was 250 mg from 200 mg of allopurinol and 1.8 gm of uridine. They also demonstrated that the allopurinol riboside does not inhibit the xanthine oxidase reaction.

VIII. Formation of Nucleoside Derivatives

The current state of development of the formation of flavoring nucleotides in Japan has made it possible to supply enough of the various bases, nucleosides, and nucleotides. Consequently, extensive studies have been done on the microbial preparation of various nucleotide derivatives from these compounds.

A. Nucleoside Polyphosphates

The phosphorylation of AMP to ATP by yeast extracts was first investigated by Lutwak-Mann and Mann (146). Ostern et al. (126) found that adenosine and ATP are phosphorylated to ADP and ATP through glycolysis by yeast.

Tochikura et al. (147) demonstrated that AMP is phosphorylated to ADP and ATP in a high yield using ground or acetone-dried cells of baker's yeast in the presence of glucose and a high concentration of inorganic phosphate. The reaction system contained 5'-AMP, 15 μmoles; glucose, 167 μmoles; and potassium phosphate buffer (pH 7.0) 333 μmoles; and ground or acetone-dried cells, 50 mg/ml. Respectively, 14% and 72% of the added AMP were phosphorylated to ADP and ATP, after a 3-hour incubation with ground cells. Acetone-dried cells of baker's yeast also phosphorylated adenosine to form ATP. The concentration of inorganic phosphorus was the most important factor for this phosphorylation of ATP. The optimum phosphate concentration for the efficient phosphorylation of AMP varied between 0.33 and 0.25 M. At lower phosphate concentrations IMP, inosine, and hypoxanthine were produced while under higher ones (0.5 and 0.66 M), the formation of ADP and ATP was depressed and a large amount of adenosine was accumulated.

The phosphorylation of GMP, CMP, and UMP to their corresponding nucleoside triphosphates was also observed under fermentative conditions similar to those described above. A series of these reactions is the basis of the fermentative formations of sugar nucleotides and cytidine derivatives.

In contrast, Tanaka et al. (148) have reported that 2.16 mg/ml of AMP, 1.59 mg/ml of ADP and 1.57 mg/ml of ATP were accumulated in a culture filtrate of B. ammoniagenes. When the culture was 2 days old, 2 mg/ml of adenine was fed to a medium containing 10% glucose, 0.6% urea, 1.0% KH_2PO_4, 1.0% K_2HPO_4, 1.0% $MgSO_4 \cdot 7H_2O$, 1.0% yeast extract, 0.01% $CaCl_2$, and 30 µg of biotin per liter. GMP, GDP, and GTP were also accumulated in the culture filtrate of the microorganism under the same conditions except that guanine was fed in place of adenine.

Recently, Murao et al. (149,150) reported a new ATP phosphotransferase produced by Streptomyces adephospholyticus isolated from soil. The enzyme was purified from the culture filtrate to a homogeneous state using disc gel electrophoresis. The enzyme transfers the pyrophosphate group of ATP to adenine nucleotides, guanine nucleotides, and inosine nucleotides and is a pyrophosphotransferase.

The reaction process, ATP + GDP → AMP + G4P occurs in this enzymic reaction. The reaction products were isolated and the structure of G4P was proved to be identical with that of ppGpp (guanosine tetraphosphate), which is of interest in connection with the stringent response of bacteria. The enzyme was concluded to be ATP:nucleotide pyrophosphotransferase, with the pyrophosphate group of ATP transferred to position 3′ of the nucleotides. See also the results of Haseltine et al. (151).

B. Sugar Nucleotides and Cytidine Derivatives

Since Leloir first discovered UDP-glucose in 1950, many sugar nucleotides containing different bases and sugars or sugar derivatives have been isolated from natural sources, as well as having been enzymically synthesized. Sugar nucleotides are known to be formed by the catalysis of sugar nucleotide pyrophorylase from the corresponding nucleoside triphosphate and sugar phosphate, and they are known to play an important role in the biosynthesis of polysaccharides and the transformation of carbohydrates (152). Sugar nucleotides, such as GDP-mannose, UDP-glucose, and UDP-N-acetylglucosamine, have been prepared in low yields by extraction from yeast cells and by chemical synthesis. Therefore, these compounds are extremely expensive in spite of their importance in the study of carbohydrate biochemistry.

Cytidine 5′-diphosphate (CDP)-choline was first found by Kennedy and Weiss (153,154) to be an important intermediate in the biosynthetic pathway of lipids. Recently, CDP-choline has attracted attention in the medical field, because it has had good curative effects on the clouding of consciousness due to brain injury and on the acute toxicosis caused by sleeping drugs as well as on nervous and muscular diseases caused by head and muscular cerebrovascular accidents. CDP-choline is now prepared industrially by chemical procedures in Japan.

CDP-ethanolamine, together with CDP-choline, was first proved by Kennedy and Weiss (*153,154*) to be an important intermediate in the biosynthetic pathway of lipids. Tochikura *et al.* (*155–165*) and Kawaguchi *et al.* (*166–168*) have developed a preparative method for sugar nucleotides and these cytidine derivatives as shown in Table II. Almost the same conditions as for ATP formation were used, except for the use of corresponding compounds and nucleotides.

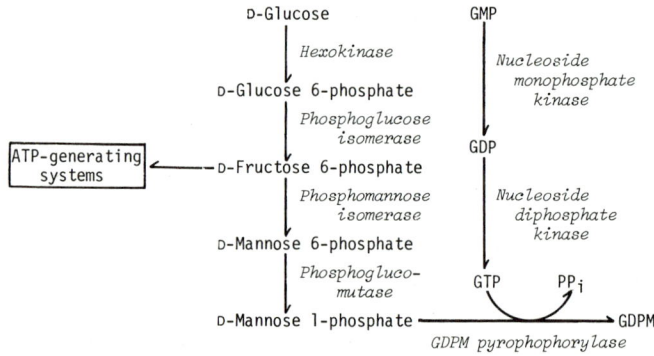

Fig. 8. The presumptive pathway for the fermentative production of guanosine 5′-diphosphate mannose (GDPM), guanosine 5′-di- and triphosphates (GDP and GTP) from guanosine 5′-monophosphate (GMP) by air-dried cells of baker's yeast. PPi, pyrophosphate.

Kawaguchi *et al.* (*166*) posited a mechanism for the synthesis of GDP-mannose, GDP, and GTP from GMP and glucose using air-dried cells of baker's yeast as shown in Fig. 8. The synthesis of GDP and GTP may be catalyzed by nucleoside mono- and diphosphate kinases. Mannose 1-phosphate, which is derived from glucose by hexokinase and phosphomannomutase, may react with GTP to form GDP-mannose and inorganic pyrophosphate due to the catalysis of GDP-mannose pyrophosphorylase.

Kawai and Tochikura (*159*) have shown the mechanism for UDP-galactose accumulation and estimated the enzyme activities involved in the galactose metabolism of *Torulopsis candida*. Remarkable activities have been reported for galactokinase and galactose 1-phosphate uridylyltransferase, whereas UDP-galactose pyrophosphorylase activity is weak. UDP-galactose formation proceeds, in the cell-free extract, along a coupling

TABLE II
FORMATION OF SUGAR NUCLEOTIDES AND CYTIDINE DERIVATIVES

Product[a]	Substrate[a]	Microorganism	Yield (max) Mg/ml	Molar (%)	References
GDP-mannose	GMP-Na₂ Glucose	Baker's yeast, air-dried cells	5.4	45	155,166
GDP-glucose	GTP-(N₄H₃) Glucose 1-phosphate	Hansenula jadinii air-dried cells	10.8	90	167
		Streptomyces sp., air-dried cells	8.5	70	168
UDP-glucose	UMP-Na₂ Glucose	Baker's yeast, ground cells	6.7	60	156
		Brewer's yeast, acetone-dried cells	7.7	70	156
		Torulopsis candida, air-dried cells	10.0	90	156,157
UDP-galactose	UMP-Na₂ Galactose	Torulopsis candida, air-dried cells	51.5	70	157–159
UDP-N-acetyl-glucosamine	UMP-Na₂ Fructose Glucosamine-HCl	Baker's yeast, air-dried cells	4.1	40	160
		Debaryomyces sp., air-dried cells	6.0	40	160
UDP-N-acetyl-galactosamine	UDP-N-acetylglucosamine	Bacillus subtilis, cell extract	2.0	35	161
CDP-choline	CMP-Na₂ (CTP) Glucose Phosphorylcholine-Ca	Brewer's yeast, air-dried cells	16.7–17.7	70	162,163
		Candida sp., freeze-dried cells	8.0	80	169
		Bacillus cereus Intact or lyophilized cells	7.1	70	170
		Growing cells	3.6	80	170
		Rhodotorula mucilaginosa, intact or lyophilized cells	8.6	85	170
	CMP-Na₂ Glucose Choline chloride	Hansenula jadinii, air-dried cells	6.4	65	164
CDP-ethanolamine	CMP-Na₂ Glucose Phosphorylethanolamine	Brewer's yeast, air-dried cells	2.8	60	164a,165

[a] CMP, CTP: cytidine 5′-mono-, triphosphates; GMP, GDP, GTP: guanosine 5′-mono-, di-, and triphosphates; UMP, UDP, uridine 5′-mono-, diphosphates.

reaction catalyzed by UDP-glucose pyrophosphorylase and galactose-1-P uridylyltransferase where the UDP-glucose or glucose-1-P acts as a catalyst. The UDP-galactose accumulation under fermentative conditions is believed to be the concerted inhibition of UDP-galactose 4-epimerase activity by 5'-UMP and the galactose present as fermentation substrates, as shown in Fig. 9.

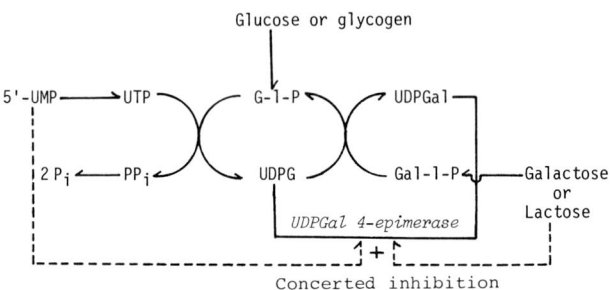

Fig. 9. The mechanism of uridine 5'-diphosphate galactose (UDPGal) fermentation by *Torulopsis candida*.

Shirota et al. (169) reported the formation of CDP-choline from CMP by dried cells of *Candida* sp. under conditions similar to those of Tochikura et al. In contrast, Miyauchi et al. (170) have investigated the formation of CDP-choline, using intact cells of microorganisms, with a reaction system containing CMP or CTP, phosphorylcholine, glucose, $MgSO_4$ and potassium phosphate buffer. They found that *B. cereus* and *Rhodotorula mucilaginosa* effectively accumulates CDP-choline in good yields from CTP or CMP.

C. FAD AND NAD

Several methods for extracting flavin adenine nucleotide (FAD), a coenzyme form of riboflavin, from yeast cells have been reported (171–173). Sakai et al. (174) selected microorganisms capable of accumulating FAD in their culture filtrates and investigated the conditions for the accumulation of FAD with *Sarcina lutea*. They obtained an accumulation of 27.1 μg and 54.7 μg of FAD in the culture filtrate of *S. lutea* by adding, respectively, flavin mononucleotide (FMN) plus AMP, and FMN plus adenine to the basal medium.

Watanabe et al. (175,176) have also attempted to obtain an efficient mutant for the industrial production of FAD from FMN and adenine. *S. lutea* was treated with N-methyl-N'-nitro-N-nitrosoguanidine, and a purine-requiring and adenosine deaminaseless mutant was obtained. After 5 days of cultivation, this strain produced about 1 mg/ml of FAD in the presence of adenine and FMN, which is more than 3-fold that produced by the parent strain.

NAD has usually been prepared from hot-water extracts of NAD-rich baker's yeast and is purified by several procedures. Takebe and Kitahara (*177*) found that *Lactobacillus plantarum* II, *L. acidophilus,* and *Streptococcus faecalis* accumulated $NAD(H_2)$ 7.26 mg, 7.80 mg, and 7.00 mg per gram of dried cells, respectively. Nakayama *et al.* (*177a*) found that *Brevibacterium ammoniagenes* accumulated NAD, AMP, ADP, and ATP in its culture filtrate on the addition of adenine and nicotinamide to a 2-day-old culture. The amount of NAD in the culture filtrate reached 2.3 mg/ml after cultivation for 5 days. When either nicotinic acid or nicotinamide without adenine was added to the medium, a large amount of nicotinic acid mononucleotide (2.3 mg/ml) and a small amount of

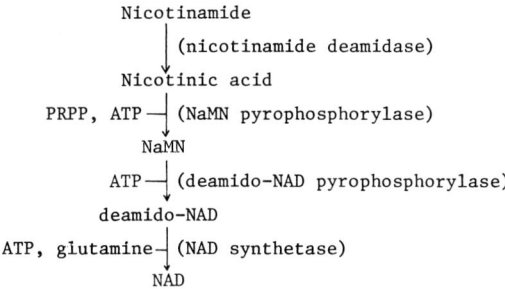

FIG. 10. Proposed pathway of nicotinamide adenine nucleotide (NAD) biosynthesis from nicotinic acid or nicotinamide in *Brevibacterium ammoniagenes*. NaMN, nicotinic acid mononucleotide.

NAD were accumulated. But, when an adenine other than nicotinic acid or nicotinamide was added, the accumulation of NAD increased and was accompanied by the accumulation of AMP, ADP, and ATP. The amount of NAD or nicotinic acid mononucleotide accumulated was several milligrams per milliliter under that for the optimum culture conditions. The biosynthetic pathway of NAD in these organisms has been proposed to be the salvage biosynthetic pathway via the nicotinic acid mononucleotide, as shown in Fig. 10.

Hayano *et al.* (*178*) attempted to increase the content of NADs (Nicotinamide Nucleotide Coenzymes) using *Candida utilis*. The optimal cultural conditions for accumulation of NADs in the cells were as follows:

(1) medium: glucose 3 gm, urea 0.4 gm, KH_2PO_4 0.15 gm, $CaCl_2 \cdot 2H_2O$ 0.032 gm, and biotin 50 ng in 100 ml of desalted water, pH 4.8.

(2) cultural conditions: temp., 30°C; oxygen supplying, Kd 2.5×10^{-6} gm · mole/min/ml/atm.

After the complete consumption of glucose, 4 days culture under above conditions, the content of total NADs reached to approximately 10 mg per gram of dry cells.

Sakai et al. (178a) have reported a method for enhancing the NAD content of baker's yeast by incubating the yeast with precursors. When commercial baker's yeast was secondarily cultured for 72 hours with shaking in a medium containing 0.3% adenine, 0.6% nicotinamide, and 0.2 M K_2HPO_4, pH 4.5; its NAD content markedly increased. Under optimum conditions, the NAD content of the yeast cells was about 12 mg/gm of the dry cells (corresponding to 2.0 mg/ml of the medium), which is about 20 times the initial content. The phosphorylation of NAD to NADP is catalyzed by NAD kinase; this reaction is known to be the sole pathway for NADP biosynthesis.

Tochikura et al. (179,180) and Kuwahara et al. (181,182) observed that when NAD and organic phosphates, such as *p*-nitrophenyl phosphate (*p*-NPP) or nucleoside monophosphates, were incubated with intact cells or a cell-free extract of *Proteus mirabilis* under an acidic condition, NADP, an NADP-analog and NAD-diphosphate were formed in the reaction mixture. The NADP analog is phosphorylated at the 2' or 3' positions in the nicotinamide ribose moiety and NAD-diphosphate (NAD-phosphate or NAD-analog phosphate) is phosphorylated at the nicotinamide ribose of NADP. These new phosphorylations of NAD are the result of the phosphotransferring reaction from organic phosphate to NAD. The NAD-analog and NAD-diphosphate showed little or no coenzyme activity for the several dehydrogenases tested.

IX. Production of Coenzyme A

CoA is an important biochemical reagent, prepared by extraction from yeasts (183,184) containing a small amount of CoA. Because of its low yield and the intricacy of the preparative process, CoA has been a very expensive biochemical reagent.

Recently Kuno et al. (185) found that *Pseudomonas alkanolytica*, an *n*-paraffin assimilative bacterium, contains a large amount of CoA. They extracted CoA from its cells with hot water and purified it by Dowex 1 × 2 (Cl⁻) column chromatography, active charcoal treatment and DEAE-cellulose column chromatography. Under suitable conditions, 80 mg of CoA per liter of the culture was obtained; a 13% yield of CoA (purity 90%).

A. The Biosynthesis of CoA and Its Isolation

The biosynthesis of CoA from pantothenic acid (PaA) has been mainly investigated by Brown et al. (186,187). Brown proposed the biosynthetic pathway shown in Fig. 11. PaA is first phosphorylated to 4'-phospho-

PaA (P-PaA) by PaA kinase (reaction 1, Fig. 11), then 4′-phosphopantothenoylcysteine (P-PaCySH) synthetase catalyzes the combination of P-PaA with cysteine to form P-PaCySH (reaction 2). This is decarboxylated to 4′-phosphopantetheine (P-PaSH) by P-PaCySH decarboxylase (reaction 3). The P-PaSH formed is combined with ATP to form 3′-dephospho-CoA (DP-CoA) (reaction 4) which is finally phosphorylated to CoA by DP-CoA kinase (reaction 5). Abiko et al. (*188–192*) have confirmed the validity of Brown's pathway using rat liver.

In addition, Pierpoint and Hughes (*193*) have found that 10% of the PaA added is incorporated into CoA when PaA is added to a glucose-salt solution containing intact cells of *Lactobacillus arabinosus* cultured on a PaA-deficient medium. Shimizu and Abiko (*188*) and Soyama (*194*) have reported that CoA is biosynthesized *in vitro* from PaA, pantethine (PaSS), or P-PaSH, using rat liver. These results indicate the possibility that CoA might be effectively biosynthesized from PaA, cysteine, and ATP under appropriate conditions.

As described in Section VII,A, AMP is phosphorylated to ATP in a high yield by ground or acetone-dried cells of baker's yeast. Ogata et al. (*195,196*) have observed that 100–150 μg/ml of CoA is synthesized in a reaction mixture containing: glucose, 167 μmoles; ATP, 15 μmoles; PaA-Na, 10 μmoles; cysteine, 10 μmoles; potassium phosphate buffer (pH 7.0), 200 μmoles; and dried cells of baker's yeast, 100 mg/ml. A relatively high activity for the accumulation of CoA (20–100 μg/ml) has also been found in some strains of yeast, e.g., *Saccharomyces sake*, *S. lactis*, brewer's yeast, *Torulopsis sake*, *T. candida*, and *Schizosaccharomyces liquefaciens*.

Ogata et al. (*197*) and Shimizu et al. (*198–200*) selected CoA-forming activities of air-dried cells of numerous bacteria with the same condition used for yeasts, except that AMP was replaced by ATP. *Sarcina*, *Corynebacterium*, and *Brevibacterium* accumulated large amounts of CoA.

The amount of CoA synthesized from PaA-Na and from PaSH after an 8-hour incubation of *B. ammoniagenes*, the most favorable strain for CoA accumulation, was approximately 1.9 mg/ml and 2.6 mg/ml, respectively, with the addition of sodium lauryl benzene sulfonate (SLBS). They have also cultivated *B. ammoniagenes* for 3 days in a medium similar to that which Tanaka et al. used for ATP formation, except that 0.2% AMP was added followed by the addition of PaA (0.04 gm), cysteine (0.04 gm) and cetyl pyridinium chloride (CPC) (0.02 gm) per 20 ml of culture broth. On further cultivating the culture broth for 2–4 days, 1–2 mg/ml of CoA was accumulated. Under favorable conditions, the accumulation of CoA increased to 5–6 mg/ml and a disulfide type of CoA (CoAS-SCoA) was the main product.

A large-scale technique for the isolation and purification of CoA from a culture filtrate has also been devised. A culture filtrate (1000 ml)

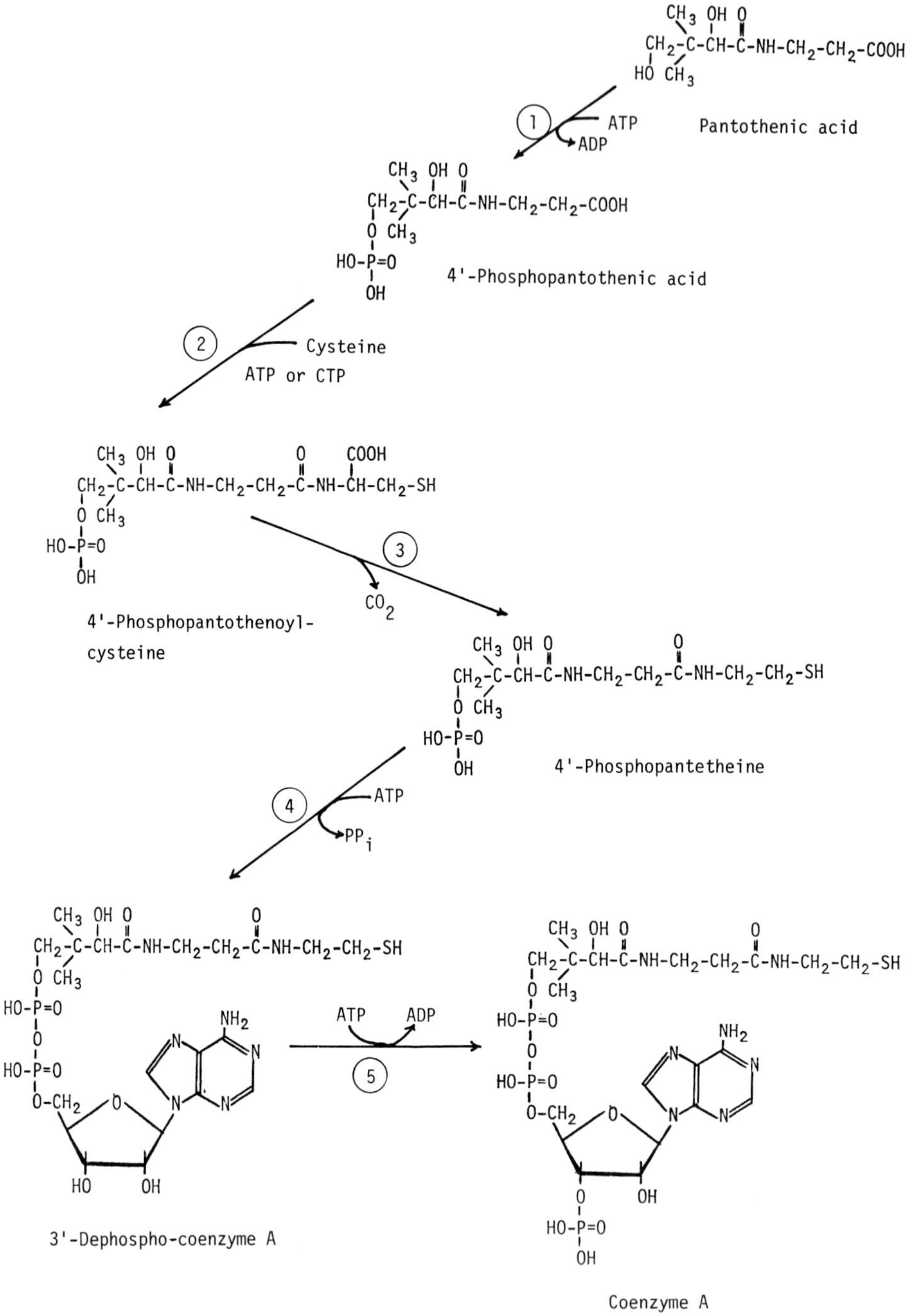

FIG. 11. The synthetic pathway of coenzyme A proposed by Brown.

containing about 2.5 gm of CoA (disulfide type) is passed through a Duolite S-30 column to remove brown pigment, then the column is washed with deionized water. The effluent is then charged on an active charcoal column to adsorb CoA. CoA is eluted with 40% acetone containing 0.028% ammonia, then is taken up on Dowex 1×2 (Cl⁻). CoA is eluted with an aqueous solution containing HCl and LiCl. The pH of the CoA fraction is adjusted to 4.5 with LiOH, and the whole is concentrated to precipitate the lithium salt of CoA. By this procedure 1007 mg of CoA (disulfide) was obtained. The purity of this CoA was 83–87%, based on a value obtained by the phosphotransacetylase method. This process is summarized in Fig. 12.

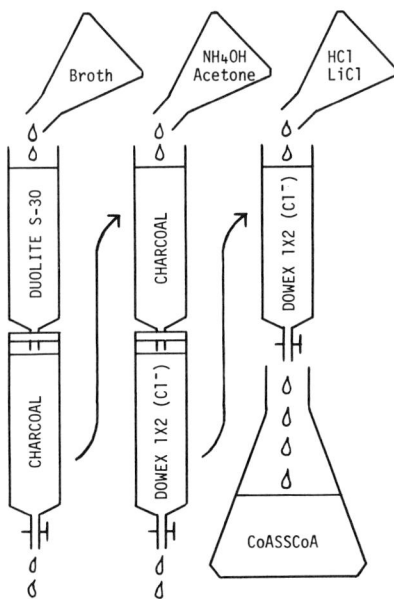

FIG. 12. Separation of coenzyme A (CoA) from the culture filtrate.

The disulfide was converted to the SH-type with 2-mercaptoethanol. The CoA obtained had 84% purity as determined by its adenosine content. Another purified sample was obtained from the SH-type by treatment with Dowex 1×2 (Cl⁻). The purity of this CoA was 97% as determined by its adenosine content. The final yield of CoA·Li from the culture broth was 34%.

Recently, Nishimura et al. (200a) reported that *Sarcina lutea* IAM 1099 accumulated 600 μg/ml of CoA in its cultured broth containing PaA, cysteine and adenine. CoA was readily isolated in high purity by the use of charcoal, DEAE-cellulose, Sephadex G-25, and Dowex 50. Yields of isolated CoA were over 33% from the culture broth.

B. THE BIOSYNTHESIS OF INTERMEDIATES OF CoA

Shimizu et al. (201,202) investigated the productivity of intermediates of CoA biosynthesis, P-PaA, P-PaSH, and DP-CoA and the metabolic regulation of CoA using these intermediates. P-PaA was obtained using the procedures for the CoA incubation of dried cells or the cultivation of *Brevibacterium ammoniagenes* without cysteine. Various nucleotides, e.g., ATP, ITP, GTP, UTP, were effective as phosphoryl donors; P-PaSH was obtained with the procedure used to produce CoA, except that CTP took the place of ATP.

Kakinuma and Igarashi (28) reported that the 3′-nucleotidase of *Bacillus subtilis* could dephosphorylate the 3′-phosphoryl residue of CoA. Later Kurooka et al. (203) prepared DP-CoA from commercial CoA using this enzyme. Shimizu et al. (204) also prepared DP-CoA by treating the CoA in the supernatant of the dried cell method with crude 3′-nucleotidase.

Shimizu et al. (204,205) investigated the metabolic regulation of CoA biosynthesis by *Brevibacterium ammoniagenes* in order to remove the negative feedback regulation by the end product. The phosphorylation of PaA was remarkably inhibited by the presence of CoA, but the coupling reaction of P-PaA and cysteine and the CoA formation from P-PaA were not significantly influenced by the presence of CoA. The amount of CoA biosynthesized from P-PaA was twice that from PaA.

These results suggest that the CoA biosynthesis of *B. ammoniagenes* is regulated at the point of the phosphorylation of PaA—the first step of the biosynthetic pathway—as pointed out by Karasawa et al. (206) and Abiko et al. (207) in rat liver. This was confirmed with the purified P-PaA kinase of *B. ammoniagenes*.

X. Conclusion

In the past 15 years, preparation methods for 5′-nucleotides, especially the flavor-enhancing purine nucleotides obtained by hydrolyzing RNA, have been established and expanded to include fermentation methods and numerous techniques for producing nucleoside derivatives. Recently, widespread investigations of the industrial production of single-cell protein of bacteria or yeast from n-paraffin or C_1-compounds, i.e., methane and methanol, have been carried out. If efficient production methods can be established, RNA and nucleotides may be more plentiful than at present.

Many antibiotics have been found to contain nucleosides in their structures. Also the recent discoveries of cAMP and guanosine tetraphosphate shed light on the physiological role of nucleic acid-related compounds.

Investigations of production of nucleic acid-related substances need to be carried out in the areas of biochemical reagents, food and fodder additives, and also in the pharmaceutical field.

Acknowledgments

The author is indebted to the members of his laboratory, Drs. Y. Tani, F. Kawai, Y. Izumi, and S. Shimizu, for their helpful assistance in the preparation of this manuscript.

References

1. Kuninaka, A. (1960). *Nippon Nogei Kagaku Kaishi* **34**, 489.
2. Nakao, Y., and Ogata, K. (1963). *Agr. Biol. Chem.* **27**, 491.
3. Kuninaka, A. (1961). *Protein, Nucl. Acid, Enzyme (Tokyo)* **6**, 403.
4. Kuninaka, A., Kibi, M., and Sakaguchi, K. (1964). *Food Technol.* **19**, 29.
5. Ogata, K. (1963). *Amino Acid Nucl. Acid (Tokyo)* **8**, 1.
6. Demain, A. L. (1968). *Progr. Ind. Microbiol.* **8**, 35.
7. Ogata, K. (1971). *In* "Biochemical and Industrial Aspects of Fermentation," pp. 37–59. Kodansha, Tokyo.
8. Kuninaka, A., Otsuka, S., Kobayashi, Y., and Sakaguchi, K. (1959). *Bull. Agr. Chem. Soc. Jap.* **23**, 239.
9. Kuninaka, A., Kibi, M., Yoshino, H., and Sakaguchi, K. (1961). *Agr. Biol. Chem.* **25**, 693.
10. Ogata, K., Nakao, Y., Igarashi, S., Omura, E., Sugino, Y., Yoneda, M., and Suhara, I. (1963). *Agr. Biol. Chem.* **27**, 110.
11. Nakao, Y., and Ogata, K. (1963). *Agr. Biol. Chem.* **27**, 199 and 291.
12. Saruno, R., Takahara, H., and Fujimoto, M. (1964). *J. Ferment. Technol.* **42**, 475.
13. Tone, H., Sasaki, K., Sayama, Y., Takata, M., and Ozaki, A. (1964). *Amino Acid Nucl. Acid (Tokyo)* **10**, 135.
14. Mouri, T., Hashida, W., Shiga, I., and Teramoto, S. (1968). *Amino Acid Nucl. Acid (Tokyo)* **17**, 178.
15. Fujimura, Y., Hasegawa, Y., Kaneko, Y., and Doi, S. (1967). *Agr. Biol. Chem.* **31**, 92.
16. Nakagiri, Y., Maekawa, Y., Kihara, R., and Miwa, M. (1968). *J. Ferment. Technol.* **46**, 605 and 610.
17. Kuninaka, A. (1961). *Food. Technol.* **4**, 9.
18. Aida, K., Chung, S., Suzuki, I., and Yagi, T. (1965). *Agr. Biol. Chem.* **29**, 508.
19. Fujishima, T., and Yoshino, H. (1967). *Amino Acid Nucl. Acid (Tokyo)* **16**, 45.
20. Yoneda, M. (1964). *J. Biochem. (Tokyo)* **55**, 475 and 481.
21. Kuninaka, A., Fujishima, T., and Fujimoto, M. (1967). *Amino Acid Nucl. Acid (Tokyo)* **16**, 281.
22. Fujimoto, M., Kuninaka, A., and Yoshino, H. (1967). *Agr. Biol. Chem.* **33**, 1517.
22a. Fujimoto, M., Kuninaka, A., and Yoshino, H. (1974). *Agr. Biol. Chem.* **38**, 777 and 785.
23. Kuninaka, A. (1957). *J. Gen. Appl. Microbiol.* **3**, 55.
24. Nakao, Y., and Ogata, K. (1963). *Agr. Biol. Chem.* **27**, 116.
25. Nakao, Y., and Ogata, K. (1963). *Agr. Biol. Chem.* **27**, 499.
26. Igarashi, S., and Kakinuma, A. (1962). *Agr. Biol. Chem.* **26**, 218.

27. Igarashi, S. (1962). *Agr. Biol. Chem.* **26**, 221.
28. Kakinuma, A., and Igarashi, S. (1964). *Agr. Biol. Chem.* **28**, 131.
29. Nakao, Y., and Ogata, K. (1963). *Agr. Biol. Chem.* **27**, 199.
30. Nakao, Y., Imada, A., Wada, T., and Ogata, K. (1964). *Agr. Biol. Chem.* **28**, 151.
31. Tsukada, Y., and Sugimori, T. (1964). *Agr. Biol. Chem.* **24**, 479 and 484.
32. Watanabe, S., Osawa, T., and Yamamoto, S. (1968). *J. Ferment. Technol.* **46**, 538.
33. Kitano, K., Akiyama, S., and Fukuda, H. (1970). *J. Ferment. Technol.* **48**, 14.
34. Okabayashi, T., and Masuo, E. (1960). *Chem. Pharm. Bull.* **8**, 1084 and 1089.
35. Ogata, K., Imada, A., and Nakao, Y. (1962). *Agr. Biol. Chem.* **26**, 586.
36. Okabayashi, T., Ide, M., and Yoshimoto, A. (1963). *Arch. Biochem. Biophys.* **100**, 158.
37. Arima, K., Fukami, T., Fujiwara, M., Yokota, M., and Tamura, G. (1963). *Nippon Nogei Kagaku Kaishi* **37**, 453.
38. Furuya, A., Abe, S., and Kinoshita, S. (1963). *Amino Acid Nucl. Acid (Tokyo)* **8**, 100.
39. Furuya, A., Araki, K., Nohara, M., Abe, S., and Kinoshita, S. (1964). *Amino Acid Nucl. Acid (Tokyo)* **9**, 24.
40. Nakayama, K., Nara, T., Suzuki, T., Sato, Z., Misawa, M., and Kinoshita, (1963). *Amino Acid Nucl. Acid (Tokyo)* **8**, 81.
41. Okabayashi, T., Ide, M., and Yoshimoto, A. (1963). *J. Bacteriol.* **86**, 930.
42. Makman, R. S., and Sutherland, E. W. (1963). *Fed. Proc., Fed. Amer. Soc. Exp. Biol.* **22**, 470.
43. Ide, M., Yoshimoto, A., and Okabayashi, T. (1967). *J. Bacteriol.* **94**, 317.
44. Okabayashi, T., Ide, M., and Yoshimoto, A. (1964). *Amino Acid Nucl. Acid (Tokyo)* **10**, 117.
45. Makman, R. S., and Sutherland, E. W. (1965). *J. Biol. Chem.* **240**, 1309.
46. Ishiyama, J., Yokotsuka, T., and Saito, N. (1974). *Agr. Biol. Chem.* **38**, 507.
47. Tomita, F., and Suzuki, T. (1974). *Agr. Biol. Chem.* **38**, 71.
48. Buchanan, J. M. (1960). In "The Nucleic Acids" (E. Chargaff and J. N. Davidson, eds.), Vol. 3, p. 303. Academic Press, New York.
49. Moat, A. G., and Friedman, H. (1960). *Bacteriol. Rev.* **24**, 309.
50. Magasanik, B., and Karibian, D. (1960). *J. Biol. Chem.* **235**, 2672.
51. Magasanik, B. (1957). *Annu. Rev. Microbiol.* **11**, 221.
52. Nara, T., Misawa, M., and Kinoshita, S. (1967). *Agr. Biol. Chem.* **31**, 1351.
53. Nara, T., Misawa, M., and Kinoshita, S. (1968). *Agr. Biol. Chem.* **32**, 561.
54. Furuya, A., Abe, S., and Kinoshita, S. (1968). *Appl. Microbiol.* **16**, 981.
55. Furuya, A., Abe, S., and Kinoshita, S. (1969). *Appl. Microbiol.* **18**, 977.
56. Misawa, M., Nara, T., and Kinoshita, S. (1969). *Agr. Biol. Chem.* **33**, 514.
57. Misawa, M., Nara, T., Udagawa, K., Abe, S., and Kinoshita, S. (1969). *Agr. Biol. Chem.* **33**, 370.
58. Skodova, H., and Skoda, J. (1969). *Appl. Microbiol.* **17**, 188.
59. Furuya, A., Abe, S., and Kinoshita, S. (1970). *Appl. Microbiol.* **20**, 263.
60. Kato, F., Furuya, A., and Abe, S. (1971). *Agr. Biol. Chem.* **35**, 1061.
61. Furuya, A., Abe, S., and Kinoshita, S. (1971). *Biotechnol. Bioeng.* **13**, 229.
62. Furuya, A., Okachi, R., Takayama, K., and Abe, S. (1973). *Biotechnol. Bioeng.* **15**, 795.
63. Uchida, K., Kuninaka, A., Yoshino, H., and Kibi, M. (1961). *Agr. Biol. Chem.* **25**, 804.

64. Hara, T., Koaze, Y., Yamada, Y., and Kojima, M. (1962). *Agr. Biol. Chem.* **26,** 61 and 747.
65. Kojima, M., Koaze, Y., and Hara, T. (1962). *Agr. Biol. Chem.* **26,** 656 and 758.
66. Koaze, Y., Yamada, Y., Kojima, M., and Hara, T. (1962). *Agr. Biol. Chem.* **26,** 754.
67. Koaze, Y., Yamada, Y., Kojima, M., Sato, K., and Aoyama, Y. (1962). *Agr. Biol. Chem.* **26,** 747.
68. Shiro, T., Yamanoi, A., Konishi, S., Okumura, S., and Takahashi, M. (1962). *Agr. Biol. Chem.* **26,** 785.
69. Aoki, R., Momose, H., Kondo, Y., Muramatsu, N., and Tsuchiya, Y. (1963). *J. Gen. Appl. Microbiol.* **9,** 387.
70. Aoki, R. (1963). *J. Gen. Appl. Microbiol.* **9,** 397.
71. Aoki, R., Kondo, Y., and Momose, H. (1963). *J. Gen. Appl. Microbiol.* **9,** 403.
72. Yamanoi, A., Konishi, S., and Shiro, T. (1967). *J. Gen. Appl. Microbiol.* **13,** 365.
73. Aoki, R., Kondo, Y., Hirose, Y., and Okada, H. (1968). *J. Gen. Appl. Microbiol.* **14,** 411.
74. Yamanoi, A., and Shiro, T. (1968). *J. Gen. Appl. Microbiol.* **14,** 1.
75. Fujiwara, M., Nakamura, H., Chu, S. Y., Yamamoto, T., Fukami, T., Tamura, G., and Arima, K. (1963). *Amino Acid Nucl. Acid (Tokyo)* **8,** 110.
76. Nara, T., Misawa, M., Nakayama, K., and Kinoshita, S. (1963). *Amino Acid Nucl. Acid (Tokyo)* **8,** 94.
77. Fujimoto, N., Morozumi, M., Midorikawa, U., Miyakawa, S., and Uchida, K. (1965). *Agr. Biol. Chem.* **29,** 918.
78. Fujimoto, M., and Uchida, K. (1965). *Agr. Biol. Chem.* **29,** 249.
79. Yamanoi, A., Hirose, Y., Aoki, M., and Shiro, T. (1965). *J. Ferment. Technol.* **11,** 339.
80. Yamanoi, A., Hirose, Y., Shiro, T., and Katsuya, N. (1965). *J. Gen. Appl. Microbiol.* **11,** 269.
81. Yamanoi, A., Hirose, Y., and Shiro, T. (1966). *J. Gen. Appl. Microbiol.* **12,** 299.
82. Fujimoto, M., and Uchida, K. (1965). *Agr. Biol. Chem.* **29,** 1150.
83. Fujimoto, M., Uchida, K., Suzuki, M., and Yoshino, H. (1966). *Agr. Biol. Chem.* **30,** 605.
84. Fujiwara, M., Bunovsky, V., Nakamura, H., Tamura, G., and Arima, K. (1967). *J. Gen. Appl. Microbiol.* **13,** 1.
85. Konishi, S., and Shiro, T. (1968). *Agr. Biol. Chem.* **32,** 396.
86. Konishi, S., Kubota, K., Aoki, R., and Shiro, T. (1968). *Amino Acid Nucl. Acid (Tokyo)* **18,** 15.
87. Momose, H., and Shiio, I. (1969). *J. Gen. Appl. Microbiol.* **15,** 399.
88. Komatsu, K., Saijo, A., Kodaira, R., and Ohsawa, H. (1970). *Amino Acid Nucl. Acid (Tokyo)* **22,** 54.
89. Kanamitsu, O. (1970). *Agr. Biol. Chem.* **34,** 1424.
90. Akiya, T., Midorikawa, Y., Kuninaka, A., Yoshino, H., and Ikeda, Y. (1972). *Agr. Biol. Chem.* **36,** 227.
91. Ishii, K., and Shiio, I. (172). *Agr. Biol. Chem.* **36,** 1511.
92. Midorikawa, Y., Akiya, T., Kuninaka, A., and Yoshino, H. (1972). *Agr. Biol. Chem.* **36,** 1529.
93. Midorikawa, Y., Akiya, T., Kuninaka, A., and Yoshino, H. (1973). *Agr. Biol. Chem.* **37,** 1595.

94. Kinoshita, K., Nishiyama, T., Tsuri, H., Konishi, S., Shiro, T., and Okada, H. (1968). *Amino Acid Nucl. Acid (Tokyo)* **17**, 150.
95. Shirafuji, H., Imada, A., Yashima, S., and Yoneda, M. (1968). *Agr. Biol. Chem.* **32**, 69.
96. Hirano, A., Akimoto, T., and Osawa, T. (1967). *Amino Acid Nucl. Acid (Tokyo)* **16**, 84.
97. Nogami, I., Kida, M Iijima, T., and Yoneda, M. (1968). *Agr. Biol. Chem.* **32**, 144.
98. Sasajima, K., Nogami, I., and Yoneda, M. (1970). *Agr. Biol. Chem.* **34**, 381.
99. Sasajima, K., and Yoneda, M. (1971). *Agr. Biol. Chem.* **34**, 381.
99a. Sasajima, K., and Yoneda, M. (1974). *Agr. Biol. Chem.* **38**, 1305.
100. Haneda, K., Hirano, R., Kodaira, R., and Ohuchi, S. (1971). *Agr. Biol. Chem.* **35**, 1906.
101. Haneda, K., Komatsu, K., Kodaira, R., and Ohsawa, S. (1972). *Amino Acid Nucl. Acid (Tokyo)* **26**, 15.
102. Komatsu, K., Haneda, K., Hirano, A., Kodaira, R., and Ohsawa, H. (1972). *J. Gen. Appl. Microbiol.* **18**, 19.
103. Komatsu, K., and Kodaira, R. (1973). *J. Gen. Appl. Microbiol.* **19**, 263.
104. Misawa, M., Nara, T., Udagawa, K., Abe, S., and Kinoshita, S. (1964). *Agr. Biol. Chem.* **28**, 688.
105. Misawa, M., Nara, T., and Kinoshita, S. (1964). *Agr. Biol. Chem.* **28**, 692.
106. Nakayama, K., Suzuki, T., Sato, Z., and Kinoshita, S. (1964). *J. Gen. Appl. Microbiol.* **10**, 133.
107. Sato, Z., Nakayama, K., Tanaka, H., and Kinoshita, S. (1965). *Agr. Biol. Chem.* **29**, 412.
108. Nakayama, K., Sato, Z., Tanaka, H., and Kinoshita, S. (1965). *Nippon Nogei Kagaku Kaishi* **39**, 118.
109. Demain, A. L., Jackson, M., Vitali, R. A., Hendlin, D., and Jacob, T. A. (1965). *Appl. Microbiol.* **13**, 757.
110. Demain, A. L., Jackson, M., Vitali, R. A., Hendlin, D., and Jacob, T. A. (1966). *Appl. Microbiol.* **14**, 821.
111. Iguchi, T., and Takeda, I. (1966). *Agr. Biol. Chem.* **30**, 709.
112. Iguchi, T., Watanabe, T., and Takeda, I. (1967). *Agr. Biol. Chem.* **31**, 569, 574, and 888.
113. Kawamoto, I., Nara, T., Misawa, M., and Kinoshita, S. (1970). *Agr. Biol. Chem.* **34**, 1142.
114. Nakayama, K., Suzuki, T., Sato, Z., and Kinoshita, S. (1963). *Amino Acid Nucl. Acid (Tokyo)* **8**, 88.
115. Ozaki, M., Tagawa, S., and Kimura, T. (1972). *Amino Acid Nucl. Acid (Tokyo)* **26**, 24.
116. Watanabe, A., Tani, K., and Sasaki, Y. (1968). *Amino Acid Nucl. Acid (Tokyo)* **18**, 9.
116a. Hanka, L. J. (1960). *J. Bacteriol.* **80**, 30.
117. Michelson, M., Drell, W., and Mitchell, H. K. (1951). *Proc. Nat. Acad. Sci. U.S.* **37**, 396.
118. Machida, H., Kuninaka, A., and Yoshino, H. (1970). *Agr. Biol. Chem.* **34**, 1129.
119. Machida, H., and Kuninaka, A. (1969). *Agr. Biol. Chem.* **33**, 868.
120. Tomita, F., and Suzuki, T. (1972). *Agr. Biol. Chem.* **33**, 133.
121. Smithies, W. R., and Gibbons, W. E. (1955). *Can. J. Microbiol.* **1**, 614.
122. Demain, A. L., Burg, R. W., and Hendlin, D. (1965). *J. Bacteriol.* **89**, 640.
123. Rudin, L., and Albertsson, P. (1967). *Biochim. Biophys. Acta* **134**, 37.

124. Nara, T., Misawa, M., Komuro, T., and Kinoshita, S. (1967). *Agr. Biol. Chem.* **31**, 1224.
125. Nara, T., Kawamoto, I., Misawa, M., and Kinoshita, S. (1968). *Agr. Biol. Chem.* **32**, 956.
126. Ostern, P., Baraunowski, J., and Terszakowec, J. (1938). *Hoppe-Seyler's Z. Physiol. Chem.* **251**, 258.
127. Ostern, P., Baranunowski, J., and Huble, S. (1938). *Hoppe-Seyler's Z. Physiol. Chem.* **255**, 104.
128. Brawermann, G., and Chargaff, E. (1955). *Biochim. Biophys. Acta* **16**, 524.
129. Katagiri, H., Yamada, H., Mitsugi, K., and Takahashi, M. (1963). *Agr. Biol. Chem.* **27**, 469.
130. Katagiri, H., Yamada, H., Mitsugi, K., and Tsunoda, T. (1964). *Agr. Biol. Chem.* **28**, 577.
131. Mitsugi, K., Nakazawa, E., Takahashi, M., and Yamada, H. (1964). *Agr. Biol. Chem.* **28**, 571, 849, and 859.
132. Mitsugi, K., Komagata, K., Takahashi, M., Iizuka, H., and Katagiri, H. (1964). *Agr. Biol. Chem.* **28**, 586.
133. Mitsugi, K. (1964). *Agr. Biol. Chem.* **28**, 658 and 668.
134. Mitsugi, K., Kamimura, A., Nakazawa, E., and Okumura, S. (1964). *Agr. Biol. Chem.* **28**, 828.
135. Mitsugi, K., Nakazawa, E., and Okumura, S. (1964). *Agr. Biol. Chem.* **28**, 838.
136. Mitsugi, K., Nakazawa, E., and Okumura, S., Takahashi, M., and Yamada, H. (1965). *Agr. Biol. Chem.* **29**, 1051.
137. Mitsugi, K., Okumura, S., Shiro, T., and Takahashi, M. (1965). *Agr. Biol. Chem.* **29**, 1104.
138. Mitsugi, K., Kamimura, A., Kimura, M., and Okumura, S. (1965). *Agr. Biol. Chem.* **29**, 1109.
139. Hikino, H., Suzuki, N., and Takemoto, T. (1969). *Tetrahedron Lett.* **50**, 5069.
140. Yamazaki, A., Kumashiro, I., and Takenishi, T. (1969). *J. Org. Chem.* **32**, 3032.
141. Sakai, T., Tochikura, T., and Ogata, K. (1965). *Agr. Biol. Chem.* **29**, 742.
142. Sakai, T., Tochikura, T., and Ogata, K. (1966). *Agr. Biol. Chem.* **30**, 245.
143. Sakai, T., Yorifuji, T., Tochikura, T., and Ogata, K. (1967). *Agr. Biol. Chem.* **31**, 525 and 535.
144. Tanaka, H., and Nakazawa, K. (1972). *Agr. Biol. Chem.* **36**, 1405.
145. Sakai, T., Uchio, K., Ichimoto, I., and Omata, S. (1974). *Agr. Biol. Chem.* **38**, 433.
146. Lutwak-Mann, C., and Mann, T. (1935). *Biochem. Z.* **281**, 140.
147. Tochikura, T., Kuwahara, M., Yagi, S., Okamoto, H., Tominaga, Y., Kano, T., and Ogata, K. (1967). *J. Ferment. Technol.* **45**, 511.
148. Tanaka, H., Sato, Z., Nakayama, K., and Kinoshita, S. (1968). *Agr. Biol. Chem.* **32**, 721.
149. Murao, S., and Nishino, T. (1973). *Agr. Biol. Chem.* **37**, 2929.
150. Murao, S., Nishino, T., and Hamagishi, Y. (1974). *Agr. Biol. Chem.* **38**, 887.
151. Haseltine, W. A., Block, R., Gilbert, W., and Weber, K. (1972). *Nature (London)* **238**, 381.
152. Ginsburg, V. (1964). *Advan. Enzymol.* **26**, 35.
153. Kennedy, E. P., and Weiss, S. B. (1956). *J. Biol. Chem.* **222**, 193.
154. Kennedy, E. P., and Weiss, S. B. (1956). *J. Biol. Chem.* **222**, 185.
155. Tochikura, T., Kawaguchi, K., Kano, T., and Ogata, K. (1969). *J. Ferment. Technol.* **47**, 564.

156. Tochikura, T., Kawai, H., Tobe, S., Kawaguchi, K., Osugi, M., and Ogata, K. (1968). *J. Ferment. Technol.* **46**, 957.
157. Tochikura, T., Kawai, H., Tobe, S., Kawaguchi, K., and Ogata, K. (1970). *Amino Acid Nucl. Acid (Tokyo)* **22**, 144.
158. Tochikura, T., Kawaguchi, K., Kawai, H., Mugibayashi, Y., and Ogata, K. (1968). *J. Ferment. Technol.* **46**, 970.
159. Kawai, H., and Tochikura, T. (1971). *Agr. Biol. Chem.* **35**, 1578 and 1587.
160. Tochikura, T., Kawai, H., and Gotan, T. (1971). *Agr. Biol. Chem.* **35**, 163.
161. Kawai, H., Yamamoto, K., Kimura, A., and Tochikura, T. (1973). *Agr. Biol. Chem.* **37**, 1741.
162. Tochikura, T., Kimura, A., Kawai, H., Tachiki, T., and Gotan, T. (1970). *J. Ferment. Technol.* **48**, 763, 769.
163. Kimura, A., Morita, M., and Tochikura, T. (1971). *Agr. Biol. Chem.* **35**, 1955.
164. Tochikura, T., Kariya, Y., and Kimura, A. (1974). *J. Ferment. Technol.* **52**, 637.
164a. Tochikura, T., Kimura, A., Kawai, H., and Gotan, T. (1971). *J. Ferment. Technol.* **49**, 1005.
165. Tochikura, T., Kimura, A., Kawai, H., and Gotan, T. (1972). *J. Ferment. Technol.* **50**, 178.
166. Kawaguchi, K., Ogata, K., and Tochikura, T. (1970). *Agr. Biol. Chem.* **34**, 908.
167. Kawaguchi, K., Tanida, S., Mugibyayashi, Y., Tani, Y., and Ogata, K. (1971). *J. Ferment. Technol.* **49**, 195.
168. Kawaguchi, K., Tanida, S., Matsuda, K., Tani, Y., and Ogata, K. (1973). *Agr. Biol. Chem.* **37**, 75.
169. Shirota, S., Watanabe, S., and Takeda, I. (1971). *Agr. Biol. Chem.* **35**, 325.
170. Miyauchi, K., Uchida, K., and Yoshino, H. (1972). *Amino Acid Nucl. Acid (Tokyo)* **25**, 47.
171. Siliprandi, N., and Bianchi, N. (1954). *Biochim. Biophys. Acta* **16**, 424.
172. Cerletti, P., and Siliprandi, N. (1958). *Arch. Biochem. Biophys.* **76**, 214.
173. Tsukihara, K., Minoura, K., and Izumiya, M. (1960). *J. Vitaminol. (Kyoto)* **6**, 68.
174. Sakai, T., Watanabe, T., and Chibata, I. (1973). *Agr. Biol. Chem.* **37**, 849.
175. Sakai, T., Watanabe, T., and Chibata, I. (1973). *Agr. Biol. Chem.* **37**, 2885.
176. Watanabe, T., Uchida, T., Kato, J., and Chibata, I. (1974). *Appl. Microbiol.* **27**, 531.
177. Takebe, I., and Kitahara, K. (1963). *J. Gen. Appl. Microbiol.* **9**, 31.
177a. Nakayama, K., Sato, Z., Tanaka, H., and Kinoshita, S. (1968). *Agr. Biol. Chem.* **32**, 1331.
178. Hayano, K., Takebe, I., and Kitahara, K. (1964). *Amino Acid Nucl. Acid (Tokyo)* **9**, 37.
178a. Sakai, T., Uchida, T., and Chibata, I. (1973). *Agr. Biol. Chem.* **37**, 1041 and 1049.
179. Tochikura, T., Kuwahara, M., Komatsubara, S., Fujisaki, M., Suga, A., and Ogata, K. (1969). *Agr. Biol. Chem.* **33**, 840.
180. Tochikura, T., Kuwahara, M., Tachiki, T., Komatsubara, S., and Ogata, K. (1969). *Agr. Biol. Chem.* **33**, 848.
181. Kuwahara, M., Tachiki, T., Tochikura, T., and Ogata, K. (1970). *Agr. Biol. Chem.* **34**, 984.
182. Kuwahara, M., Tachiki, T., Tochikura, T., and Ogata, K. (1972). *Agr. Biol. Chem.* **36**, 745.

183. Beinert, H., von Korff, R. W., Green, D. E., Buyske, D. A., Handschumacher, R. E., Higgins, H., and Strong, F. M. (1953). *J. Biol. Chem.* **200**, 385.
184. Reece, H. C., Donald, M. B., and Crook, E. M. (1959). *J. Biochem. Micribiol. Technol. Eng.* **1**, 217.
185. Kuno, M., Kikuchi, M., Nakao, Y., and Yamatodani, S. (1973). *Agr. Biol. Chem.* **37**, 313.
186. Ward, G. B., Brown, G. M., and Snell, E. E. (1955). *J. Biol. Chem.* **213**, 869.
187. Brown, G. M. (1959). *J. Biol. Chem.* **234**, 370.
188. Shimizu, M., and Abiko, Y. (1965). *Chem. Pharm. Bull.* **13**, 189.
189. Abiko, Y. (1967). *J. Biochem. (Tokyo)* **61**, 290.
190. Abiko, Y. (1967). *J. Biochem. (Tokyo)* **61**, 300.
191. Abiko, Y., Suzuki, T., and Shimizu, M. (1967). *J. Biochem. (Tokyo)* **61**, 309.
192. Suzuki, T., Abiko, Y., and Shimizu, M. (1967). *J. Biochem. (Tokyo)* **62**, 642.
193. Pierpoint, W. S., and Hughes, D. E. (1954). *Biochem. J.* **56**, 130.
194. Soyama, K. (1966). *Vitamins* **34**, 214.
195. Ogata, K., Tani, Y., Shimizu, S., and Uno, K. (1972). *Agr. Biol. Chem.* **36**, 93.
196. Ogata, K., Shimizu, S., and Tani, Y. (1972). *Agr. Biol. Chem.* **36**, 84.
197. Ogata, K., Shimizu, S., and Tani, Y. (1972). *Agr. Biol. Chem.* **36**, 1757.
198. Shimizu, S., Tani, Y., and Ogata, K. (1972). *Agr. Biol. Chem.* **36**, 370.
199. Shimizu, S., Miyata, K., Tani, Y., and Ogata, K. (1972). *Biochim. Biophys. Acta* **279**, 583.
200. Shimizu, S., Miyata, K., Tani, Y., and Ogata, K. (1973). *Agr. Biol. Chem.* **37**, 607 and 615.
200a. Nishimura, N., Shibatani, T., Kakimoto, T., and Chibata, I. (1974). *Appl. Microbiol.* **28**, 117.
201. Shimizu, S., Satsuma, S., Kubo, K., Tani, Y., and Ogata, K. (1973). *Agr. Biol. Chem.* **37**, 857.
202. Shimizu, S., Kubo, K., Satsuma, S., Tani, Y., and Ogata, K. (1974). *J. Ferment. Technol.* **52**, 114.
203. Kurooka, S., Hosoki, K., and Yoshioka, Y. (1967). *Pharm. Bull.* **15**, 944.
204. Shimizu, S., Kubo, K., Tani, Y., and Ogata, K. (1973). *Agr. Biol. Chem.* **37**, 2863.
205. Shimizu, S., Kubo, K., Morioka, H., Tani, Y., and Ogata, K. (1974). *Agr. Biol. Chem.* **38**, 1015.
206. Karasawa, T., Yoshida, K., Furukawa, K., and Hosoki, K. (1972). *J. Biochem. (Tokyo)* **71**, 1065.
207. Abiko, Y., Ashida, S., and Shimizu, M. (1972). *Biochim. Biophys. Acta* **268**, 364.

Synthesis of L-Tyrosine-Related Amino Acids by β-Tyrosinase[1]

HIDEAKI YAMADA AND HIDEHIKO KUMAGAI

*The Research Institute for Food Science,
Kyoto University, Uji Kyoto, Japan*

I. Introduction 249
II. Physicochemical Properties of β-Tyrosinase 250
    A. Purification Procedure 250
    B. Stability 253
    C. Molecular Weight 253
    D. Cofactor Requirements 254
    E. Catalytic Properties 255
III. Reaction Mechanism 258
    A. Mechanism of the α,β-Elimination, β-Replacement, and Racemization Reactions 258
    B. Mechanism of the Synthetic Reaction 264
    C. Modification of the Essential Histidyl Residue of β-Tyrosinase 266
    D. Interaction of β-Tyrosinase with Pyridoxal Phosphate Analogs 268
    E. Stereochemistry of the β-Replacement Reaction 270
IV. Immobilization of β-Tyrosinase on Sepharose 273
V. Enzymic Preparation of L-Tyrosine and L-Dopa 274
    A. Selection of Microorganisms Having High β-Tyrosinase Activity 274
    B. Culture Conditions for the Preparation of Cells Containing High β-Tyrosinase Activity 275
    C. Reaction Conditions for the Synthesis of L-Tyrosine and L-Dopa 278
VI. Conclusions 285
References 285

I. Introduction

Kakihara and Ichihara (1953) reported that phenol is produced from L-tyrosine in bacterial cultures, not through stepwise degradation, but through primary fission of the side chain. The tyrosine-inducible enzyme responsible for this conversion was subsequently named β-tyrosinase by Uchida *et al.* (1953). The enzyme was partially purified by Uchida *et al.* (1953) and by Yoshimatsu (1957) from cells of *Bacterium coli phenologenes* grown in a medium containing L-tyrosine. Yoshimatsu (1957) demonstrated that β-tyrosinase catalyzes the stoichiometric conversion of L-tyrosine into phenol, pyruvate, and ammonia (Eq. 1) and requires pyridoxal phosphate as a cofactor. A similar enzymic reaction

---

[1] Systematic name: Tyrosine phenol-lyase (deaminating) (EC 4.1.99.2).

has been described by Brot et al. (1965), who used a partially purified enzyme preparation from *Clostridium tetanomorphum*.

Apparently homogeneous preparations of the enzyme were prepared in our laboratory from cells of *Escherichia intermedia* (Kumagai et al., 1970a) and *Erwinia herbicola* (Kumagai et al., 1972) grown on media supplemented with L-tyrosine (Kumagai et al., 1970d). We found that crystalline preparations of the enzyme catalyze a series of α,β-elimination [Eqs. (1–3)], β-replacement [Eqs. (4) and (5)], and racemization [Eq. (6)] reactions (Kumagai et al., 1969, 1970c; Ueno et al., 1970). The reverse of the α,β-elimination reaction to synthesize L-tyrosine or 3,4-dihydroxyphenyl-L-alanine (L-dopa) [Eqs. (7) and (8)] was also catalyzed by crystalline preparations of the enzyme (Yamada et al., 1972b).

$$L(D)\text{-Tyrosine} + H_2O \rightarrow \text{pyruvic acid} + NH_3 + \text{phenol} \quad (1)$$
$$L(D)\text{-Serine} \rightarrow \text{pyruvic acid} + NH_3 \quad (2)$$
$$L(D)\text{-Tyrosine} + \text{pyrocatechol} \rightarrow L\text{-dopa} + \text{phenol} \quad (3)$$
$$L(D)\text{-Serine} + \text{phenol} \rightarrow L\text{-tyrosine} + H_2O \quad (4)$$
$$L(D)\text{-Serine} + \text{pyrocatechol} \rightarrow L\text{-dopa} + H_2O \quad (5)$$
$$L(D)\text{-Alanine} \rightarrow DL\text{-alanine} \quad (6)$$
$$\text{Pyruvic acid} + NH_3 + \text{phenol} \rightarrow L\text{-tyrosine} + H_2O \quad (7)$$
$$\text{Pyruvic acid} + NH_3 + \text{pyrocatechol} \rightarrow L\text{-dopa} + H_2O \quad (8)$$

In recent studies, we have proved that this enzyme catalyzes the synthesis of L-tyrosine or its related amino acids from DL-serine and phenol or its related compounds or from pyruvic acid, ammonia, and phenol or its related compounds, in significantly high yields (Enei et al., 1971, 1972a; Yamada et al., 1972a). We here describe the enzymic methods for the preparation of L-tyrosine or its related amino acids.

II. Physicochemical Properties of β-Tyrosinase

A. Purification Procedure

Escherichia intermedia was taken from an agar slant of the basal medium (pH 7.0), which consisted of 0.5% peptone, 0.5% meat extract, 0.2% NaCl, and 0.05% yeast extract in tap water (pH 7.0), and was inoculated into a subculture (500 ml of the basal medium in a 2-liter flask). Incubation was carried out at 30°C for 24 hours with reciprocal shaking. The subculture was, in turn, inoculated into a 30-liter fermentor containing 20 liters of the basal medium supplemented with 0.2% L-tyrosine. Incubation was carried out at 30°C for 14 hours with aeration (15 liters per minute). Mature cells were harvested by continuous-flow centrifugation, then were washed with distilled water and suspended in 0.01 M potassium phosphate buffer, pH 6.0, containing 0.005 M mercaptoethanol. Approximately, 10 gm of cells (wet weight) were obtained per liter of medium.

All operations described below were carried out at 0–5°C.

Step 1. Preparation of the cell extract. The cell paste was suspended in 0.01 M potassium phosphate buffer, pH 6.0, containing 0.005 M mercaptoethanol, to give a suspension of about 1 gm/5 ml. The suspension was divided into 500-ml portions, and each portion was subjected to ultrasonic oscillation (20 Kc) for 30 minutes. Cells and debris were removed by centrifugation.

Step 2. Ammonium sulfate fractionation. Solid ammonium sulfate was added to the cell extract to 0.30 saturation. After standing overnight, the precipitate was removed by centrifugation and discarded. The ammonium sulfate concentration was then increased to 0.70 saturation by the addition of solid ammonium sulfate. After standing overnight, the precipitate was collected by centrifugation and dissolved in 0.01 M potassium phosphate buffer, pH 7.0, containing 0.005 M mercaptoethanol. The solution was dialyzed for 36 hours against the same buffer.

Step 3. Protamine sulfate treatment. One-tenth the volume of a 6% protamine sulfate solution was added to the dialyzate, then the precipitate formed was removed by centrifugation.

Step 4. DEAE-Sephadex column chromatography. The enzyme solution was subjected to DEAE-Sephadex column chromatography. The solution was passed through a column of DEAE-Sephadex A-50, equilibrated with 0.01 M potassium phosphate buffer, pH 7.0 and containing 0.005 M mercaptoethanol. The column was washed with 0.1 M potassium phosphate buffer, pH 7.0, containing 0.005 M mercaptoethanol, which removed much of the inactive protein. The enzyme was subsequently eluted with 0.1 M potassium phosphate buffer, pH 7.0, containing 0.005 M mercaptoethanol and 0.1 M potassium chloride. Active fractions were combined and concentrated by adding solid ammonium sulfate to 0.70 saturation. The precipitate obtained by centrifugation was dissolved in 0.01 M potassium phosphate buffer, pH 6.0, containing 0.005 M mercaptoethanol, then was dialyzed for 48 hours against the same buffer.

Step 5. Hydroxyapatite column chromatography. The enzyme solution was subjected to hydroxyapatite column chromatography. The solution was placed on a column of hydroxyapatite, equilibrated with 0.01 M potassium phosphate buffer, pH 6.0, containing 0.005 M mercaptoethanol. The column was washed with 0.03 M potassium phosphate buffer, pH 6.0, containing 0.005 M mercaptoethanol. The enzyme was subsequently eluted with 0.1 M potassium phosphate buffer, pH 6.0, containing 0.005 M mercaptoethanol. Active fractions were combined and concentrated by adding solid ammonium sulfate to 0.70 saturation. The precipitate was collected by centrifugation, then was dissolved in a minimum amount of 0.01 M potassium phosphate buffer, pH 6.0, containing 0.005 M mercaptoethanol.

Step 6. Sephadex G-150 gel filtration. The enzyme solution was introduced into a column of Sephadex G-150, equilibrated with 0.01 M potassium phosphate buffer, pH 6.0, containing 0.005 M mercaptoethanol, after which it was filtered with the same buffer. Active fractions were combined and concentrated by adding solid ammonium sulfate to 0.70 saturation. The precipitate obtained by centrifugation was dissolved in a minimum amount of 0.01 M potassium phosphate buffer, pH 6.0, containing 0.005 M mercaptoethanol.

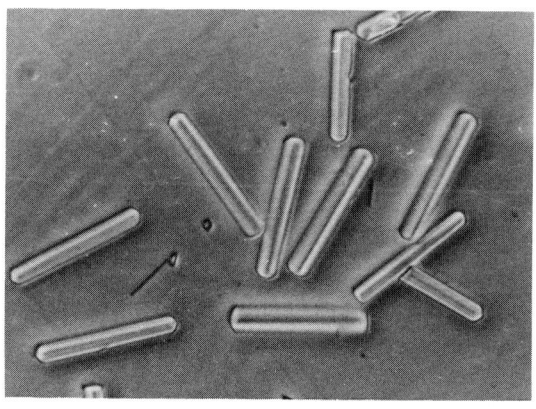

FIG. 1. Photomicrograph of crystalline β-tyrosinase from *Escherichia intermedia*. ×150.

Step 7. Crystallization. Solid fine powdered ammonium sulfate was cautiously added to the purified enzyme solution until it became slightly turbid, then the mixture was placed in an ice bath. Crystallization began after about 2 hours and was virtually complete within 1 week. A photomicrograph of the crystalline enzyme, which appears as hexagonal rods, is shown in Fig. 1 (Kumagai *et al.*, 1970a). For recrystallization, the crystals were dissolved in a minimum amount of 0.01 M potassium phosphate buffer, pH 6.0, containing 0.005 M mercaptoethanol. Solid ammonium sulfate was added as described above. Approximately a 110-fold purification was achieved with an overall yield of 4.4%.

β-Tyrosinase was also purified and crystallized from cells of *Erwinia herbicola* grown in a medium supplemented with 0.2% L-tyrosine; a procedure used was similar to that described above. Figure 2 (Kumagai *et al.*, 1972) shows a photomicrograph of the crystalline enzyme prepared from *Erwinia herbicola*, which appears as dodecahedrons.

FIG. 2. Photomicrograph of crystalline β-tyrosinase from *Erwinia herbicola*. ×150.

B. Stability

Crude and partially purified preparations of the β-tyrosinase were preserved at 5°C in 0.01 M potassium phosphate buffer, pH 6.0, containing 0.005 M mercaptoethanol. Variations in pH in the range of pH 6.0–7.0 had little effect on enzyme stability. Outside this range, however, the rate of inactivation increased rapidly. Solutions of highly purified enzyme were less stable than those of partially purified preparations. For example, a sample of the enzyme from *Escherichia intermedia*, with a specific activity of 1.94, lost 6–10% of its initial activity after standing for 2 days at 5°C in 0.01 M potassium phosphate buffer, pH 6.0, containing 0.005 M mercaptoethanol.

The crystalline enzyme preparations could be stored at 0–5°C as suspensions in 0.01 M potassium phosphate buffer, pH 6.0, containing 0.005 M mercaptoethanol and 60% saturated ammonium sulfate. The specific and total activity of the enzyme remained constant for periods of more than 1 month. Preservation in this fashion constituted the best storage procedure.

On heating, the crystalline enzyme was rather stable up to 40°C, but above 50°C, it was rapidly inactivated.

C. Molecular Weight

The specific activity of the enzyme, achieved after the second crystallization, was not altered upon further recrystallization. Recrystallized enzyme preparations gave a single band on acrylamide gel electrophoresis carried out at pH 8.3. The recrystallized enzyme sedimented as a single symmetric peak in the ultracentrifuge in 0.01 M potassium phosphate buffer, pH 6.0. The sedimentation coefficients in water at 20°C, extrapolated to zero protein concentration ($s^0_{20,w}$) were 7.77 S and 8.24 S for the enzymes from *Escherichia intermedia* and *Erwinia herbicola*, respectively. Diffusion coefficients ($D_{20,w}$) of 4.42 and 3.08 $\times$ 10^{-7} cm^2 sec^{-1} were respectively determined for 7.8 mg of protein from *Escherichia intermedia* and for 2.3 mg of protein from *Erwinia herbicola*. Assuming a partial specific volume of 0.75, the molecular weight of the enzyme from *Escherichia intermedia* was calculated as approximately 170,000 (Kumagai *et al.*, 1970a) and that of *Erwinia herbicola* as 259,000 (Kumagai *et al.*, 1972).

D. Cofactor Requirements

1. *Pyridoxal Phosphate Requirement* (Kumagai et al., 1970b, 1972)

Crystalline β-tyrosinases prepared by the above procedure showed negligible activity when assayed in the absence of added pyridoxal phosphate; therefore, they entirely represent the apoenzyme. Half the maximum enzymic activity was obtained at the respective concentrations of 1.3 $\times$ 10^{-6} M and 1.5 $\times$ 10^{-6} M for the enzymes from *Escherichia intermedia* and *Erwinia herbicola*. The apoenzyme showed no appreciable absorbance at 340 and 430 nm. On association with pyridoxal phosphate, pronounced absorption maxima appeared at 340 and 430 nm, as shown in Fig. 3. Absorption in the 410–440 nm region is characteristic of many pyridoxal phosphate enzymes and has been attributed to a hydrogen-bounded azomethine of pyridoxal phosphate (Metzler, 1957). Unlike other pyridoxal phosphate enzymes, the absorption maxima of holo-β-tyrosinase were not appreciately shifted by variations in pH between 6.0 and 9.0.

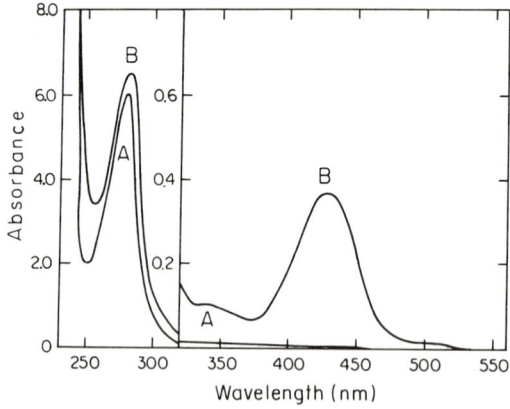

Fig. 3. Absorption spectra of β-tyrosinase in the presence (curve B) and in the absence (curve A) of pyridoxal phosphate. From Kumagai *et al.* (1970b).

2. Cation Requirement (Kumagai et al., 1970b)

Ichihara et al. (1956) have studied in some detail the cation requirements for the maximum activity of β-tyrosinase, with a crude enzyme preparation. The crystalline enzymes from *Escherichia intermedia* and *Erwinia herbicola* showed a similar requirement for K^+ and NH_4^+ for their maximum activities. However, Na^+ was inactive. When the K^+ was replaced by NH_4^+, essentially no change in the spectrum of holo β-tyrosinase was observed. However, if the K^+ was replaced by Na^+, a small but pronounced decrease in absorption at 430 nm was observed.

3. Amount of Pyridoxal Phosphate Bound (Kumagai et al., 1970b, 1972)

The amount of pyridoxal phosphate bound by the apoenzyme was determined after dialysis of the crystalline enzymes from *Escherichia intermedia* and *Erwinia herbicola* against 0.05 M potassium phosphate buffer, pH 8.0, containing 3×10^{-5} M pyridoxal phosphate. After 36 hours, when control experiments without protein showed complete equilibration, the concentration of pyridoxal phosphate inside and outside the dialysis sac was determined. An excess concentration of pyridoxal phosphate was found within the dialysis sac, which corresponded to the binding of 2 moles of pyridoxal phosphate by 1 mole of the apoenzyme.

The amount of pyridoxal phosphate bound was also determined by spectrophotometric titration of the apoenzyme at the wavelengths of 370 and 450 nm, with graded amounts of pyridoxal phosphate. The increment in absorbance at 450 nm and the decrement at 370 nm were read at 25°C. Those two determinations have end points corresponding to the binding of 2 moles of pyridoxal phosphate by 1 mole of the apoenzyme.

The results obtained from these two procedures clearly indicate that there are two pyridoxal phosphate binding sites per mole of enzyme. Table I summarizes some physicochemical properties of the β-tyrosinase from *Escherichia intermedia* and *Erwinia herbicola*.

E. Catalytic Properties

Using the crystalline β-tyrosinase from *Escherichia intermedia*, the catalytic properties of the enzyme were investigated. The crystalline enzyme catalyzed a series of α,β-elimination, β-replacement, and racemization reactions [Eqs. (1–8)].

1. α,β-Elimination Reaction (Kumagai et al., 1970a, 1972)

β-Tyrosinase catalyzed the conversion of L-tyrosine into phenol, pyruvate, and ammonia, in the presence of added pyridoxal phosphate.

TABLE I
Physicochemical Properties of β-Tyrosinase

	Values	
Property	Escherichia intermedia	Erwinia herbicola
$E_{1\,cm}^{1\%}$ at 280 nm	8.37	8.08
$E_{260\,nm}/E_{280\,nm}$	0.54	0.55
$s_{20,w}^{0}$	7.77 S	8.24 S
$D_{20,w}$	4.42^a	3.08^a
Molecular weight		
Sedimentation-diffusion	170,000 (±20,000)	259,000
Sephadex G-200 filtration	182,000	
Number of subunit	4	
Molecular weight of subunit		
Sedimentation-diffusion	55,000	
SDS-gel electrophoresis	50,000	
PLP[c] bound to enzyme	2^b	2^b

[a] $\times 10^{-7}$ cm² sec⁻¹.
[b] Moles per mole of enzyme.
[c] PLP, pyridoxal phosphate.

D-Tyrosine, in comparison with L-tyrosine, showed approximately 25% reactivity under the conditions used. Pyruvate formation was also observed with β-chloro-L-alanine, S-methyl-L-cysteine, L- and D-serine, and L- and D-cysteine.

L-Alanine inhibited pyruvate formation from L-tyrosine, S-methyl-L-cysteine, and L-serine. The inhibition was shown to be competitive for these substrates. Phenol, pyrocatechol, and resorcinol also inhibited pyruvate formation, but the inhibition was a mixed type. The α,β-elimination reactions proceed optimally around pH 8.2.

2. β-Replacement Reaction (Kumagai et al., 1969; Ueno et al., 1970)

β-Tyrosinase catalyzed the synthesis of L-tyrosine from L- or D-serine and phenol in the presence of pyridoxal phosphate. β-Chloro-L-alanine, S-methyl-L-cysteine, and L-cysteine replaced L- or D-serine in the synthesis of L-tyrosine. When pyrocatechol, resorcinol, pyrogallol, and hydroxyhydroquinone were added to the reaction mixture in place of phenol, L-dopa, 2,4-dihydroxyphenyl-L-alanine (2,4-L-dopa), 2,3,4-trihydroxyphenyl-L-alanine (2,3,4,-L-topa), (Rapp et al., 1974), and 2,4,5-trihydroxyphenyl-L-alanine (2,4,5-L-topa) (Kumagai et al., 1975) were, respectively, synthesized. The β-replacement reactions proceed optimally at pH 8.0–8.5.

The K_m values for its substrates, the K_i values for its inhibitors and the maximum velocities of the α,β-elimination and replacement reactions catalyzed by β-tyrosinase were determined from Lineweaver–Burk plots and are summarized in Table II.

TABLE II
COMPARATIVE SUBSTRATE AFFINITIES AND ACTIVITIES OF β-TYROSINASE IN CATALYSIS OF DIFFERENT REACTIONS

Compound	Role	Product measured	K_m (mM)	K_i (mM)	V_{max} (μmoles/ min/mg)
α,β-Elimination reactions					
L-Tyrosine	Substrate	Pyruvate	0.23	—	1.9
L-Serine	Substrate	Pyruvate	34	—	0.35
S-Methyl-L-cysteine	Substrate	Pyruvate	1.8	—	1.2
β-Chloro-L-alanine	Substrate	Pyruvate	4.5	—	18.2
L-Alanine	Inhibitor	Pyruvate	—	6.5	—
L-Phenylalanine	Inhibitor	Pyruvate	—	2.0	—
Phenol	Inhibitor	Pyruvate	—	0.04	—
Pyrocatechol	Inhibitor	Pyruvate	—	0.46	—
Resorcinol	Inhibitor	Pyruvate	—	0.16	—
β-Replacement reactions					
L-Serine	Cosubstrate	L-Tyrosine	35	—	0.33
S-Methyl-L-cysteine	Cosubstrate	L-Tyrosine	1.8	—	0.82
β-Chloro-L-alanine	Cosubstrate	L-Tyrosine	4.5	—	1.4
Phenol	Cosubstrate	L-Tyrosine	1.2	—	—

3. Racemization Reaction (Kumagai et al., 1970c)

The conversion of D- or L-alanine into the racemate is catalyzed by β-tyrosinase in the presence of pyridoxal phosphate. The K_m value for L-alanine was 2.6×10^{-2} M and the maximum velocity of the racemization reaction was 0.05 μmole/min per milligram of protein. The optimum pH for this racemization reaction is between 7.2 and 7.5.

4. Synthetic Reaction (Yamada et al., 1972b)

The synthesis of L-tyrosine from phenol, pyruvate, and ammonia is catalyzed by β-tyrosinase in the presence of pyridoxal phosphate. This synthesis proceeded as a function of the concentration of phenol, pyruvate, and ammonia. Michaelis–Menten kinetics were observed with pyruvate and ammonia, but phenol showed strong substrate inhibition at high concentrations. The apparent K_m values for phenol and pyruvate were calculated as 4.4×10^{-3} M and 5.1×10^{-3} M, respectively. More accurate K_m values were obtained by using three substrate kinetics (Cleland, 1963). Double reciprocal plots of the extrapolated reaction velocity at infinite concentrations of one substrate and the variable concentrations of the remaining two substrates are shown in Fig. 4 (Kumagai et al., 1973c). A parallel set of curves was obtained when pyruvate was at the infinite concentration. The K_m values for phenol, pyruvate, and ammonia, calculated from Fig. 4, were 1.1 mM, 12 mM, and 20 mM, respectively, and the maximum velocity was 3.3 μmoles/min per milligram of protein, which is about 1.5 times higher than that of L-tyrosine degradation by the reaction of Eq. (1).

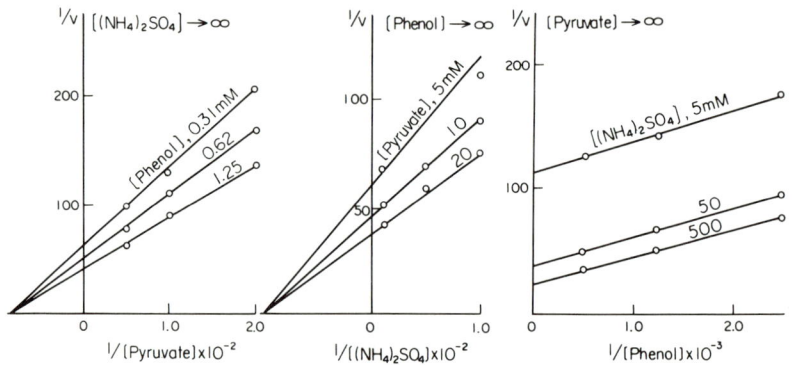

FIG. 4. Kinetic attempts to establish the order of addition of phenol, pyruvate, and ammonia to β-tyrosinase during synthesis of L-tyrosine.

When pyrocatechol and resorcinol were, respectively, added to the reaction mixture in place of phenol L-dopa and 2,4-L-dopa were synthesized. Cresols (m- and o-) and chlorophenols (m- and o-) also replaced phenol to synthesize, respectively, methyl-L-tyrosine (2- and 3-) and chloro-L-tyrosine (2- and 3-) (Kumagai et al., 1974b). Syntheses of L-tyrosine and its related amino acids by the reversal of the α,β-elimination reaction catalyzed by β-tyrosinase are summarized in Table III.

III. Reaction Mechanism

A. Mechanism of the α,β-Elimination, β-Replacement, and Racemization Reactions

The preceding description shows that crystalline preparations of β-tyrosinase from *Escherichia intermedia* (and *Erwinia herbicola*) catalyze a variety of α,β-elimination, β-replacement, racemization, and synthetic reactions. These reactions are partially explainable by adopting the general mechanism for pyridoxal-dependent reactions proposed by Braunstein and Shemyakin (1953) and by Metzler et al. (1954). This mecha-

TABLE III
RELATIVE VELOCITY OF SYNTHESIS OF L-TYROSINE-RELATED AMINO
ACIDS FROM PYRUVATE, AMMONIA, AND PHENOL DERIVATIVES
BY β-TYROSINASE

Phenol derivative	L-Amino acid[a] synthesized	Relative velocity of synthesis
HO–⟨phenyl⟩	HO–⟨phenyl⟩–R	100
3-CH₃, HO–⟨phenyl⟩	3-CH₃, HO–⟨phenyl⟩–R	45.4
2-CH₃, HO–⟨phenyl⟩	2-CH₃, HO–⟨phenyl⟩–R	52.7
3-Cl, HO–⟨phenyl⟩	3-Cl, HO–⟨phenyl⟩–R	47.3
2-Cl, HO–⟨phenyl⟩	2-Cl, HO–⟨phenyl⟩–R	32.7
3-HO, HO–⟨phenyl⟩	3-HO, HO–⟨phenyl⟩–R	60.0
2-OH, HO–⟨phenyl⟩	2-OH, HO–⟨phenyl⟩–R	58.2

[a] R represents L-alanyl moiety.

nism is shown in Fig. 5, with a modification in which pyridoxal phosphate in the enzyme is shown as an internal azomethine. The enzyme-bound α-aminoacrylate (EA) is the key intermediate in this proposed mechanism. The α,β-elimination reaction is considered to proceed via steps:

$$E + S \rightleftharpoons ES \xrightarrow{-H^+} EX \xrightarrow{-R^-} EA \xrightarrow{+H_2O} E + \text{pyruvate} + NH_3$$

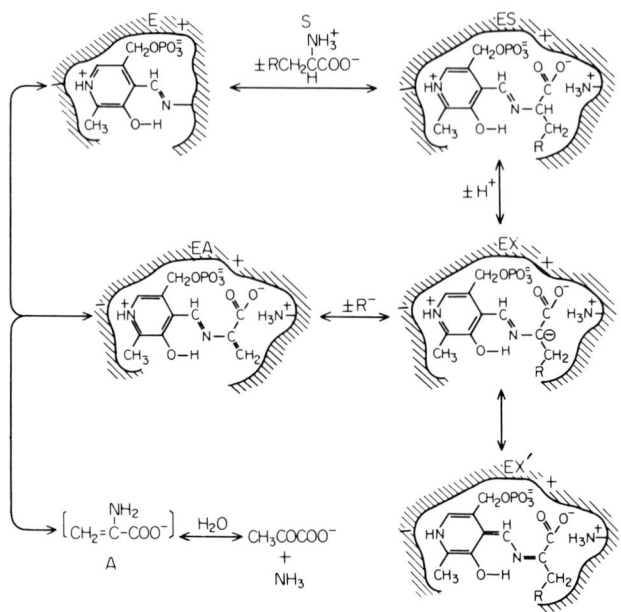

FIG. 5. Schematic representation of the mechanism for the reactions catalyzed by β-tyrosinase. E, ES, etc., represent species of the enzyme and enzyme–substrate complexes.

the β-replacement reaction via steps:

$$E + S \rightleftharpoons ES \xrightarrow{-H^+} EX \xrightarrow{-R^-} EA \xrightarrow{+R'^-} EX' \xrightarrow{+H^+} ES' \to E + S'$$

and the racemization reaction via steps:

$$E + \text{L-S} \rightleftharpoons E \cdot \text{L-S} \xrightarrow{-H^+} EX \xrightarrow{+H^+} E \cdot \text{DL-S} \to E + \text{DL-S}$$

The synthetic reaction of L-tyrosine seems to proceed via the reversal steps of α,β-elimination. Morino and Snell (1967a,b) have made detailed studies on the mechanism of the tryptophanase which catalyzes the α,β-elimination and β-replacement reactions of L-tryptophan as it does with β-tyrosinase.

In many of the pyridoxal phosphate-dependent enzymes examined to date, the azomethine linkages between pyridoxal phosphate and the apoenzymes have involved the ε-amino group of the lysine residues of the apoenzymes (Fisher et al., 1958; Hughes et al., 1962; Schirch and Mason, 1962; Dempsey and Snell, 1963). Reduction of holo-β-tyrosinase by sodium borohydride resulted in a complete loss in activity, indicating the presence of the azomethine linkage between the coenzyme and the apoenzyme. ε-Pyridoxyllysine was isolated from the hydrolyzate of the reduced enzyme in amounts corresponding to 2 moles of pyridoxal phosphate (Kumagai et al., 1970b). The results show that each of the 2 moles of pyridoxal phosphate present in the holoenzyme participates in an azomethine linkage similar to that of E (Fig. 5) and that the amino group involved is the ϵ-amino group of a lysine residue in each case.

The first step in the proposed reaction sequence for catalysis by β-tyrosinase (Fig. 5) is interaction of the holoenzyme with substrate by a transaldimination reaction that replaces the azomethine linkage between coenzyme and apoenzyme (E, Fig. 5) with a similar bond between coenzyme and substrate (ES). Inactivation of the enzyme by borohydride reduction was partially blocked in the presence of the substrate amino acids (Kumagai et al., 1973a), especially L-alanine, indicating the formation of the ES complex shown in Fig. 5. L-Alanine is not a substrate for the α,β-elimination and β-replacement reactions but is a competitive inhibitor for these reactions.

Many pyridoxal phosphate enzymes have been shown to interact with their substrates to form enzyme–substrate complexes with absorption characteristics distinctly different from those of the enzyme itself (Jenkins and Sizer, 1957; Shukuya and Schwert, 1960; Jenkins, 1961a,b; Morino and Snell, 1967b). The interaction of L-tyrosine with β-tyrosinase also produced spectral changes. The addition of L-tyrosine to the holoenzyme resulted in the appearance of a new absorption peak between 495 and 507 nm with a shoulder near 450 nm. The appearance of the peak around 500 nm was always associated with a decrease in the peak at 430 nm, as observed from the holoenzyme spectrum. The new peak around 500 nm disappeared as L-tyrosine was degraded to phenol, pyruvate, and ammonia. L-Alanine also produced an intense absorption band at 500 nm. However, in contrast to a similar peak formed with L-tyrosine, this band did not disappear even after long standing (Fig. 6) (Kumagai et al., 1970b, 1972). D-Alanine produced the very same change as in the absorption spectrum with L-alanine (Kumagai et al., 1970c).

Although L-alanine is not a substrate for the α,β-elimination and β-replacement reactions, it does act as a competitive inhibitor for these reactions and as a substrate for the racemization reaction. L-Alanine cannot undergo β-elimination to form an EA species (Fig. 5), but can proceed through ES to an EX species. Absorption at 500 nm is unlikely for the ES species, and can be tentatively ascribed to the deprotonated intermediate, EX, or to a species in equilibrium with this intermediate (Morino and Snell, 1967b).

Proton exchange at the α-position of L-alanine should occur, if the absorption in the 500 nm region is the result of the formation of an EX complex, as concluded above. The β-tyrosinase-dependent incorpora-

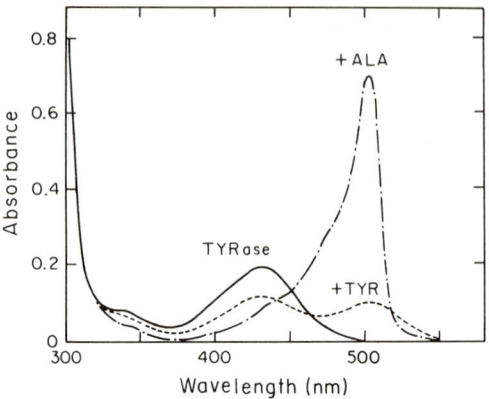

Fig. 6. Absorption spectra of holo-β-tyrosine (TYRase) in the presence of L-tyrosine (+TYR) and L-alanine (+ALA). From Kumagai et al. (1970b).

tion of tritium into L-alanine was observed, in fact, to be a function of incubation time. The exchange of the α-hydrogen atom of L-alanine was confirmed by following the nuclear magnetic resonance spectra (Fig. 7) (Kumagai et al., 1973a) of L-alanine during prolonged incubation

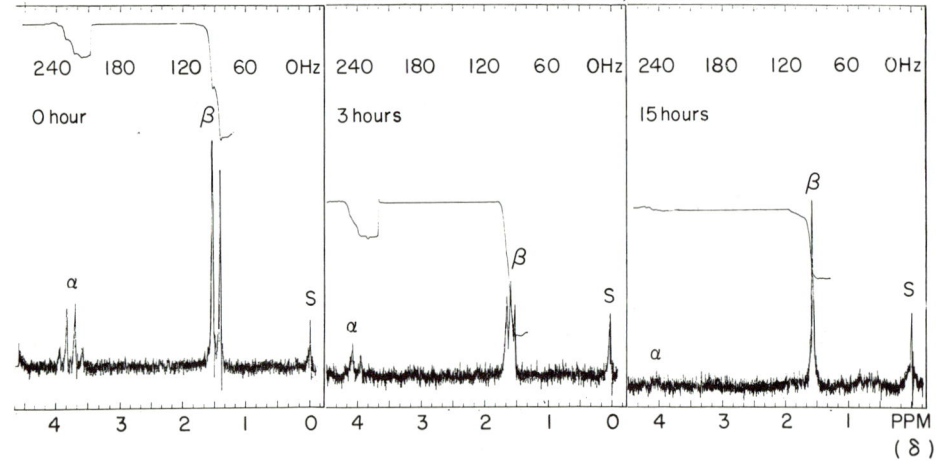

Fig. 7. Nuclear magnetic resonance spectra of L-alanine during reaction with β-tyrosinase in D_2O.

with β-tyrosinase in D_2O. Peaks at δ3.5–4.0, attributable to the α-proton, disappear almost completely as this proton is replaced by deuterium. While peaks at δ1.5–1.75, which represent the β-protons, remain and become a single peak with no change in integral magnitude. This supports catalysis of the labilization of the α-hydrogen atom of L-alanine and is consistent with the assumption that the species absorbing at 500 nm represents a deprotonated enzyme–L-alanine complex, EX (Fig. 5). No pyruvate was found in the reaction mixture containing L-alanine, β-tyrosinase, and pyridoxal phosphate, indicating that neither β-elimination nor transamination occurred. D-Alanine was detected in the reaction mixture, and its formation was proportional to the reaction time and enzyme concentration. D-Alanine was also converted to the racemic mixture by the enzyme, and the ratio of conversion for both isomers never exceeded 50% (Fig. 8). The addition of D-alanine to holo-β-tyrosinase

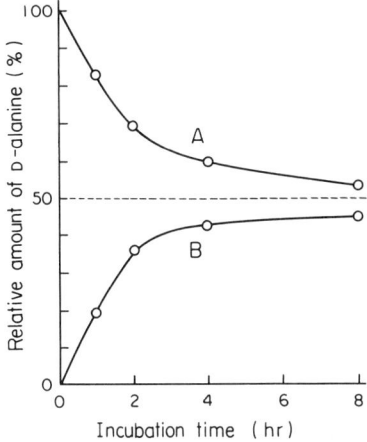

FIG. 8. Racemization of L- (curve A) or D-alanine (curve B) by β-tyrosinase. From Kumagai et al. (1970c).

resulted in an absorption at 500 nm similar to that observed by the addition of L-alanine. These results show that the proton exchange reaction of L-alanine which proceeds via steps

$$E + S \rightleftharpoons ES \xrightarrow{-H^+} EX \xrightarrow{+H^+} ES \rightarrow E + S$$

is nonstereospecific. The stereochemical random incorporation of proton into the complex of enzyme and deprotonated L-alanine

$$(EX \xrightarrow{+H^+} ES)$$

results in racemization of the substrate.

In the reaction of the enzyme with a substrate which has an electronegative substituent in the β-position, the elimination of the β-substituent from EX to form an enzyme–aminoacrylate complex EA takes place subsequent to the elimination of the α-hydrogen. The enzyme bound α-aminoacrylate is the key common intermediate in both α,β-elimination and β-replacement reactions. In the β-elimination reactions, α-aminoacrylate is hydrolyzed to pyruvate and ammonia. In the β-replacement reactions, the enzyme-bound aminoacrylate reacts with R'H, in a reaction which is essentially the reverse of α,β-elimination, and new amino acids replaced with the R' residue at the β-position are liberated via reversal steps of the α,β-elimination.

$$EX' \xrightarrow{+H^+} ES' \rightarrow E + S'$$

B. Mechanism of the Synthetic Reaction

The α,β-elimination reaction of L-tyrosine catalyzed by β-tyrosinase is reversible at high concentrations of pyruvate and ammonia (Yamada et al., 1972b). L-Tyrosine or certain of its analogs, e.g., 3,4-dihydroxyphenyl-L-alanine, can be synthesized by this reaction from pyruvate, ammonia, and phenol or an appropriate derivative such as pyrocatechol (Kumagai et al., 1974b). Degradation of tryptophan to indole, pyruvate, and ammonia by tryptophanase was also reversible (Nakazawa et al., 1972a,b; Watanabe and Snell, 1972; Yoshida et al., 1974).

The absorption of holo-β-tyrosinase at 430 nm was increased by adding ammonia to the concentration required for the synthetic reactions (Fig. 9) (Kumagai et al., 1973c). This indicates the formation of an enzyme-bound, imine complex of pyridoxal phosphate and ammonia. The results of double-reciprocal plots of the extrapolated reaction velocity at infinite concentrations of one substrate and variable concentrations of the re-

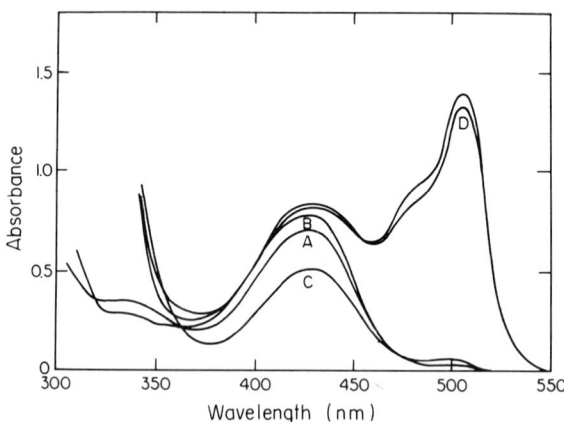

Fig. 9. Absorption spectra of holo-β-tyrosinase in the presence of ammonia, pyruvate, and phenol. Curve A: holo-β-tyrosinase; curve B: holo-β-tyrosinase plus ammonia; curve C: holo-β-tyrosinase plus ammonia and pyruvate; curve D: holo-β-tyrosinase plus ammonia, pyruvate, and phenol.

maining two substrates (kinetics for three substrates, Fig. 4) are consistent with the ordered Ter-Uni mechanism of Cleland (1970), indicating that pyruvate may be the second substrate to combine with the enzyme. A spectral study in the presence of a high concentration of ammonia suggested that ammonia is the first substrate to interact with the enzyme in the synthetic reaction, and that phenol should be the third.

The scheme for the mechanism of β-tyrosinase-catalyzed reactions is modified as shown in Fig. 10, taking into consideration the results

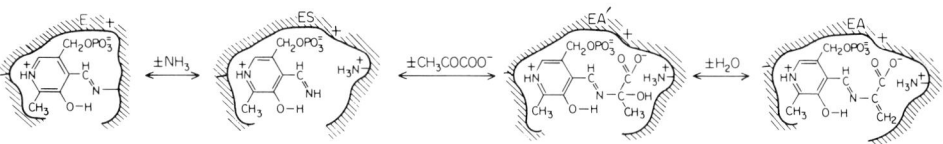

Fig. 10. Schematic representation of the mechanism for the synthesis of enzyme–α-aminoacrylate complex catalyzed by β-tyrosinase. E, ES, etc., represent species of the enzyme and enzyme–substrate complexes.

obtained from the reversal reaction. These results support a mechanism for β-tyrosinase-catalyzed reactions in which α-aminoacrylate, which functions as a common enzyme-bound intermediate in both synthetic and degradative reactions, is not released into the medium during these reactions.

Proton exchange at the C-3 of pyruvate should occur during incubation of the enzyme with ammonia and pyruvate, if enzyme-bound α-aminoacrylate is formed from the two substrates. The β-tyrosinase-dependent incorporation of tritium into pyruvate does, in fact, occur (Kumagai et al., 1973c). This was confirmed by following the nuclear magnetic resonance spectra (Fig. 11) (Kumagai et al., 1973c) of pyruvate during incubation with holo-β-tyrosinase and NH_3 in D_2O. Peaks at $\delta 2.5$–2.6, attributable to three hydrogens at the C-3 of pyruvate, disappear almost completely as these hydrogens are replaced by deuterium. Peaks at $\delta 3.5$–3.6, attributable to the three hydrogens of methanol, present as the internal standard, did not change in integral magnitude during incubation. This proton exchange reaction of pyruvate showed the same optimum pH as in the synthetic reaction of L-tyrosine. This suggests that formation of the α-aminoacrylate intermediate determines the rate of the overall synthetic reaction. To confirm this, holo-β-tyrosinase, pyruvate, and ammonia were preincubated in the deuterium medium for various periods, then phenol was added to the reaction mixture to synthesize L-tyrosine. Figure 12 (Kumagai et al., 1973c) shows the nuclear magnetic resonance spectra enzymically synthesized L-tyrosine in the deuterium medium. Signals around $\delta 3.4$ are assigned to the proton at

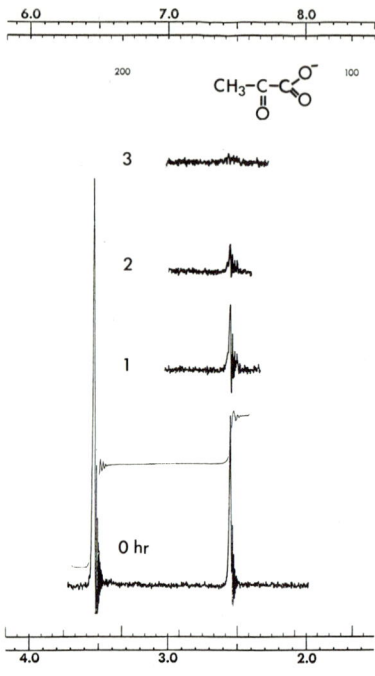

Fig. 11. Nuclear magnetic resonance spectra of pyruvate during reaction with β-tyrosinase and ammonia in D_2O.

C-2, at $δ2.8$ to C-3 and at $δ6.8$ to aromatic hydrogens. No peaks attributable to the hydrogens at C-3 were observed in the spectrum of the L-tyrosine synthesized in deuterium after 4 hours of preincubation without phenol. However, the spectrum of L-tyrosine synthesized without preincubation shows that only the α-proton is exchanged with the deuterium. The ratio of the integral magnitude of the protons of C-3 to the aromatic protons is 2:4. These results are consistent with the supposition that the formation of α-aminoacrylate is the rate-limiting step for the L-tyrosine synthetic reaction.

C. Modification of the Essential Histidyl Residue of β-Tyrosinase

The first catalytic step in all phases of the α,β-elimination reaction is the removal of the α-hydrogen from the substrate amino acid to give an intermediate (EX; see Fig. 5). Although this step is greatly facilitated by electron withdrawal through the conjugated system of pyridoxal, it probably requires the presence of a basic group on the protein to

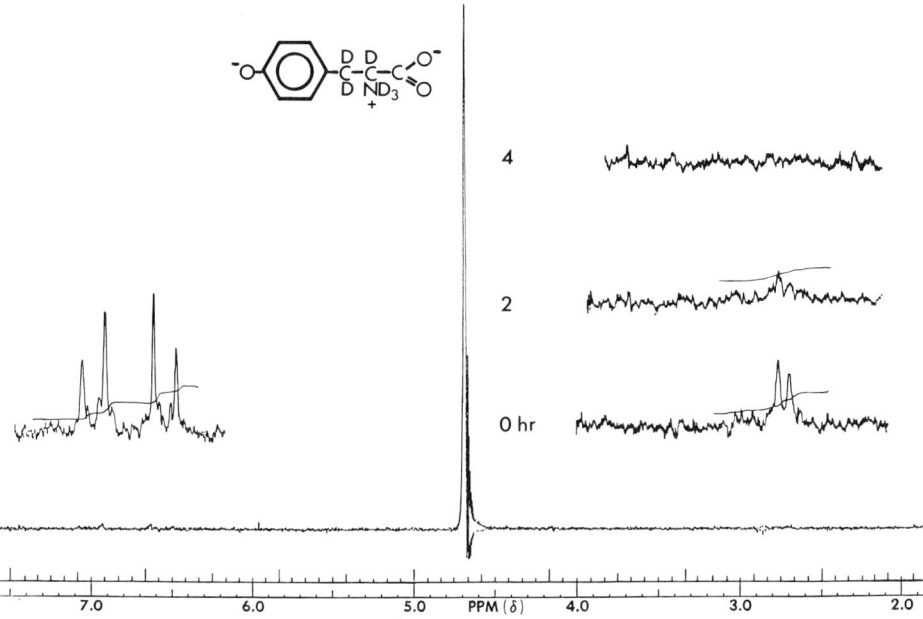

FIG. 12. Nuclear magnetic resonance spectra of L-tyrosine synthesized by β-tyrosinase.

accept the hydrogen (Thanassi et al., 1965; Peterson and Martinez-Carrion, 1970; Miles and Kumagai, 1974).

The β-tyrosinase of *Escherichia intermedia* is inactivated by treatment with diethylpyrocarbonate (Fig. 13) (Mühlard et al., 1967; Ovadi et

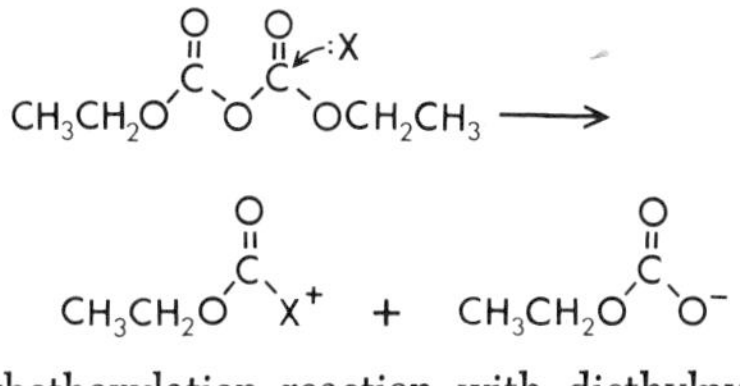

FIG. 13. Carbethoxylation reaction with diethylpyrocarbonate.

al., 1967; Melchior and Fahrney, 1970) at pH 6.0 and 4°C (Kumagai et al., 1974a). Spectrophotometric studies have shown that this inactivation is stoichiometric with a modification of two histidyl residues per molecule of the enzyme. That this inactivation is largely reversed by treatment with hydroxylamine indicates that the inactivation is mainly due to modification of the histidyl residues (Melchior and Fahrney, 1970). No changes in sulfhydryl content or in the aromatic amino acids were observed as a result of this modification. The modified enzyme retains most of its ability to form a nearly normal complex with its coenzyme, pyridoxal phosphate. This has been shown by studies of its absorption, by the determination of pyridoxal phosphate and by the reduction of the holoenzyme with sodium borohydride. The modified enzyme also appears to form an initial Schiff base intermediate with L-alanine.

The modified holoenzyme fails to catalyze the exchange of the α-hydrogen of L-alanine with the tritium from tritiated water. This is consistent with a catalytic role for the modified histidyl residues at the active site of the enzyme, which is the removal of the α-hydrogen of the substrate.

D. Interaction of β-Tyrosinase with Pyridoxal Phosphate Analogs

The interaction of apo-β-tyrosinase with two pyridoxal phosphate analogs, pyridoxal phosphate N-oxide (Fig. 14A) and 2′-hydroxy pyri-

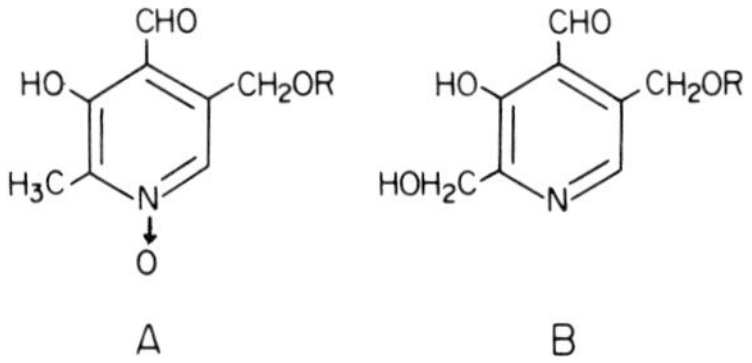

Fig. 14. Structure of pyridoxal phosphate N-oxide (A) and 2′-hydroxy pyridoxal phosphate (B).

doxal phosphate (Fig. 14B), and the effect of this modification on the catalytic activities of the resulting holoenzyme (Kumagai et al., 1973b) are described in this section.

The incorporation of an oxygen atom to the pyridine nitrogen of pyridoxal phosphate has been reported to decrease its electron-attracting ability and to lower its catalytic activities in several enzyme systems and model reactions (Fukui et al., 1969a, 1970a; Ivanov and Karpeisky, 1969; Masugi et al., 1973). The 2-methyl group of pyridoxal phosphate is not known to play a catalytic role in nonenzymic model reactions nor in the reaction catalyzed by pyridoxal phosphate proteins, but it does play an important spatial role in the hydrophobic interaction with the coenzyme binding site of the apoenzyme (Morino and Snell, 1967c). Replacement of the 2-methyl group with a hydroxymethyl group resulted in a significant change in the interaction of the pyridoxal-phosphate analog with pyridoxal phosphate enzymes (Fukui et al., 1969b, 1970b).

β-Tyrosinases reconstituted with pyridoxal phosphate N-oxide and 2′-hydroxy pyridoxal phosphate have catalyzed the α,β-elimination reaction of L-tyrosine. Table IV shows comparative coenzyme activities of pyridoxal phosphate and its analogs for β-tyrosinase. The $K_{coenzyme}$ (K_{co}) values for pyridoxal phosphate N-oxide and 2′-hydroxy pyridoxal phosphate, i.e., the concentrations of the analogs required for the half-maximum activity of the reconstituted enzyme, were determined in each instance from double-reciprocal plots of the initial rate of pyruvate formation vs. the coenzyme concentration at 30°C. The affinity of pyridoxal phosphate N-oxide for apo-β-tyrosinase was lower than that of pyridoxal phosphate. The extrapolated maximum velocity (v) of the pyridoxal phosphate N-oxide-activated enzyme was 80% that of the native enzyme.

TABLE IV
COENZYME ACTIVITIES OF PYRIDOXAL PHOSPHATE (PLP) ANALOGS FOR β-TYROSINASE[a]

Compound	K_m (M)	V_{max} (μmoles/min/mg)	$K_{coenzyme}$ (M)
PLP	4.0×10^{-4}	2.2	2.2×10^{-6}
PLPN→O	2.2×10^{-4}	1.8	6.3×10^{-6}
ω-OH-PLP	3.2×10^{-4}	1.3	1.3×10^{-5}

[a] Determined with L-tyrosine as substrate.

The low coenzyme activity of the pyridoxal phosphate N-oxide has been observed in other pyridoxal phosphate enzymes (Fukui et al., 1969a; Fonda, 1971). The K_{co} value of 2′-hydroxy pyridoxal phosphate for the apoenzyme was markedly larger than that of pyridoxal phosphate. The v value catalyzed by the 2′-hydroxy pyridoxal phosphate-bound enzyme was about 60% of that catalyzed by the native enzyme. Replacement of the 2-methyl group of the pyridoxal phosphate with a hydroxymethyl group decreased the coenzyme activity of pyridoxal phosphate for the apo-β-tyrosinase. This behavior of the analog is similar to that observed with *Escherichia coli* tryptophanase (Masugi et al., 1973).

Holo-β-tyrosinase shows two absorption bands with maxima at 340 and 430 nm. When pyridoxal phosphate is replaced by pyridoxal phosphate N-oxide, a major peak at 430 nm and a shoulder at 340 nm appeared. The 2′-hydroxy pyridoxal phosphate-bound β-tyrosinase showed two similar bands with the pyridoxal phosphate enzyme, but the intensities of both peaks were higher than those observed with the native enzyme.

The absorption peak near 430 nm for the reconstituted enzymes suggests that both the analogs, pyridoxal phosphate N-oxide and 2′-hydroxy pyridoxal phosphate, bind with the apo-β-tyrosinase through an azomethine linkage as does the native coenzyme, pyridoxal phosphate.

On the addition of substrates or competitive inhibitors, such as L-alanine, β-tyrosinase exhibits an absorption maximum near 500 nm. The absorption at 500 nm was observed on the addition of L-alanine to the pyridoxal phosphate N-oxide-bound enzyme, but the intensity was very low in comparison with the pyridoxal phosphate enzyme. One possible explanation is that the decrease in the electronegativity of the analog stabilizes an ES complex of the aldimine type and does not favor the electron shift necessary for the subsequent deprotonation reaction (Fukui et al., 1969a). β-Tyrosinase reconstituted with 2′-hydroxy pyridoxal phosphate has shown a quite similar peak at 500 nm with the pyridoxal phosphate-bound enzyme, on the addition of L-alanine.

E. Stereochemistry of the β-Replacement Reaction

β-Tyrosinase catalyzes the replacement reaction of the phenol residue of L-tyrosine by resorcinol to form L-(—)-2,4-dihydroxyphenylalanine (2,4-L-dopa). To investigate the steric course of this replacement reaction, two diastereomers of tyrosine, L-(—)-(2S,3R)-[3-^{2}H]tyrosine and DL-(±)-(2S,3S):(2R,3R)-[2,3-^{2}H$_2$]tyrosine (Kirby and Michael, 1971; Sawada et al., 1973), were converted by β-tyrosinase to the two corresponding 2,4-L-dopas (Sawada et al., 1974). The configurations at the C-3 of the products were determined by comparing their NMR spectra with that of the L-(—)-(2S,3R)-[3-^{2}H]-2,4-L-dopa of known configuration (Lambooy, 1954).

Three different results, racemization, retention, and inversion, are considerable in the configuration at the C-3 of products from the β-replacement reaction. First, supposing the stereospecific removal of the p-phenolyl moiety from the C-3 of tyrosine [(II) in Fig. 15], the prochirality of the original substrate should be kept in the α-aminoacrylate complex (Va) which has a σ-plane around the C-3 atom. Conversely, by the nonstereospecific liberation of the aryl anion, both the methylene protons on C-3 should become equivalent and racemization would take place. Second, if the resonance equilibrium, (Va) and (Vb), is present in the α-aminoacrylate intermediate, tentative formation of a carbonium ion at C-3 would possibly make random the prochirality at C-3 by free rotation between the C-2 and C-3 pivot bonds. The third problem is stereospecificity due to the addition of the aryl anion species to the C-3 of the α-aminoacrylate intermediate. Under the lesser contribution of this resonance and the steadier presence of the intermediate (Va), localization of the double bond and, thus, the prohibition of free rotation between the C-2 and C-3 bonds might induce chirality retention or

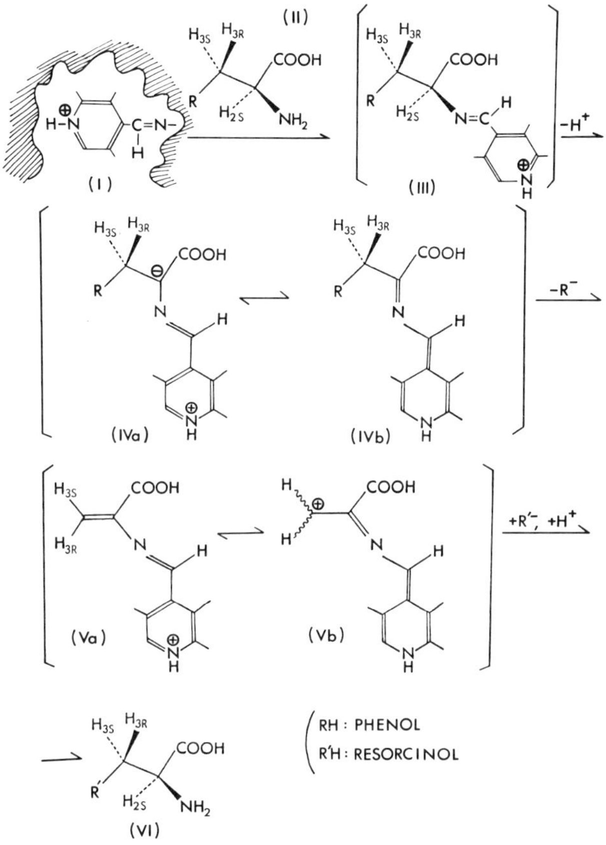

FIG. 15. Schematic representation of the stereochemistry for the β-replacement reactions catalyzed by β-tyrosinase. From Sawada et al. (1974).

inversion in the product (VI), by the subsequent stereospecific and nucleophilic attack of the aryl anion species to the C-3 of the α-aminoacrylate intermediate.

The optical activity and nuclear magnetic resonance data of both the deuterated substrates and products are shown in Table V. The 2,4-L-dopa formed enzymically from L-(—)-(2S,3R)-[3-^{2}H]tyrosine has the absolute configuration of (2S,3R), and the product from another isomeric substrate has the configuration of (2S,3S). These results show that the overall stereochemistry in the elimination of phenol and the addition of resorcinol catalyzed by β-tyrosinase is the retention of configuration at the C-3 position of the substrate; the stereospecific removal of the p-phenol moiety from C-3 or tyrosine, the presence of the α-aminoacrylate intermediate which has no resonance equilibrium as in the scheme

TABLE V
Optical Activity and Nuclear Magnetic Resonance (NMR) Data of Deuterated Substrates and Products for β-Replacement Reaction Catalyzed by β-Tyrosinase

Substrate	NMR in D_2O			Product	NMR in D_2O		
	Chemical shift (ppm)	Coupling constant (cps)	$(\alpha)_D^{25}$ in 5 N HCl		Chemical shift (ppm)	Coupling constant (cps)	$(\alpha)_D^{25}$ in H_2O
Tyrosine	$H_{2S} = 3.37$ q $H_{3R} = 2.64$ q $H_{3S} = 2.82$ q	$J_{2S,3R} = 7.0$ $J_{2S,3S} = 5.0$ $J_{3R,3S} = 13.5$	-10.0	2,4-Dopa	$H_{2S} = 4.02$ q $H_{3R} = 3.00$ q $H_{3S} = 3.25$ q	$J_{2S,3R} = 8.0$ $J_{2S,3S} = 4.5$ $J_{3R,3S} = 14.2$	-13.5 $c = 2.0$
L-(−)-(2S)-	$H_{2S} = 3.37$ d $H_{3S} = 2.80$ d	$J_{2S,3S} = 5.0$	-9.8 $c = 1.1$	L-(−)-(2S)-	$H_{2S} = 4.07$ d $H_{3S} = 3.20$ d	$J_{2S,3S} = 4.5$	-14.0 $c = 2.0$
L-(−)-(2S,3R)-[3-²H]-				L-(−)-(2S,3R)-[3-²H]-			
L-(−)-(2S,3S)-[3-²H] and its enantiomer	$H_{3R} = 2.67$ bs		Racemate	L-(−)-(2S,3S)-[3-²H]-	$H_{2S} = 4.01$ d $H_{3R} = 2.95$ d	$J_{2S,3R} = 8.0$	-14.5 $c = 1.7$

(Fig. 15), and the addition of the resorcinol anion to the intermediate from the direction in which the phenol moiety is removed.

After our work had been completed, two results of related enzymic studies appeared: (1) Skye et al. (1974) reported that the synthesis of tryptophan from serine and indole, catalyzed by tryptophan synthetase, occurs with retention of configuration at C-3 of serine; and (2) Fuganti et al. (1974) reported that the synthesis of tyrosine from serine and phenol, catalyzed by β-tyrosinase takes place with retention of configuration at C-3 of serine. Thus, the growing family of pyridoxal phosphate-dependent enzymic reactions which carry out α,β-elimination reactions and their reversals through an enzyme-bound α-aminoacrylate–pyridoxal phosphate complex appear to share a common stereochemical course. This leaves no doubt that the functional or binding group in the enzyme interacts with the α-aminoacrylate complex to keep it relatively planar, as in (Va), maintaining the prochirality at C-3 and the distinction between H_{3S} and H_{3R} throughout the reaction.

IV. Immobilization of β-Tyrosinase on Sepharose

Much attention has recently been focused on the immobilization of various enzymes, which should provide new approaches to elucidating the relation between the structure and function of enzymes, as well as giving new data for the practical application of enzymes. A number of papers have dealt with the preparation of insoluble enzyme derivatives (Katchalski et al., 1971). However, on the immobilization of enzymes requiring cofactors, only a few reports have been published.

The insolubilization of β-tyrosinase was first investigated by Dinelli (1972). He reported that β-tyrosinase is efficiently insolubilized by using a system which consists of entrapping the enzyme within the pores of wet spun synthetic fibers.

Apo- and holo-β-tyrosinases from *Escherichia intermedia* were covalently immobilized on CNBr-activated Sepharose (Fukui et al., 1975a) in a way similar to that described by Axén and Ernbach (1971). The immobilization efficiency of each was about 60%. The immobilized enzyme obtained from the apoenzyme showed about 40% of the catalytic activity of the free enzyme, and that from holoenzyme had somewhat lower activity.

The immobilization of β-tyrosinase with Sepharose-bound pyridoxal phosphate, which has been effective for tetrameric tryptophanase (Ikeda and Fukui, 1973), was not favorable.

By the use of immobilized apo-β-tyrosinase on CNBr-activated Sepharose, its catalytic properties were examined. The immobilized enzyme reconstituted by pyridoxal phosphate lost its activity gradually when used repeatedly in a batch system without the addition of pyridoxal phosphate, but the initial activity was restored by a supplement of pyridoxal phosphate to the reaction system.

These results indicate that the enzyme protein immobilized on Sepharose would be fairly stable and that the coenzyme would be readily resolved from the holoenzyme during repeated batch reactions. The apparent K_m values of the immobilized enzyme for the substrate and the coenzyme were determined, in comparison with those of the free enzyme. The affinity of the enzyme for L-tyrosine was not altered upon immobilization, whereas its affinity for pyridoxal phosphate was somewhat decreased. The steric environment around the active site of the enzyme may be slightly, but not seriously, influenced by immobilization.

Immobilized β-tyrosinase also lost its activity when used continuously in a flow system without added pyridoxal phosphate. The decrease in activity obeyed good first-order kinetics ($K = 1.76 \times 10^{-4}$ sec^{-1}). Since this decrease is considered to be a result of the dissociation of pyridoxal phosphate from the immobilized holoenzyme, the K value obtained may represent the rate constant of coenzyme dissociation. In the continuous-flow reaction system, the dissociated coenzyme should be quickly removed from the vicinity of the enzyme molecule. Accordingly, a high exact value would be obtained for the rate constant of coenzyme dissociation in the presence of substrate.

The above results indicate that the immobilized β-tyrosinase on Sepharose may be useful for the continuous synthesis of L-tyrosine or its related amino acids by the flow method (Fukui et al., 1975b). Detailed investigations on the immobilized enzyme are in progress in our laboratories.

V. Enzymic Preparation of L-Tyrosine and L-Dopa

A. Selection of Microorganisms Having High β-Tyrosinase Activity

To select microorganisms which produce β-tyrosinase at high levels, the activity of this enzyme in microorganisms was investigated with 646 strains of bacteria, 140 strains of yeast, 138 strains of fungi, and 117 strains of actinomyces (Enei et al., 1972b). β-Tyrosinase activity was determined by measuring the amount of L-tyrosine or L-dopa synthesized through the β-replacement reaction between L-serine and phenol or pyrocatechol, respectively. Cells of microorganisms were directly added to the reaction mixture as enzyme for practical applications. β-Tyrosinase activity occurred widely in a variety of bacteria, most of which belong to the Enterobacteriaceae; especially to the genera *Escher-*

ichia, Aerobacter, Proteus, and Erwinia. No activity was observed in yeasts, fungi, and actinomyces. Interestingly, a large number of bacterial strains isolated from rice and corn plants showed high β-tyrosinase activity. Cells of Erwinia herbicola ATCC 21434, which showed the highest activity, were selected as a promising source of enzyme for the preparation of L-tyrosine and L-dopa.

B. Culture Conditions for the Preparation of Cells Containing High β-Tyrosinase Activity

1. Effects of Aeration, pH, and Temperature on the Formation of β-Tyrosinase

To establish the optimum culture conditions under which cells of Erwinia herbicola would grow and under which β-tyrosinase would be sufficiently accumulated in growing cells, the effects of aeration, pH, and temperature were investigated (Enei et al., 1973a). Cells were cultivated by varying the volume of the culture medium in a 500-ml shaking flask, under controlled pH and temperatures throughout the culture period.

The optimum volume of the medium for the formation of β-tyrosinase was 40–60 ml in a 500-ml shaking flask, but cell growth was rather repressed in this volume. The optimum pH for enzyme formation was in the range of 7.0–7.5. The pH was adjusted by adding 3 N KOH and 10% acetic acid during cultivation. The optimum temperature for enzyme formation was in the range of 27–32°C.

2. Time Course of the Formation of β-Tyrosinase

Changes in β-tyrosinase activity during cultivation were studied with Erwinia herbicola (Enei et al., 1972b). Cells were grown at 31.5°C for 32 hours with reciprocal shaking, in a medium containing 0.2% tyrosine, 0.2% KH_2PO_4, 0.1% $MgSO_4 \cdot 7H_2O$, 1.0% yeast extract, 0.5% polypeptone, and 0.5% meat extract (pH 7.5). At 4-hour intervals, the tyrosine concentration in the medium, cell growth and the enzyme activity of the cells were determined. The maximum formation of the enzyme was observed in growing cells in the early stationary phase (cultivation for 26–30 hours); thereafter, the enzyme gradually disappeared with the consumption of L-tyrosine that had been added to the medium. It seems that the enzyme may be adaptive in nature and that it is responsible for cell growth on a medium containing L-tyrosine as the sole carbon source (Kumagai et al., 1970d).

3. Effect of Nutrients on the Formation of β-Tyrosinase

To establish a suitable medium for cell growth, as well as for the accumulation of β-tyrosinase in growing cells, the effect of nutrients was studied with *Erwinia herbicola* (Enei et al., 1973a). Cells were cultivated at 28°C for 28 hours in a basal medium consisting of 0.2% L-tyrosine, 0.2% K_2HPO_4, and 0.1% $MgSO_4 \cdot 7H_2O$ (pH 7.5). Various amounts of the nutrients were added to the basal medium.

Additions of yeast extract, meat extract, polypeptone, and the hydrolyzate of soybean protein to the basal medium enhanced cell growth as well as the formation of β-tyrosinase. Of the amino acids tested, DL-methionine, DL-alanine, and glycine stimulated enzyme formation. When these amino acids were added together to the medium, a remarkable increase in enzyme formation was observed. However, L-serine, L-ornithine, and L-cystine rather inhibited enzyme formation. The addition of pyridoxine to the medium enhanced enzyme formation. Since β-tyrosinase is a pyridoxal phosphate-requiring enzyme, it is likely that pyridoxine may be utilized as a precursor of the coenzyme for enzyme formation.

The formation of inducible enzyme is known to be inhibited when the bacteria has an excellent alternative energy source. This is called catabolite repression since enzyme induction is inhibited by catabolic products of the good energy source (Epstein and Beckwith, 1968). On formation of β-tyrosinase by *Erwinia herbicola*, a similar inhibitory effect was observed on adding glucose, pyruvate, and α-ketoglutarate to the medium at high concentrations. Glycerol was a rather suitable carbon source for cell growth as well as for the accumulation of the enzyme in growing cells. A marked increase in enzyme formation was especially observed when glycerol was added together with succinate, fumarate or malate.

Among the metallic ions tested, the Fe^{2+} ion was the only one effective for the formation of β-tyrosinase.

4. Effect of L-Tyrosine and Its Related Amino Acids on the Induction of β-Tyrosinase

β-Tyrosinase has been shown to be an inducible enzyme and the addition of L-tyrosine to the medium to be essential for formation of the enzyme (Kumagai et al., 1970d; Enei et al., 1973a). However, when large amounts of L-tyrosine were added, inhibition of enzyme formation and repression of cell growth were observed. The phenomena seem to be caused by the phenol which is liberated from the L-tyrosine. In fact, phenol added to the medium containing L-tyrosine inhibited enzyme formation as well as cell growth. This inhibition was restored by removal of the phenol from the medium with an ion-exchange resin during cultivation (Enei et al., 1973a).

Various analogs of L-tyrosine were investigated in the search for an inducer of β-tyrosinase which does not decompose during cultivation, one similar to the β-thiogalactosides used to induce β-galactosidase (Monod and Cohn, 1952). Among the analogs tested, D-tyrosine induced this enzyme to almost the same extent as did L-tyrosine. D-Tyrosine is, however, known to be the substrate of β-tyrosinase and was degraded to phenol, pyruvate, and ammonia as was L-tyrosine. o-Methoxyphenyl-L-alanine, 3,4-dimethoxyphenyl-L-alanine, and m-fluorophenyl-L-alanine, which are not substrates for β-tyrosinase, induced β-tyrosinase, but to a much lower degree than did L-tyrosine (Table VI). No effective, non-

TABLE VI
Effect of Tyrosine Analogs as Inducer on the Formation of β-Tyrosinase

L-Tyrosine analog[a]	Cultivation		Activity of L-dopa synthesis (gm/dl/hour)
	Final pH	Growth[b]	
L-Tyrosine	8.35	0.240	0.119
D-Tyrosine	8.35	0.220	0.105
Tyramine	8.20	0.145	0.020
p-Hydroxy-L-phenylglycine	8.30	0.190	0.020
p-Methoxy-L-phenylalanine	8.30	0.195	0.053
p-Methoxy-L-phenylglycine	8.30	0.185	0.017
p-Hydroxycinnamic acid	8.25	0.285	0.035
L-Dopa	8.10	0.360	0.033
3,4-Dimethoxy-L-phenylalanine	8.40	0.280	0.058
L-Phenylalanine	8.35	0.300	0.038
D-Phenylalanine	8.40	0.280	0.025
Phenylpyruvic acid	8.40	0.320	0.035
L-Phenylglycine	8.30	0.280	0.026
Cinnamic acid	7.50	0.000	0.000
m-Fluoro-L-phenylalanine	8.30	0.290	0.062

[a] The analogs were added to the medium at the concentration of 0.2%.
[b] Optical density at 562 nm of the culture broth after it was diluted 26 times.

decomposable inducer was found among the L-tyrosine analogs tested (Enei et al., 1973a).

L-Phenylalanine, D-phenylalanine, and phenylpyruvate did not induce this enzyme by themselves, but they did show a synergistic effect on the induction of the enzyme by L-tyrosine (Table VII) (Enei et al., 1973a). L- and D-phenylalanines have been shown to be competitive inhibitors of β-tyrosinase (Kumagai et al., 1972).

TABLE VII
EFFECT OF SYNERGISTIC MATERIALS ON THE INDUCTION
OF β-TYROSINASE BY L-TYROSINE

Synergistic material[a]	Cultivation		Activity of L-dopa synthesis (gm/dl/hour)
	Final pH	Growth[b]	
—	8.35	0.250	0.125
L-Phenylalanine	8.30	0.390	0.625
D-Phenylalanine	8.25	0.330	0.493
Phenylpyruvic acid	8.40	0.320	0.600
D-Tyrosine	8.05	0.220	0.080
L-Phenylglycine	8.20	0.210	0.095
m-Fluoro-L-phenylalanine	8.30	0.220	0.100
p-Methoxy-L-phenylalanine	8.25	0.230	0.105

[a] The materials were added to the medium at the concentration of 0.1%.
[b] Optical density at 562 nm of the culture broth after it was diluted 26 times.

5. Culture Conditions for the Preparation of Cells

Cells of *Erwinia herbicola* with high β-tyrosinase activity were prepared by growing them at 28°C for 28 hours in a medium containing 0.2% L-tyrosine, 0.2% KH_2PO_4, 0.1% $MgSO_4 \cdot 7H_2O$, 2 ppm Fe^{2+} ($FeSO_4 \cdot 7H_2O$), 0.01% pyridoxine HCl, 0.6% glycerol, 0.5% succinic acid, 0.1% DL-methionine, 0.2% DL-alanine, 0.05% glycine, 0.1% L-phenylalanine, and 12 ml of hydrolyzed soybean protein in 100 ml of tap water, with the pH controlled at 7.5 throughout cultivation (Enei et al., 1973a). Under these conditions, β-tyrosinase is efficiently accumulated in the cells of *Erwinia herbicola* and makes up about 10% of the total soluble cellular protein.

C. REACTION CONDITIONS FOR THE SYNTHESIS OF L-TYROSINE AND L-DOPA

1. Synthesis of L-Tyrosine or L-Dopa from DL-Serine and Phenol or Pyrocatechol

Reaction conditions for the synthesis of L-tyrosine or L-dopa through the β-replacement reaction, respectively, between DL-serine and phenol or pyrocatechol, were studied using cells of *Erwinia herbicola* with high β-tyrosinase activity (Enei et al., 1971, 1972c, 1973b). The optimum pH for this reaction was in the range of 8.0–8.5, and the optimum temperature range was between 37° and 40°C for the synthesis of L-tyrosine and between 22° and 25°C for that of L-dopa.

L-Dopa is known to be rapidly oxidized by atmospheric oxygen under alkaline conditions. Some heavy metals also accelerate this oxidation. To protect the L-dopa synthesized in the reaction mixture from oxidation, reducing and chelating agents were added. The addition of sulfides to the mixture increased the amount of L-dopa synthesized. When EDTA was added to the mixture, together with sulfides, a further increase in the amount of L-dopa was observed.

Phenol or pyrocatechol at high concentrations showed strong substrate inhibition in the β-tyrosinase-catalyzed reactions (Yamada et al., 1972b; Kumagai et al., 1973c). The effect of the concentration of phenol or pyrocatechol and DL-serine, respectively, on the synthesis of L-tyrosine or L-dopa, is shown in Fig. 16. To synthesize L-tyrosine, the optimum

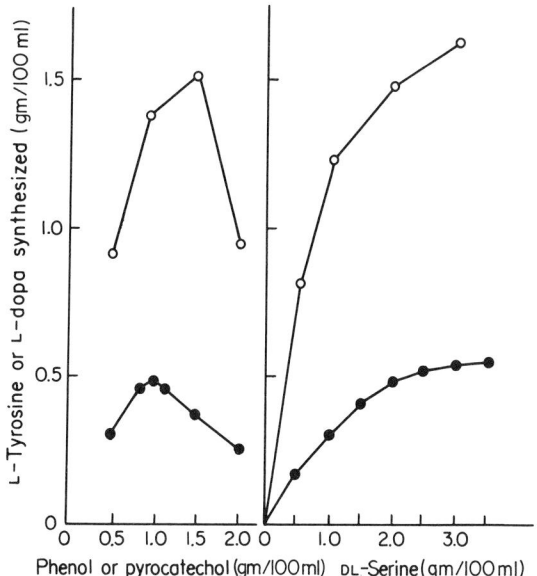

Fig. 16. Effect of substrate concentration on the synthesis of L-tyrosine (O———O) and L-dopa (●———●).

concentration of phenol was around 1.5%, and a sufficient concentration of DL-serine was above 3.0%. The optimum concentration of pyrocatechol in the synthesis of L-dopa was about 1.0%, and a sufficient concentration of DL-serine was around 2.0%.

The maximum yield of L-tyrosine or L-dopa was obtained under the following reaction conditions. The synthesis of L-tyrosine was carried out at 37°C for 16 hours in a reaction mixture containing 4 gm of DL-serine, 1 gm of ammonium acetate, 1 gm of phenol, and cells harvested from 100 ml of cultured broth, in a total volume of 100 ml. The pH of the reaction mixture was controlled at 8.0 by adding ammonia. At 2-hour intervals during incubation, phenol was added to maintain the initial concentration. Under these conditions, 5.35 gm of L-tyrosine was synthesized (Fig. 17). The synthesis of L-dopa was carried out at 22°C for 48 hours in a reaction mixture containing 2 gm of DL-serine, 1 gm of ammonium acetate, 0.7 gm of pyrocatechol, 0.2 gm of sodium sulfite, 0.1 gm of EDTA, and cells harvested from 100 ml of broth in a total volume of 100 ml. The pH of the reaction mixture was controlled at

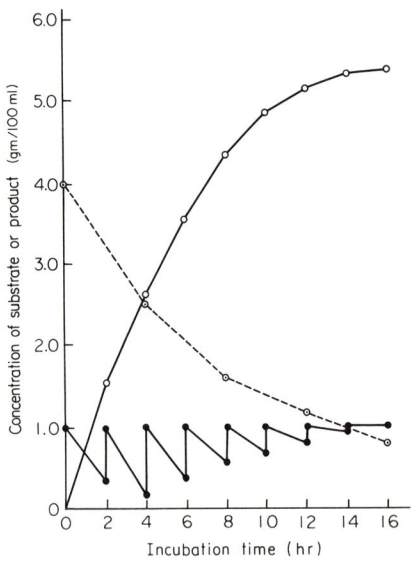

FIG. 17. Synthesis of L-tyrosine (○——○) from DL-serine (⊙ - - - ⊙) and phenol (●——●) by the cells of *Erwinia herbicola*.

8.0 by adding ammonia. At 4-hour intervals, pyrocatechol was added to maintain the initial concentration, then after incubation for 8 and 16 hours, 1 gm of DL-serine was added. Under these conditions, 5.10 gm of L-dopa was synthesized (Fig. 18).

Since the solubilities of L-tyrosine and L-dopa in water were 0.045 gm/100 ml (at 25°C) and 0.5 gm/100 ml (at 20°C), respectively, the L-tyrosine and L-dopa synthesized were precipitated in the reaction mixture (see Figs. 19 and 21).

2. *Synthesis of L-Tyrosine or L-Dopa from Pyruvate, Ammonia, and Phenol or Pyrocatechol*

Reaction conditions for the synthesis of L-tyrosine or L-dopa from pyruvate, ammonia, and phenol or pyrocatechol, respectively, through

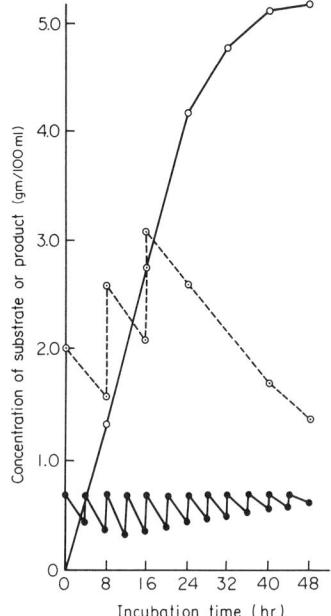

FIG. 18. Synthesis of L-dopa (O——O) from DL-serine (⊙---⊙) and pyrocatechol (●——●) by the cells of *Erwinia herbicola*.

the reversal of the α,β-elimination reaction, were studied using cells of *Erwinia herbicola* with high β-tyrosinase activity (Enei et al., 1972a, 1973c). The optimum pH for this reaction was around 8.0, and the optimum temperature was in the range of 30–37°C for the synthesis of L-tyrosine and at 16°C for that of L-dopa. Since pyruvate was unstable in the reaction mixture at a high temperature, low temperatures were favorable for the synthesis of L-tyrosine and L-dopa.

The effect of pyruvate concentration on the synthesis of L-tyrosine is shown in Table VIII. The reaction was carried out at 37°C for 10

TABLE VIII
EFFECT OF PYRUVATE CONCENTRATION ON THE SYNTHESIS OF L-TYROSINE

Sodium pyruvate added (gm)				Phenol consumed (gm)	L-Tyrosine synthesized (gm)
0	2	4	6		
(hours)					
0.4	0	0	0	3.90	5.80
0.2	0	0.2	0	3.50	6.05
0.1	0.1	0.1	0.1	2.33	3.50

hours in a reaction mixture containing varying amounts of sodium pyruvate, 5 gm of ammonium acetate, 1.0 gm of phenol, and cells harvested from 100 ml of cultured broth, in a total volume of 100 ml. The pH of the mixture was adjusted to 8.0 by the addition of ammonia. The concentration of phenol was maintained at 1.0% by adding phenol at 2-hour intervals during incubation. The maximum synthesis of L-tyrosine (6.05 gm/100 ml) was obtained when 2 gm of sodium pyruvate was added twice to the reaction mixture, at 0 and 4 hours after incubation (Fig. 19).

FIG. 19. Synthesis of L-tyrosine from pyruvate, ammonia, and phenol by the cells of *Erwinia herbicola*. The photograph was taken after incubation for 10 hours.

The effect of pyruvate concentration on L-dopa synthesis is shown in Fig. 20. The reaction was carried out at 16°C for 48 hours in a reaction mixture containing various amounts of sodium pyruvate, 5 gm of ammonium acetate, 0.6 gm of pyrocatechol, 0.2 gm of sodium sulfite, 0.1 gm of EDTA, and cells harvested from 100 ml of broth, in a total volume of 100 ml. The pH was adjusted to 8.0 by the addition of ammonia. At 2-hour intervals, sodium pyruvate and pyrocatechol were added to the reaction mixture to maintain the initial concentrations. The maximum synthesis of L-dopa (5.85 gm/100 ml) was obtained when the concentration of sodium pyruvate was kept at 0.5% (Fig. 21).

When a higher concentration of pyruvate was added to the reaction mixture during the synthesis of L-dopa, two by-products were formed (Fig. 20) (Enei *et al.*, 1973c). These by-products, X-1 and X-2, were isolated from a large-scale reaction mixture by ion-exchange resins, and

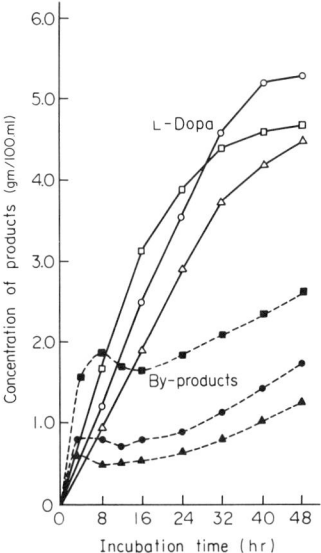

FIG. 20. Effect of pyruvate concentration on the synthesis of L-dopa and by-products by the cells of *Erwinia herbicola*. Sodium pyruvate was added to maintain the initial concentrations of 0.35 (△), 0.5 (○), and 0.8 gm (□) per 100 ml.

FIG. 21. Synthesis of L-dopa from pyruvate, ammonia, and pyrocatechol by the cells of *Erwinia herbicola*. The photograph was taken after the incubations of 0 (A) and 48 hours (B).

were identified physicochemically as 1-methyl-1,3-dicarboxy-6,7-dihydroxy-1,2,3,4-tetrahydroisoquinoline (Daxenbichler *et al.*, 1972) and 1-methyl-3-carboxy-6,7-dihydroxy-3,4-dihydroisoquinoline, respectively. The by-products were shown to be formed through a spontaneous condensation reaction (Pictet and Spengler, 1911) of the synthesized L-dopa with excess pyruvate, followed by a decarboxylation reaction (Fig. 22). Formation of the by-products in the reaction mixture increased at high concentrations of pyruvate, at high incubation temperatures and under more alkaline conditions.

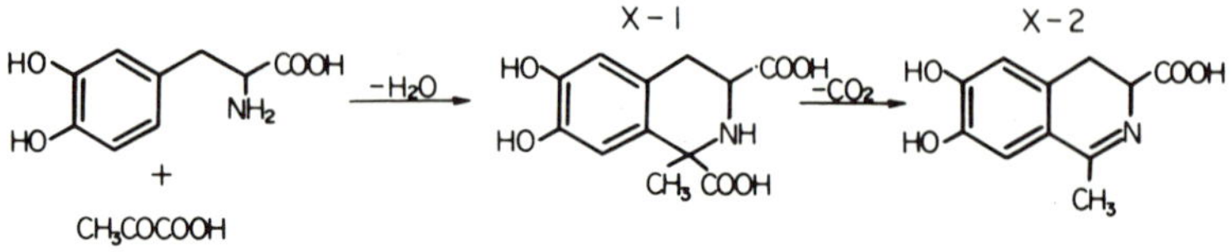

FIG. 22. Possible mechanism of the formation of by-products X-1 and X-2.

3. Synthesis of L-Dopa from Pyruvate, Ammonia, and Pyrocatechol–Borate Complex

Pyrocatechol at high concentrations showed strong substrate inhibition for the synthesis of L-dopa. When boric acid was added to this reaction mixture, inhibition was considerably restored. Boric acid is known to form a complex with pyrocatechol, thus the use of this complex for the synthesis of L-dopa was examined with no feeding of pyrocatechol and sodium pyruvate (Enei et al., 1973c).

The effect of boric acid on the synthesis of L-dopa is shown in Table IX. The reaction was carried out at 20°C for 22 hours in a reaction

TABLE IX
EFFECT OF BORIC ACID ON THE SYNTHESIS OF L-DOPA

Boric acid added (gm)	Molar ratio[a]	Synthesis of	
		L-Dopa (gm)	By-products (gm)
1.76	0.8	2.48	0.54
2.20	1.0	3.30	0.46
2.64	1.2	2.13	0.18

[a] Molar ratio of boric acid to pyrocatechol.

mixture containing boric acid at various concentrations, 4 gm of pyrocatechol, 3 gm of sodium pyruvate, 3 gm of ammonium acetate, 0.2 gm of sodium sulfite, and 0.1 gm of EDTA in a total volume of 100 ml (pH adjusted to 8.0 by ammonia). Cells harvested from 100 ml of broth were added to the mixture after the pyrocatechol had been mixed with boric acid to form the complex. The maximum synthesis of L-dopa (3.30 gm/100 ml) was obtained when pyrocatechol and boric acid were added to the reaction mixture at the molar ratio of one.

VI. Conclusions

β-Tyrosinase was shown to catalyze a variety of α,β-elimination, β-replacement, and racemization reactions and the synthesis of L-tyrosine and its related amino acids. The mechanisms for these reactions have been studied in some detail, by adopting the general mechanism proposed for pyridoxal-dependent reactions. The enzyme-bonud α-aminoacrylate is the key intermediate for these reactions.

An enzymic method for preparing L-tyrosine and its related amino acids was developed using bacterial cells with high β-tyrosinase activity. This method is simple and is one of the most economical processes to date for preparing certain L-tyrosine related amino acids, e.g., L-dopa, from the starting materials: DL-serine, pyruvic acid, and appropriate derivatives of phenol, such as pyrocatechol.

This enzymic method can also be applied to the synthesis of L-tryptophan (Nakazawa et al., 1972a,b; Yamada et al., 1974a,b) from pyruvic acid, ammonia, and indole, using bacterial cells with high tryptophanase activity.

Acknowledgments

We are deeply indebted to Professor Emeritus K. Ichihara, who discovered β-tyrosinase in 1952 at his laboratory in the Department of Biochemistry, Osaka University, Osaka. We wish to thank Professor K. Ogata of the Department of Agricultural Chemistry, Kyoto University, Kyoto for his interest and encouragement during the course of this work. We also thank Professors H. Fukami and S. Fukui and Dr. T. Ueno of Kyoto University, Kyoto; Drs. S. Okumura and H. Enei of the Central Research Laboratories, Ajinomoto Co., Ltd., Kawasaki; Professor R. K. Hill of the University of Georgia, Athens, Georgia: Associate Professor S. Sawada of Kyoto University of Education, for their splendid collaboration on all phases of this work.

References

Axén, R., and Ernbach, S. (1971). *Eur. J. Biochem.* 18, 351–360.
Braunstein, A. E., and Shemyakin, M. M. (1953). *Biokhimiya* 18, 393–411.
Brot, N., Smit, Z., and Weissbach, H. (1965). *Arch. Biochem. Biophys.* 112, 1–6.
Cleland, W. W. (1963). *Biochim. Biophys. Acta* 67, 104–137.
Cleland, W. W. (1970). In "The Enzymes" (P. D. Boyer, ed.), 3rd ed., Vol. 2, pp. 13–18. Academic Press, New York.
Daxenbichler, M. E., Kleiman, R., Weisleder, D., VanEtten, C. H., and Carlson, K. E. (1972). *Tetrahedron Lett.* 18, 1801–1802.
Dempsey, W. B., and Snell, E. E. (1963). *Biochemistry* 2, 1414–1419.

Dinelli, D. (1972). *Process Biochem.* **7**, 9–12.
Enei, H., Matsui, H., Okumura, S., and Yamada, H. (1971). *Biochem. Biophys. Res. Commun.* **43**, 1345–1349.
Enei, H., Nakazawa, H., Matsui, H., Okumura, S., and Yamada, H. (1972a). *FEBS (Fed. Eur. Biochem. Soc.) Lett.* **21**, 39–41.
Enei, H., Matsui, H., Yamashita, K., Okumura, S., and Yamada, H. (1972b). *Agr. Biol. Chem.* **36**, 1861–1868.
Enei, H., Matsui, H., Okumura, S., and Yamada, H. (1972c). *Agr. Biol. Chem.* **36**, 1869–1876.
Enei, H., Yamashita, K., Okumura, S., and Yamada, H. (1973a). *Agr. Biol. Chem.* **37**, 485–492.
Enei, H., Matsui, H., Nakazawa, H., Okumura, S., and Yamada, H. (1973b). *Agr. Biol. Chem.* **37**, 493–499.
Enei, H., Nakazawa, H., Okumura, S., and Yamada, H. (1973c). *Agr. Biol. Chem.* **37**, 725–735.
Epstein, W., and Beckwith, J. (1968). *Annu. Rev. Biochem.* **37**, 411–436.
Fisher, E. H., Kent, A. B., Snyder, E. R., and Krebs, E. G. (1958). *J. Amer. Chem. Soc.* **80**, 2906–2907.
Fonda, M. L. (1971). *J. Biol. Chem.* **246**, 2230–2240.
Fuganti, C., Ghiringhelli, D., Giangrasso, D., Grasselli, P., and Santopietro, A. (1974). *Pap., 9th IUPAC Symp. Chem. Natur. Prod.*
Fukui, S., Ohishi, N., and Shimizu, S. (1969a). *Arch. Biochem. Biophys.* **130**, 584–593.
Fukui, S., Ohishi, N., Nakai, Y., and Masugi, F. (1969b). *Arch. Biochem. Biophys.* **132**, 1–7.
Fukui, S., Ikeda, S., Fujimura, M., Yamada, H., and Kumagai, H. (1975a). *Eur. J. Biochem.* **51**, 155–165.
Fukui, S., Ikeda, S., Fujimura, M., Yamada, H., and Kumagai, H. (1975b). *Eur. J. Appl. Microbiol.* **1**, 25–39.
Fukui, S., Ohishi, N., and Shimizu, S. (1970a). In "Methods in Enzymology" (D. B. McCormick and L. D. Wright, eds.), Vol. 18A, pp. 598–603. Academic Press, New York.
Fukui, S., Nakai, Y., and Masugi, F. (1970b). In "Methods in Enzymology" (D. B. McCormick and L. D. Wright, eds.), Vol. 18A, pp. 603–606. Academic Press, New York.
Hughes, R. C., Jenkins, W. T., and Fisher, E. H. (1962). *Proc. Nat. Acad. Sci. U.S.* **48**, 1615–1618.
Ichihara, K., Yoshimatsu, H., and Sakamoto, Y. (1956). *J. Biochem. (Tokyo)* **43**, 803–810.
Ikeda, S., and Fukui, S. (1973). *Biochem. Biophys. Res. Commun.* **52**, 482–488.
Ivanov, V. I., and Karpeisky, M. Ya. (1969). *Advan. Enzymol.* **32**, 21–53.
Jenkins, W. T. (1961a). *J. Biol. Chem.* **236**, 1121–1125.
Jenkins, W. T. (1961b). *Fed. Proc., Fed. Amer. Soc. Exp. Biol.* **20**, 978–981.
Jenkins, W. T., and Sizer, I. W. (1957). *J. Amer. Chem. Soc.* **79**, 2655–2656.
Kakihara, Y., and Ichihara, K. (1953). *Med. J. Osaka Univ.* **3**, 497–507.
Katchalski, E., Silman, I., Goldman, R. (1971). *Advan. Enzymol.* **34**, 445–536.
Kirby, G. W., and Michael, J. (1971). *Chem. Commun.* **187**, 415–416.
Kumagai, H., Matsui, H., Ohkishi, H., Ogata, K., Yamada, H., Ueno, T., and Fukami, H. (1969). *Biochem. Biophys. Res. Commun.* **34**, 266–270.
Kumagai, H., Yamada, H., Matsui, H., Ohkishi, H., and Ogata, K. (1970a). *J. Biol. Chem.* **245**, 1767–1772.

Kumagai, H., Yamada, H., Matsui, H., Ohkishi, H., and Ogata, K. (1970b). *J. Biol. Chem.* **245**, 1773–1777.
Kumagai, H., Kashima, N., and Yamada, H. (1970c). *Biochem. Biophys. Res. Commun.* **39**, 796–801.
Kumagai, H., Matsui, H., and Yamada, H. (1970d). *Agr. Biol. Chem.* **34**, 1259–1261.
Kumagai, H., Kashima, N., Yamada, H., Enei, H., and Okumura, S. (1972). *Agr. Biol. Chem.* **36**, 472–482.
Kumagai, H., Ohkishi, H., Kashima, N., and Yamada, H. (1973a). *Amino Acid Nucl. Acid (Tokyo)* **27**, 11–18.
Kumagai, H., Yamada, H., Masugi, F., and Fukui, S. (1973b). *Biochim. Biophys. Acta* **327**, 510–514.
Kumagai, H., Utagawa, T., and Yamada, H. (1973c). *Pap., Annu. Meet. Jap. Biochem. Soc.*
Kumagai, H., Utagawa, T., and Yamada, H. (1974a). *J. Biol. Chem.* **250**, 1661–1667.
Kumagai, H., Utagawa, T., and Yamada, H. (1974b). *Pap., Annu. Meet. Jap. Biochem. Soc.*
Kumagai, H., Kitamura, S., Yamada, H., Ueno, T., and Fukami, H. (1975). *Pap., Meet. Agric. Chem. Soc. Jap., Kyoto*, February.
Lambooy, J. P. (1954). *J. Amer. Chem. Soc.* **76**, 133–138.
Masugi, F., Natori, Y., Shimizu, S., and Fukui, S. (1973). *J. Nutr. Sci. Vitaminol.* **19**, 55–70.
Melchior, W. B., and Fahrney, D. (1970). *Biochemistry* **9**, 251–258.
Metzler, D. E. (1957). *J. Amer. Chem. Soc.* **79**, 485–490.
Metzler, D. E., Ikawa, M., and Snell, E. E. (1954). *J. Amer. Chem. Soc.* **76**, 648–652.
Miles, E. W., and Kumagai, H. (1974). *J. Biol. Chem.* **249**, 2483–2491.
Monod, J., and Cohn, M. (1952). *Advan. Enzymol.* **3**, 67–119.
Morino, Y., and Snell, E. E. (1967a). *J. Biol. Chem.* **242**, 2793–2799.
Morino, Y., and Snell, E. E. (1967b). *J. Biol. Chem.* **242**, 2800–2809.
Morino, Y., and Snell, E. E. (1967c). *Proc. Nat. Acad. Sci. U.S.* **57**, 1692–1699.
Mühlard, A., Higyi, G., and Toth, G. (1967). *Acta Biochim. Biophys. Acad. Sci. Hung.* **2**, 19–29.
Nakazawa, H., Enei, H., Okumura, S., and Yamada, H. (1972a). *FEBS (Fed. Eur. Biochem. Soc.) Lett.* **25**, 43–45.
Nakazawa, H., Enei, H., Okumura, S., and Yamada, H. (1972b). *Agr. Biol. Chem.* **36**, 2523–2528.
Ovadi, J., Libor, S., and Elüdi, P. (1967). *Acta Biochim. Biophys.* **2**, 455–458.
Peterson, D. L., and Martinez-Carrion, M. (1970). *J. Biol. Chem.* **245**, 806–813.
Pictet, J., and Spengler, W. (1911). *Ber. Deut. Chem. Ges.* **44**, 2030–2036.
Rapp, P., Kumagai, H., Yamada, H., Ueno, T., and Fukami, H. (1974). *Pap., Meet. Agr. Chem. Soc. Jap.*
Sawada, S., Kumagai, H., Yamada, H., Hill, R. K., Mugibayashi, Y., and Ogata, K. (1973). *Biochim. Biophys. Acta* **315**, 204–207.
Sawada, S., Kumagai, H., Yamada, H., and Hill, R. K. (1974). *J. Amer. Chem. Soc.* (in press).
Schirch, L., and Mason, M. (1962). *J. Biol. Chem.* **238**, 1032–1037.
Shukuya, R., and Schwert, G. W. (1960). *J. Biol. Chem.* **235**, 1653–1657.
Skye, G. E., Potts, R., and Floss, G. H. (1974). *J. Amer. Chem. Soc.* **96**, 1593–1595.
Thanassi, J. W., Butler, A. R., and Bruce, T. C. (1965). *Biochemistry* **4**, 1463–1472.
Uchida, M., Taketomo, Y., Kakihara, Y., and Ichihara, K. (1953). *Med. J. Osaka Univ.* **3**, 509–519.

Ueno, T., Fukami, H., Ohkishi, H., Kumagai, H., and Yamada, H. (1970). *Biochim. Biophys. Acta* **206**, 476–479.
Watanabe, T., and Snell, E. E. (1972). *Proc. Nat. Acad. Sci. U.S.* **69**, 1086–1090.
Yamada, H., Kumagai, H., Enei, H., Matsui, H., and Okumura, S. (1972a). *Proc. IFS, Ferment. Technol. Today, 4th, 1972* pp. 445–454.
Yamada, H., Kumagai, H., Kashima, N., Torii, H., Enei, H., and Okumura, S. (1972b). *Biochem. Biophys. Res. Commun.* **46**, 796–801.
Yamada, H., Nakazawa, H., Yoshida, H., and Kumagai, H. (1974a). *Pap., Intersect. Congr. IAMS, 1st.*
Yamada, H., Yoshida, H., Nakazawa, H., and Kumagai, H. (1974b). *Acta Vitaminol. Enzymol.* (in press).
Yoshida, H., Kumagai, H., and Yamada, H. (1974). *Agr. Biol. Chem.* **38**, 463–464.
Yoshimatsu, H. (1957). *Med. J. Osaka Univ.* **9**, 727–730.

Effects of Toxicants on the Morphology and Fine Structure of Fungi

Donald V. Richmond

Long Ashton Research Station, University of Bristol, Bristol, England

I. Introduction 289
II. Morphological Changes Induced by Toxicants 291
   A. Early Reports 291
   B. Effects of Systemic Fungicides 292
   C. Effects of Chlorinated Nitrobenzenes and Related Compounds 295
   D. Effects of Antibiotics 297
III. Effects on Dimorphic Fungi 301
   A. Animal Pathogens 302
   B. Plant Pathogens 303
   C. *Mucor* 304
IV. Effects of Toxicants on Fine Structure 306
   A. Captan 308
   B. Benomyl 309
   C. 2-Aminobutane 309
   D. Oxycarboxin 311
   E. Griseofulvin 313
   F. Cycloheximide 314
   G. Chloramphenicol 314
   H. Ethidium Bromide 314
   I. Other Compounds 315
References 316

I. Introduction

The fungi are so different from other organisms in their nutrition, vegetative structure, and cell wall composition that they are frequently classed in a kingdom or subkingdom of their own. They feed entirely by absorption, not by photosynthesis or ingestion (Whittaker, 1969), their hyphae are coenocytic, and the vast majority have cell walls composed of chitin and glucan (Bartnicki-Garcia, 1968a). The higher fungi and some lower fungi differ from green plants in synthesizing lysine by a pathway in which α-aminoadipic acid is a key intermediate. Plants, bacteria, and blue-green algae all synthesize lysine by a pathway involving α,ϵ-diaminopimelic acid (DAP); animals are unable to synthesize lysine (Vogel *et al.*, 1970). The pathways of lysine biosynthesis are closely connected with the presence of cellulose or chitin in the cell walls; the Oomycetes, which have cells walls containing cellulose, use the DAP pathway (Lé John, 1971).

Despite the features fungi have in common with one another, they are an extremely diverse group. Some workers so emphasize the differences between the major groups of fungi as to suggest that they are essentially unrelated (Cain, 1972). Ultrastructural evidence does, however, imply that the differences between fungi and plants or animals are much greater than the differences between groups within the fungi (Moore, 1971).

The fungi are an extremely varied and successful, actively evolving, group able to live in a wide range of habitats from fresh and marine waters to extreme xerophytic conditions; when in association with algae in the lichens they are the first colonizers of bare rocks. In fact, fungi are able to grow on practically all organic materials provided the temperature and humidity are suitable. Under tropical conditions they can cause enormous damage to all kinds of equipment and apparatus. Fungi are even able to grow in sterile media containing inorganic salts without the addition of any organic compounds (Mirocha and DeVay, 1971; Tribe and Mabadeje, 1972).

The success of the fungi is partly due to their genetic adaptability and the great variety of life histories and sexual processes they have evolved; parasexuality and heterokaryosis are widespread and are the only means of genetic recombination in the Fungi Imperfecti, which have no sexual stage. As most fungi, except Oomycetes, are haploid in the vegetative stage, harmful mutations are immediately expressed and eliminated (Raper, 1968). Some species of fungi, including important plant and animal pathogens, can exist either in yeastlike or in mycelial form. The mycelial form of animal pathogens is usually the saprophytic phase, and the yeastlike form occurs in the tissue of the host. The conversion from the mycelial form into the yeastlike form is induced by increase in temperature, anaerobic conditions, sulfur compounds, metal ions, and chelating complexes (Pine, 1962).

The versatility and adaptability of fungi is also expressed by their enormous capacity for individual variation. The visual appearance of fungal colonies and the microscopic morphology of their hyphae can be profoundly altered by changes in the composition of the media on which they grow. The production of special vegetative or reproductive structures, the initiation and abundance of fruit bodies as well as the presence of pigmentation are frequently dependent on nutritional factors, such as carbon or nitrogen source, concentration of nutrients, and pH of the medium. Sexual reproduction is regulated by specific endogenous hormones. Fungal growth and spore production can be modified by volatile organic compounds (Norrman, 1968), and volatile microbial metabolites can produce abnormalities in growing fungi (Hutchinson, 1973; Fries, 1973; Moore-Landecker and Stotzky, 1973). Giant cells may be produced by nutritional changes or by products of metabolism (Burguillo et al., 1972; Wildman, 1966).

Fungicides, antibiotics, and metabolic inhibitors can also, at suitable concentrations, modify the growth and development of fungi. Some compounds act on germinating spores to produce swollen and distorted germ tubes, others change the morphology of growing hypae or convert one form of a dimorphic fungus into the other. Compounds which neither kill the fungus nor produce any obvious visual effect may induce changes detectable at the ultrastructural level. Much information is now available concerning the mechanism of antibiotic action, and antibiotics have played a major role in the elucidation of the biochemistry of many fundamental cellular processes. Despite recent advances in knowledge much less is known about the mode of action of fungicides. The main purpose of this review is to consider the contribution of morphological and ultrastructural studies to our understanding of the mechanisms of fungicidal action. Some reference is made in this review to the effects of toxicants other than fungicides, but less attention is paid here to the many studies in which antibiotics, with known modes of action, have been used to study cellular processes. Nutritional or hormonal effects are mentioned only in so far as they are necessary for an understanding of toxicant action.

II. Morphological Changes Induced by Toxicants

A. Early Reports

When a dilute suspension of fungal spores is incubated with the necessary nutrients in the presence of a suitable concentration of a protective fungicide, such as captan [N-(trichloromethylthio)-3a,4,7,7a-tetrahydrophthalimide] or dodine (dodecylguanidine acetate), individual spores are either prevented from germinating altogether or they germinate normally although germ tubes may be shorter. This suggests that these compounds are acting directly to inhibit spore germination. Some antibiotics and a few of the newer systemic fungicides have, however, little effect on spore germination but are highly active in preventing hyphal growth or cell division. When spores are incubated in the presence of some of these compounds the spores germinate but the germ tubes produced are often highly distorted.

Brian et al. (1946) described the striking effects they observed when spores of *Botrytis allii* germinated in the presence of a substance, later known as griseofulvin, which was produced by *Penicillium janczewskii*. After treatment with the antibiotic, germ tubes were stunted, hyphae were distorted and excessively branched, and, at lower concentrations, the hyphae were curled or wave shaped. Later Vincent (1947) reported, independently that very similar effects occurred when hyphae of *Asper-*

gillus niger, Penicillium roquefortii, and *Byssochlamys fulva* grew in the presence of *p*-hydroxybenzoic acid methyl ester. The higher homologs were less active than the methyl ester. Griseofulvin is much more active than the compounds described by Vincent and can be detected at concentrations as low as 6 μM while *p*-hydroxybenzoic acid methyl ester is detected at a concentration of 300 μM.

Many phenolic compounds also produce distorted hyphae at relatively high concentrations. Armitage and Verdcourt (1948) found that *Aspergillus niger* formed abnormal hyphae when grown on agar containing 25 μg of sodium pentachlorophenate per milliliter. At 30 $\mu g/ml$ spores were converted into giant cells. Giant cells were also produced by thymol at 180$\mu g/ml$, 2:4 dichlorophenol at 50 $\mu g/ml$, and 2-naphthol at 80$\mu g/ml$ (Verdcourt, 1952). More recently Strzelczyk (1968) found that low concentrations of phenyl mercury acetate in the vapor phase produced swollen spores and distorted germ tubes in *Trichothecium roseum* and *Penicillium frequentans*. Pentachlorophenol produced similar effects in *T. roseum*. The early reports of compounds producing distorted growth in fungi were summarized and discussed by Horsfall (1956) and Cochrane (1958). Horsfall (1956) considered that compounds which produced distorted germ tubes acted by inhibiting mitosis; the fungus was able to grow, but cell division could not take place and hence the hyphae became gnarled and distorted to accommodate the excess material. Although not concerned with germination it may be worth mentioning here the results of Loveless *et al.* (1954), who examined the effects of a number of compounds on the growth and cell division of *Saccharomyces cerevisiae* and *Escherichia coli*. Although 42 compounds inhibited both growth and cell division of *S. cerevisiae,* only two compounds [methylbis-(β-chloroethyl) amine hydrochloride and 2,4,6-tris(ethyleneimino)-1,3,5-*s*-triazine] had a specific effect on cell division.

Most of the early reports of fungal distortions were concerned with the effects produced by relatively high concentrations of nonspecific toxicants. The abnormal growth patterns induced by low concentrations of fungicides or antibiotics are of greater interest, as they are more likely to express direct responses of organisms to interference with specific metabolic pathways.

B. Effects of Systemic Fungicides

Until comparatively recently most fungicides in commercial use acted as protectants and were of relatively low toxicity compared with other biologically active compounds (Somers, 1962). In the last 10 years a number of highly active systemic fungicides has been introduced partly as a result of changes in the screening techniques used by chemical manufacturers. These new fungicides which have specific modes of action have had dramatic success in the control of plant diseases (Erwin, 1969). Many of these new compounds do not inhibit spore germination but induce morphological effects in growing hyphae.

1. *Benomyl and Related Compounds*

Benomyl [methyl-(1-butylcarbamoyl)benzimidazol-2-ylcarbamate] is a fungicide which controls a wide range of fungal diseases. The compound breaks down in aqueous solution to form carbendazim(methyl benzimidazole-2-ylcarbamate) (MBC) a stable compound as toxic to most fungi as the parent compound. When conidia of *Neurospora crassa* were incubated in a growth medium containing MBC (1 μg/ml), they developed enlarged and misshapen germ tubes (Clemons and Sisler, 1971). Similar distortions were formed in germ tubes of *Mycosphaerella musicola* (Stover, 1969) and *Cercospora beticola* (Solel, 1970). When conidia of *Botrytis fabae* germinated in the presence of benomyl (0.1–0.3 μg/ml), the germ tubes were swollen and distorted and they branched more frequently than did untreated controls. Multiple germ tube emergence also occurred to a greater extent than in untreated conidia (Richmond and Pring, 1971a).

The related compound thiabendazole [2-(4′thiazolyl)benzimidazole] (TBZ) produced similar effects in *Botrytis cinerea* (Fig. 1b), *Botrytis fabae* and *Neurospora crassa* (Gottlieb and Kumar, 1970). Another systemic fungicide, thiophanate [1,2-di-(3-ethoxycarbonyl-2-thioureido)benzene] is readily converted into ethyl benzimidazole-2-yl carbamate, the ethyl analog of MBC, and hence probably has the same mode of action as benomyl (Vonk and Kaars Sijpesteijn, 1971). Thiophanate produced similar distortions in *B. allii* (Fig. 1d) to those produced in *B. cinerea* by MBC (Fig. 1c) and benomyl (Fig. 1e). Figure 1a shows control spores of *B. cinerea* germinating under the same conditions but in the absence of fungicides.

All the benzimidazole and related fungicides probably have similar modes of action since strains of fungi resistant to one compound are usually resistant to other derivatives (Bartels-Schooley and MacNeill, 1971; Bollen, 1971). When MBC was added to germinating conidia of *Aspergillus nidulans,* the increase in the number of nuclei in each germ tube was inhibited at once while the increase in DNA content was inhibited progressively until 100% inhibition was achieved after 8 hours (Davidse, 1973). Hammerschlag and Sisler (1973) studied the effect of MBC on synchronous cultures of *Ustilago maydis* and *Saccharomyces cerevisiae*. Mitosis and cytokinesis stopped when MBC was added to sporidia of *U. maydis* even though they had synthesized the DNA

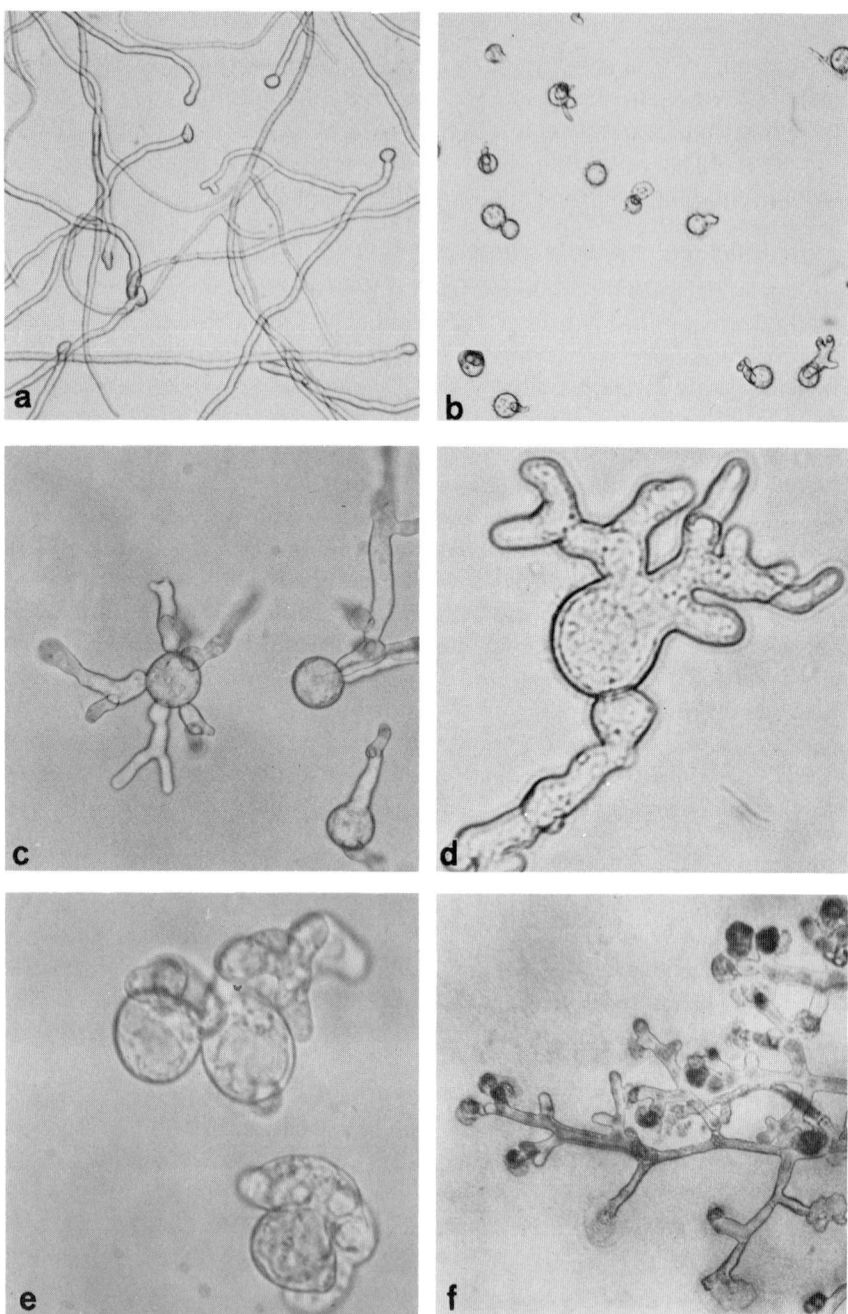

FIG. 1. (a) Conidia of *Botrytis cinerea* germinating in growth medium without added fungicide. (b) Conidia of *B. cinerea* germinating in thiabendazole (40 µg/ml). (c) Conidia of *B. cinerea* germinating in carbendazim (10 µg/ml). (d) Conidia of *B. allii* germinating in thiophanate (20 µg/ml). (e) Conidia of *B. cinerea* germinating in benomyl (10 µg/ml). (f) *Actinomucor elegans* growing on malt agar containing dicloran (4 µg/ml). Note lysis of hyphal tips. a, ×130; b, ×140; c, f, ×300; d, ×870; e, ×650.

required for replication before the addition of toxicant. Synthesis of DNA preceding the first cell division of *S. cerevisiae* was not inhibited by MBC, but mitosis and cytokinesis were prevented. Both Davidse (1973) and Hammerschlag and Sisler (1973) considered that MBC acted by inhibiting mitosis, probably by interfering with the mitotic spindle.

2. *Triarmol and Triforine*

Triarimol [α-(2,4-dichlorophenyl)-α-phenyl-5-pyrimidinemethanol] is a systemic fungicide which is effective against a range of plant pathogens including powdery mildews (Hickey and Drake, 1972). The compound had no effect on the germination of conidia of *Erysiphe graminis* on wheat or of *E. polygoni* on bean and haustoria were produced as usual in both plants but all further fungal growth was completely stopped (Brown and Hall, 1971). Elongation of the germ tubes of *Cladosporium cucumerinum* and *Aspergillus niger* was also inhibited (Houseworth et al., 1971). Sporidia of *Ustilago maydis* continued to grow in the presence of triarimol, but multiplication ceased. Instead, enlarged, branched, multicelled sporidia were formed. Each cell contained a nucleus, and hence mitosis must have continued after multiplication had stopped (Ragsdale and Sisler, 1973). Another systemic fungicide triforine [1,4-di-(2,2,2-trichloro-1-formamidoethyl)piperazine] although structurally unrelated to triarimol appears to have a similar mode of action (Sherald et al., 1973). The two compounds had similar effects on germinating conidia of *C. cucumerinum* and *Neurospora crassa*. After 28 hours, untreated spores had developed an extensive meshwork of hyphae while spores treated with triarimol or triforine had abnormally swollen germ tubes, usually no more than 1–3 cells in length. Mutants of *C. cucumerinum* resistant to triarimol were also resistant to triforine. Sherald et al. (1973) considered the primary site of action of both compounds to be inhibition of the biosynthesis of ergosterol.

The two groups of compounds discussed above produce malformations in germ tubes but do not inhibit spore germination, Koch (1971) suggested that the presence of deformations in germ tubes in normal spore germination tests could give useful indications of systemic activity. Tests on growing plants require much labor, time, and expense whereas a spore germination test is complete in 1 day. Out of 11 commercial systemic fungicides he examined, only 1 inhibited spore germination but 6 produced distorted germ tubes.

C. Effects of Chlorinated Nitrobenzenes and Related Compounds

Fungicides such as pentachloronitrobenzene (PCNB), dicloran (2,6-dichloro-4-nitroaniline) and diphenyl were included by Georgopoulos and Zaracovitis (1967) in the "aromatic hydrocarbon group." Fungicides in the group probably have similar modes of action since strains resistant to one compound frequently show cross resistance to other fungicides in the group (Priest and Wood, 1961).

Although these fungicides have no systemic properties they resemble systemic fungicides in that resistant strains arise quite readily. In fact tolerant sectors arise easily after exposure to the fungicide in the laboratory (Georgopoulos and Zaracovitis, 1967). While systemic fungicides can exert a selective pressure on the pathogen by translocation within the host plant, fungicides in this group may exert a similar pressure by their activity in the vapor phase.

Diphenyl is used for controlling citrus green mold caused by *Penicillium digitatum*. When germinating in the presence of diphenyl vapor, spores of *P. digitatum* form distorted hyphae and develop giant cells (Ramsey et al., 1944). Isolates resistant to diphenyl germinate normally. Hence germination tests *in vitro* can be used to predict sensitivity to disease control by diphenyl (Duran, 1962). Dicloran and tecnazene (2,3,5,6-tetrachloronitrobenzene) are two fungicides frequently used for the control of diseases caused by various species of *Botrytis* (Higgons, 1961).

Neither compound prevented the germination of spores of *Botrytis cinerea* in water, but germination was inhibited in the presence of agar. Germ tubes from spores grown in agar containing either of the fungicides at sublethal concentrations became swollen and branched, local lysis of hyphal tips occurred followed by release of the cell contents (Esuruoso et al., 1968). Dicloran probably acts by inhibiting some stage of protein synthesis (Weber and Ogawa, 1965). Dicloran produces similar effects in *Actinomucor elegans* growing on agar: lysis occurs at the hyphal tips (Fig. 1f).

Pentachloronitrobenzene (PCNB) produced striking cytological changes in hyphae of *Neurospora crassa* (Macris and Georgopoulos, 1973) and *Aspergillus nidulans* (Threlfall, 1972). On solid medium in the presence of PCNB hyphae of *A. nidulans* had thicker cell walls, branched less, and had fewer nuclei in each cell than untreated controls. Hyphal growth was also severely restricted.

When growth in liquid culture *A. nidulans* normally secretes an extracellular polysaccharide, and hence the culture filtrate becomes very viscous. In the presence of PCNB the production of extracellular polysaccharide was considerably reduced.

The mycelium of *N. crassa* formed compact colonies consisting of much-branched distorted hyphae when grown in liquid medium in the presence of PCNB. The morphological effects, such as branching at the hyphal tip, were similar to those produced by L-sorbose (DE Terra and Tatum, 1961, 1963). Treatment with PCNB had marked effects on the composition of the *N. crassa* cell wall. In the presence of PCNB the walls contained 49% less glucosamine and 15% more glucose than control walls. These results contrast with those obtained for *A. nidulans* where there was an increase in hexosamine and reduction in hexose in the presence of PCNB. The morphology of PCNB-resistant strains of *N. crassa* and *A. nidulans* was unaffected by the toxicant. Resistant strains of both organisms had less hexosamine than wild-type strains. The reduction of hexosamine in resistant strains is of interest since chloronitrobenzenes are effective only against those fungal species that have chitin in their cell walls (Reavill, 1954).

D. Effects of Antibiotics

1. General

Considerable attention has been paid to the effects of antibiotics on the growth and morphogenesis of fungi. Bekker et al. (1971) investigated the effects of a range of antibiotics on the proliferation of conidiophores of *Aspergillus fumigatus*. Streptomycin increased proliferation because the cells grew but were unable to divide, while griseofulvin decreased proliferation as the cells divided but could not grow. Aurantin and antitumor compounds caused reduced proliferation at low concentrations but increased proliferation at high concentrations. Nystatin, erythromycin and chloramphenicol had no effect, Bekker et al. (1971) suggested that the effects were due to shifts in metabolism toward synthesis of either DNA or RNA. Streptomycin effectively controls hop downy mildew. Sporangia of *Pseudoperonospora humuli*, the casual agent of the disease germinated in the presence of sublethal concentrations of streptomycin to produce a high proportion of multiflagellate, multinucleate zoospores (Griffin and Coley-Smith, 1971), in agreement with the findings of Bekker et al. (1971).

Baráthová et al. (1969) studied the effect of a number of antibiotics on the growth of *Botrytis cinerea* on agar. The center of a petri dish was inoculated with a suspension of *B. cinerea* conidia and after 48 hours of incubation a filter paper disc impregnated with antibiotic was placed on the agar 5 mm from the edge of the growing colonies. The morphological changes were examined after a further 24 hours. Twenty of the 31 antibiotics tested induced some form of morphological change at concentrations several times lower than the minimum inhibitory concentrations. Some antibiotics induced several types of changes while with others the effects varied with the concentration. The most frequent change was an increase in the amount of branching. The scopamycins (Hütter et al., 1965) stimulate terminal branching of *B. cinera* in a similar manner. Griseofulvin was the only antibiotic to produce the curling effect. Citrinin had the effect of narrowing the hyphae. Robinson and Park 1966) found that this antibiotic had a similar effect on the hyphae of each major group of fungi. Aspergillic acid, cyanein, and flavofungin had a broadening effect on hyphae. Other antibiotics caused swelling or bulging of hyphae followed by lysis. Morphological changes occurred only in hyphae growing in agar, aerial hyphae were unaltered. Baráthová et al. (1969) were unable to correlate the morphological effects with the modes of action of the antibiotics. Antibiotics with different mechanisms of action induced similar morphological changes. The authors suggested that the changes produced in growing hyphae were associated with changes in permeability and with the impairment of cell wall synthesis.

2. *Macrolide Antibiotics*

Cyanein, monorden, cytochalasin A, and cytochalasin D are macrolide antibiotics of fungal origin (Betina and Mičeková, 1972). All have antimicrobial properties. All 4 compounds produced similar morphological changes in *Botrytis cinerea* (Betina and Mičeková, 1973). Conidia of *B. cinerea* germinating in the presence of cytochalasins A and D had swollen and distorted germ tubes which branched more freely than those of normal conidia. The germ tubes of conidia germinating in the presence of monorden were also distorted and had numerous bulbous swellings. When added to growing hyphae, cyanein induced terminal branching whereas monorden induced lateral branching with branches perpendicular to the main hyphae. The effects of these compounds on *B. cinerea* are remarkably similar to those produced in germinating conidia of *B. fabae* by the fungicide benomyl already discussed in Section II,B,1. In particular the effect of monorden in producing much lateral branching on growing hyphae of *B. cinerea* closely resembles the effect of benomyl on *B. fabae* (Richmond and Pring, 1971a).

Cytochalasins prevent completion of cytokinesis by disorganizing microfilaments. They have no effect on mitosis (Carter, 1967; Wessells et al., 1971). Betina et al. (1972) did not study the effects of cytochalasin B on *Botrytis cinerea* since they found that this compound, unlike cytochalasins A and D, had no antifungal properties. The effect of cytochalasin B on hyphae of *Aspergillus nidulans* was investigated by Oliver (1973). Hyphae treated with cytochalasin B were more branched and were periodically constricted giving a beaded appearance. Tip branching was more frequent in treated cultures. Oliver (1973) suggested that the morphological effects were due to a failure to maintain the concentration of wall growth materials at a single point near the cell surface. The compound acted on microfilaments near the cell surface but was unable to reach deep-seated filaments within the cell.

Betina and Mičeková (1973) suggested that the morphological changes induced by antibiotics were associated with impairment of cytoplasmic membrane or cell wall synthesis. The toxicant could act at various stages in these processes, such as inhibition of synthesis of wall precursors in the cytoplasm, alteration in transport of precursors through the plasmalelmma, activation of enzymes causing softening of the cell wall, or finally inhibition of attachment of precursors to the growing cell wall.

The bulbous swellings produced by monorden in *B. cinerea* (Betina and Mičeková, 1973) and by cytochalasin B in *A. nidulans* (Oliver, 1973) closely resemble the effects produced by incorporating phytanic acid (3,7,11,15-tetramethylhexadecanoic acid) into the membranes of a fatty acid-requiring strain of *Neurospora crassa* (Brody and Allen, 1972). In this mutant strain 10–15% of the fatty acids in the phospholipid were replaced by phytanic acid. The morphological change was associated with increased osmotic fragility and may be due to interference with the secretion and assembly of cell wall components.

3. Polyenes

The polyenes are a group of macrolide antibiotics which alter the permeability of cells whose membranes contain sterols. They are active against many fungi but have no effect on bacteria or blue-green algae (Gale *et al.*, 1972). They inhibit spore germination and produce morphological changes in growing fungal cells.

Filipin produced swellings in mycelia of *Aspergillus flavus* which subsequently burst with loss of protoplasm (Gottlieb *et al.*, 1960). The effect of filipin may be an extreme example of the consequences of loss of membrane integrity as compared with the milder modifications produced by the macrolide antibiotics discussed in the previous section.

4. Cycloheximide

Cycloheximide is a glutarimide antibiotic which inhibits protein synthesis. It is effective against a number of fungal diseases (Gottlieb and Shaw, 1970). Although this antibiotic does not generally produce morphological effects in fungi, some have been reported. Abnormally large cells were produced when cells of *Saccharomyces pastorianus* were treated with cycloheximide (Gundersen and Wadstein, 1962). This effect might be due to failure to synthesize cell wall protein or the enzymes required for wall synthesis. In fact, protoplasts of *S. cerevisiae* were unable to regenerate complete cell walls in the presence of cycloheximide. The fibrillar groundwork was formed, but not the cell wall matrix

(Nečas et al., 1968; Farkaš et al., 1970). Katz and Rosenberger (1971a) studied the effect of cycloheximide on the incorporation of labeled N-acetylglucosamine into the walls of a glucosamine-requiring mutant of *Aspergillus nidulans*. Growing hyphae incorporated N-acetylglucosamine mainly at the tip. In the presence of cycloheximide, labeling occurred at subapical regions and tip incorporation was greatly reduced. A similar change occurred when the hyphae were subjected to osmotic shock. Both treatments increased branching and the numbers of septa. These effects were probably a response to the disintegration of the hyphal tip.

5. *Polyoxin D*

Polyoxin D is an antibiotic produced by *Streptomyces cacaoi* var. *asoensis*, which inhibits the growth of some filamentous fungi but has no action on bacteria or yeasts. The antibiotic prevents chitin synthesis by competitively inhibiting chitin synthetase. The germ tubes of *Cochliobolus miyabeanus* when grown in the presence of 0.1 mM polyoxin D were distorted and finally ruptured. The addition of 20% sucrose stabilized the germ tubes which continued growing to produce giant protoplastlike structures. These giant cells lysed when placed in distilled water but regenerated to produce normal mycelium when transferred to a suitable growth medium. Germ tubes of *Neurospora crassa* were distorted in the presence of polyoxin D but did not rupture (Endo et al., 1970). Polyoxin D also inhibited swelling and germ tube emergence in sporangiospores of *Mucor rouxii*. When added to growing cultures the hyphal tips burst (Bartnicki-Garcia and Lippman, 1972).

The effects of polyoxin D on fungi were very similar to those produced when a mutant strain of *Aspergillus nidulans* containing 7–15% of the normal amount of chitin grew in growth medium in the absence of an osmotic stabilizer. Hyphae were grown in the presence of N-acetylglucosamine and then transferred to either buffer solution or a growth medium containing neither osmotic stabilizer nor N-acetylglucosamine. The hyphae became distorted and lysed in growth medium but not in buffer. As the cells did not lyse in dilute buffer, wall-splitting enzymes must presumably be produced along the whole length of the hyphal walls during the growing process. In the absence of fresh chitin synthesis the growing cells, but not those in buffer alone, burst at the weak points (Katz and Rosenberger, 1971b).

Many fungi differ from green plants in having cell walls containing chitin. Polyoxin D is thus a compound of considerable practical and theoretical importance, as it is the first known antifungal agent to exploit this fundamental biochemical difference between green plants and higher fungi by specifically inhibiting chitin synthesis.

6. Other Substances

Some other substances produced effects in fungi similar to those caused by polyoxin D. Rubratoxin B obtained from cultures of *Penicillium rubrum* induced giant cells and swollen hyhal tips in *Aspergills niger* and *A. flavus* (Reiss, 1972). Germ tubes of *A. niger* and *Rhizopus nigricans* had a "beaded" appearance, and lysis occurred at the hyphal tips of *A. niger*. The effects were similar to those produced by aflatoxins (Wildman, 1966; Reiss, 1971) and may be due to an interference with wall synthesis. Large spherical cells which retained at least part of their cell walls were also obtained from various fungi after treatment with a culture filtrate from *Bacillus cereus* (Koltin and Chorin-Kirsch, 1971). Antibiotics produced by strains of actinomycetes also produced swellings and lysis of hyphal tips (Links et al., 1957; Stevenson, 1956).

Many of the compounds discussed earlier cause distortions in fungal hyphae by interfering either directly or indirectly with the synthesis of essential wall components. The cell wall compositions of the two forms of the dimorphic fungus *Mucor rouxii* are different, (Bartnicki-Garcia and Nickerson, 1962a) and the conversion of the yeastlike form into the mycelial form is accompanied by the synthesis of RNA and protein (Haidle and Storck, 1966a). This organism and other dimorphic fungi may therefore provide useful systems for studying the effects of toxicants on cellular control processes in the fungi.

III. Effects on Dimorphic Fungi

The two most studied groups of dimorphic fungi are the species of *Mucor* and the animal pathogens. A third economically important group is the dimorphic species of *Verticillium* (Wang and Bartnicki-Garcia, 1970) and *Ceratocystis* (Hofsten and Hofsten, 1958) which are plant pathogens producing wilts in a wide variety of plants.

Fungi which attack the internal organs of man and animals develop a yeastlike form which is found in infected tissues. In contrast, the dermatophytes which cause superficial infections of the skin are mycelial both in the infective stage and in culture. Some animal pathogens are converted from the saprophytic mycelial form to the parasitic yeastlike form by increasing the temperature to 37°C; others require additional nutrients as well as increase of temperature. Both yeastlike and mycelial forms of *Candida* occur together independent of temperature, and both forms may be pathogenic (Scherr and Weaver, 1953; Romano, 1966).

A number of other fungal species produce yeastlike forms under certain conditions and, in fact, no clear distinction can be made between those fungi in which a yeastlike state is normal and those which exist predominantly in the mycelial state. The Ustilaginales develop a true mycelium but also include a budding stage in their life history. Some fungi, such as *Cordyceps militaris*, produce yeastlike forms under certain culture conditions (Marks et al., 1971), and *Aspergillus parasiticus* is converted into enlarged yeastlike cells in the absence of manganese (Detroy and Ciegler, 1971). Conidia of *Neurospora crassa* form yeastlike cells when they germinate in the presence of 2-phenylethanol (PEA) (Turian et al., 1972). On the other hand, even the true yeasts can, under certain conditions, develop elongated filamentous forms (Scherr and Weaver, 1953).

A. Animal Pathogens

Paecilomyces viridis is a dimorphic fungus which causes a mycosis of chameleons. Agar cultures of the saprophytic mycelial form of this fungus were converted into the parasitic yeastlike form by cyanein, actinomycin D, azalomycin F, citrinin, griseofulvin, and five other antibiotics. Twenty-four antibiotics inhibited growth of *P. viridis* without producing any morphological changes (Betina et al., 1966; Baráthová et al., 1969). The yeastlike forms of *P. viridis* induced by cyanein and azalomycin F were morphologically similar and were reconverted into the mycelial forms when transferred to agar free of antibiotic (Betina et al., 1966; Baráthová et al., 1972). This reconversion was inhibited by 2-phenylethanol and 2,4-dinitrophenol. 5-Fluorouracil also inhibited the reconversion of azalomycin F-induced cells but had no effect on cyanein-induced cells.

The yeastlike cells induced by treatment with cyanein contained less DNA and protein but more RNA than the mycelial form. The two forms also differed quantitatively in the lipid and carbohydrate composition of their cell walls. Examination of thin sections with the electron microscope showed that the cell walls of the yeastlike form were twice as thick as those of the mycelial form (Baráthová et al., 1972; Koman et al., 1972; Baráth and Betina, 1972).

Candida albicans causes a number of human and animal infections. The organism differs from other fungal pathogens in that both yeastlike and mycelial forms can be pathogenic and both may be found in infected tissue. The yeastlike form is more virulent than the mycelial form, but the mycelial form may be more invasive and may play an important role in the early stages of the infection (Romano, 1966). The fungus secretes an autoinhibitor, PEA, which favors yeastlike growth (Lingappa et al., 1969, 1971). Scherr and Weaver (1953) suggested that since the yeastlike form was better adapted to growth in tissues, treatment with plant growth substances which converted the yeastlike form into the mycelial form *in vitro* should halt the growth of the pathogen. The compounds were ineffective, however, in controlling the disease in mice inoculated with *C. albicans*.

Tetracycline and other broad-spectrum antibiotics stimulate the conversion of the yeastlike form of *C. albicans* into the mycelial form. Treatment of mice inoculated with *C. albicans* with tetracycline increases the pathogenicity of the fungus. The antibiotic stimulates the invasiveness of the mycelial form of *C. albicans*. Later applications of tetracycline encourage the recovery of lesions caused by the yeastlike form (Strippoli and Simonetti, 1973).

B. Plant Pathogens

Verticillium albo-atrum is a fungus which causes wilt diseases of a number of economic plants. The fungus is dimorphic and grows as mycelium in stationary culture but forms budding yeastlike structures in shake culture (Malca *et al.*, 1966). The yeastlike structures are frequently described as "spores" or "conidia," and Buckley *et al.* (1969) in a study of the fine structure of *V. albo-atrum* considered that the production of "secondary conidia" bore no resemblance to budding. In view of variations in the budding process which occur in true yeasts, Wang and Bartnicki-Garcia (1970) considered the multiplication of single cells in *V. albo-atrum* to be strictly comparable to mycelial-yeastlike dimorphism in other fungi.

The yeastlike form plays an important role in the pathogenicity of the fungus. The fungus spreads throughout the plant by rapid transport of the yeastlike cells through the vascular tissue, but not by mycelial growth (Garber and Houston, 1966; Pomerlau, 1970). The pathogenic behavior of *V. albo-atrum* in plants is thus very similar to that of the animal pathogens in animals.

In view of the importance of the yeastlike form to the pathogenicity of *V. albo-atrum*, Keen *et al.* (1971) investigated the effect of a number of inhibitors of spore production on shake cultures of the fungus. Semicarbazide, phenlhydrazine, deoxyadenosine, gossypol, and 5-fluorodeoxyuridine all converted the yeastlike form into mycelium when the initial spore inoculum was less than 10^8 cells/ml. At higher inoculum concentrations negligible amounts of mycelium were formed. The compounds, except gossypol, did not reduce the total culture dry weight. Unfortunately, the application of deoxyadenosine or 5-fluorodeoxyuridine to the roots of cotton plants did not significantly reduce the symptoms of *Verticillium* wilt. The failure of the two compounds to control the disease in the field may be due to lack of uptake, transformation by the plant, or insensitivity of the yeastlike cells of the fungus in the xylem.

Ceratocytis (Ophiostoma) multiannulatum also grows as a mycelium on solid medium and in yeastlike form in shake culture (Hofsten and Hofsten, 1958). Treatment with 5-fluorodeoxyuridine or deoxyadenosine converts the yeastlike form into mycelium (Hofsten, 1964). Biehn (1972) has also found that *p*-fluorophenylalanine at 5–10 µg/ml inhibits the formation of yeastlike cells of *C. ulmi*, the causal agent of Dutch elm disease. In view of the serious losses caused by wilt diseases it would seem well worthwhile to investigate further the possibility of using compounds which inhibit the mycelium to yeast transformation to control these diseases.

C. *Mucor*

The dimorphism of *Mucor* species has been studied in considerable detail (Bartnicki-Garcia and Nickerson, 1962b; Bartnicki-Garcia, 1963). Sporangiospores of the Mucorales swell considerably before the germtubes are produced. The swelling process does not require oxygen, but those species which produce little ethanol by fermentation do require oxygen for germ tube emergence. Species which are strong fermentors germinate by budding to produce yeastlike forms in the absence of oxygen (Wood-Baker, 1955). The mycelial form of *Mucor rouxii* is converted into the yeastlike form by growth under anaerobic conditions either in the presence of CO_2 or of about 8% hexoses in the medium (Bartnicki-Garcia, 1968b). When yeastlike cells of *M. rouxii* growing in an atmosphere of CO_2 and N_2 were transferred to air they were converted to hyphae. This conversion was accompanied by protein and RNA synthesis and was inhibited by cycloheximide (Haidle and Storck, 1966a,b).

Mucor rouxii also grew in the yeastlike form under aerobic conditions in the presence of PEA (0.2% v/v) and 2–5% of a hexose. The aerobic yeastlike growth involved an increase in ethanolic fermentation and a corresponding decrease in respiration. Terenzi and Storck (1968) suggested that PEA acted as an uncoupling agent by stimulating fermentation and inhibiting oxidative phosphorylation, and they concluded that conversion to the yeastlike form occurred when ethanolic fermentation reached a critical level. The addition of PEA to 9 other species of *Mucor* and 2 of *Rhizopus* increased aerobic fermentation by all the species, but only the *Mucor* species were converted into yeastlike forms (Terenzi and Storck, 1969a).

Stable yeastlike mutants of *M. bacilliformis* which have lost the ability to grow in the mycelial form have been found to occur spontaneously with a frequency of about one in every 3000 colonies (Storck and Morrill, 1971). These mutants had no cytochrome oxidase activity and were unable to respire normally; the energy necessary for growth was obtained by fermentation. These results provided support for the contention that the conversion from the mycelial form to the yeastlike form was intimately connected with a metabolic change from respiration to aerobic fermentation. Since conidia were not produced by *Neurospora crassa* in the presence of PEA, Turian et al. (1972) suggested that there was a morphogenetic sequence in the filamentous fungi from yeastlike to mycelial to conidial forms. Each step in the sequence could be repressed stepwise by increasing restriction of oxidative metabolism.

The precise mode of action of PEA is uncertain, but it appears to inhibit DNA synthesis by preventing the initiation of replication at a fixed point of origin on the chromosome. At the high concentrations used for production of yeastlike forms, PEA may also alter membrane permeability (Gale et al., 1972). The compound, which is an autoantibiotic secreted by *Candida albicans* and is also a common constituent of essential oils (Lingappa et al., 1969, 1971), causes a variety of morphological changes in microorganisms, plants, and animals (Müller et al., 1971).

Clark-Walker (1973) has studied the effect of chloramphenicol on the morphology and respiration of *Mucor genevensis*. The effect of chloramphenicol on *M. genevenis* was not as dramatic as that of PEA. There was only a partial conversion to yeastlike form under aerobic conditions in the presence of chloramphenicol (4 mg/ml). The fungus grew partly as yeastlike forms and partly as shortened club-shaped hyphae. The antibiotic eliminated cytochromes aa_3 and b from cells and abolished cyanide-sensitive respiration. The residual respiration was about 40% of that present in controls and was insensitive to cyanide. Clark-Walker claimed that chloramphenicol eliminated functional mitochondria without producing a complete conversion to yeastlike form. He suggested that both chloramphenicol and PEA might prevent mycelial growth not by uncoupling oxidative phosphorylation, but by altering the concentrations of high-energy molecules or by indirectly interfering with production of the components required for hyphal wall formation. Chloramphenicol inhibits protein synthesis in bacteria and produces abnormal morphological forms in bacteria at sublethal concentrations (Malik, 1972). At the high concentration (4 mg/ml) used by Clark-Walker (1973) the antibiotic may have other nonspecific effects.

Another member of the Mucorales, *Mycotypha africana*, grew as yeastlike cells either in an atmosphere of nitrogen or in the presence of inhibitors of electron transport when glucose, fructose, or mannose were the carbon source. In air, yeastlike forms were produced by PEA when the medium contained more than 1% glucose (Hall and Kolankaya, 1973).

D. J. Fisher, D. V. Richmond, and R. J. Pring (unpublished observations, 1974) studied the effect of a number of antibiotics and fungicides on the germination of sporangiospores of *Mucor hiemalis, M. pusillus, M. rouxii,* and *Actinomucor elegans* in liquid shake cultures under aerobic conditions. Yeastlike forms were produced in one or more of these organisms by treatment with dicloran, PEA, ethidium bromide, and carboxin (2,3-dihydro-6-methyl-5-phenylcarbamoyl-1,4-oxathiin) at 10^{-4} M. At the same concentration, streptomycin, tyrocidine, dimethirimol (5-butyl-2-dimethylamino-4-hydroxy-6-methylpyrimidine) and oxycarboxin (2,3-dihydro-6-methyl-5-phenylcarbamoyl-1,4-oxathiin-4,4-dioxide) produced some yeastlike forms, but the compounds were partially toxic. A number of compounds, such as triphenyltin acetate, nabam (disodium ethylenbisdithiocarbamate), and captan killed the spores. Dicloran was one of the most effective compounds for inducing yeastlike growth. The yeastlike colonies of *M. hiemalis* growing on agar containing dicloran are clearly distinguishable from normal mycelial growth (Fig. 2e). The distinctive multipolar budding of the yeastlike forms of *M. hiemalis* and *M. pusillus* induced by dicloran can be clearly seen both by the light microscope and by the scanning electron microscope (Fig. 2a–d). *A. elegans* was converted into yeastlike cells by 15 µg of PEA per milliliter. This fungus was much more sensitive to PEA than other *Mucor* species, which required 1–2 mg/ml to form pure yeastlike cultures (Terenzi and Storck, 1969b).

The compounds which induced the formation of yeastlike cells had little effect on the oxygen uptake of the cells, but all produced an increase in respiratory quotient and all increased aerobic ethanolic fermentation. The yeastlike cells induced by dicloran were shown by electron microscopy to have thickened cell walls and functional mitochondria. They thus differed from the yeastlike cells of *M. genevensis* induced by chloramphenicol which did not have functional mitochondria (Clark-Walker, 1973). Further studies will be needed before we are able to understand fully all the factors involved in the process of conversion from a mycelial to a yeastlike morphology.

IV. Effects of Toxicants on Fine Structure

The fine structure of the fungi resembles that of other eukaryotic organisms. Organelles such as nuclei, mitochondria, and vacuoles occur, but plastids are absent and cell walls may or may not be present. Fungal cells are almost invariably smaller than the cells of green plants and animals (Bracker, 1967). Studies with the electron microscope can give information about the effects of toxicants on the structural components of cells. Care must be taken, however, to ensure that the cells examined

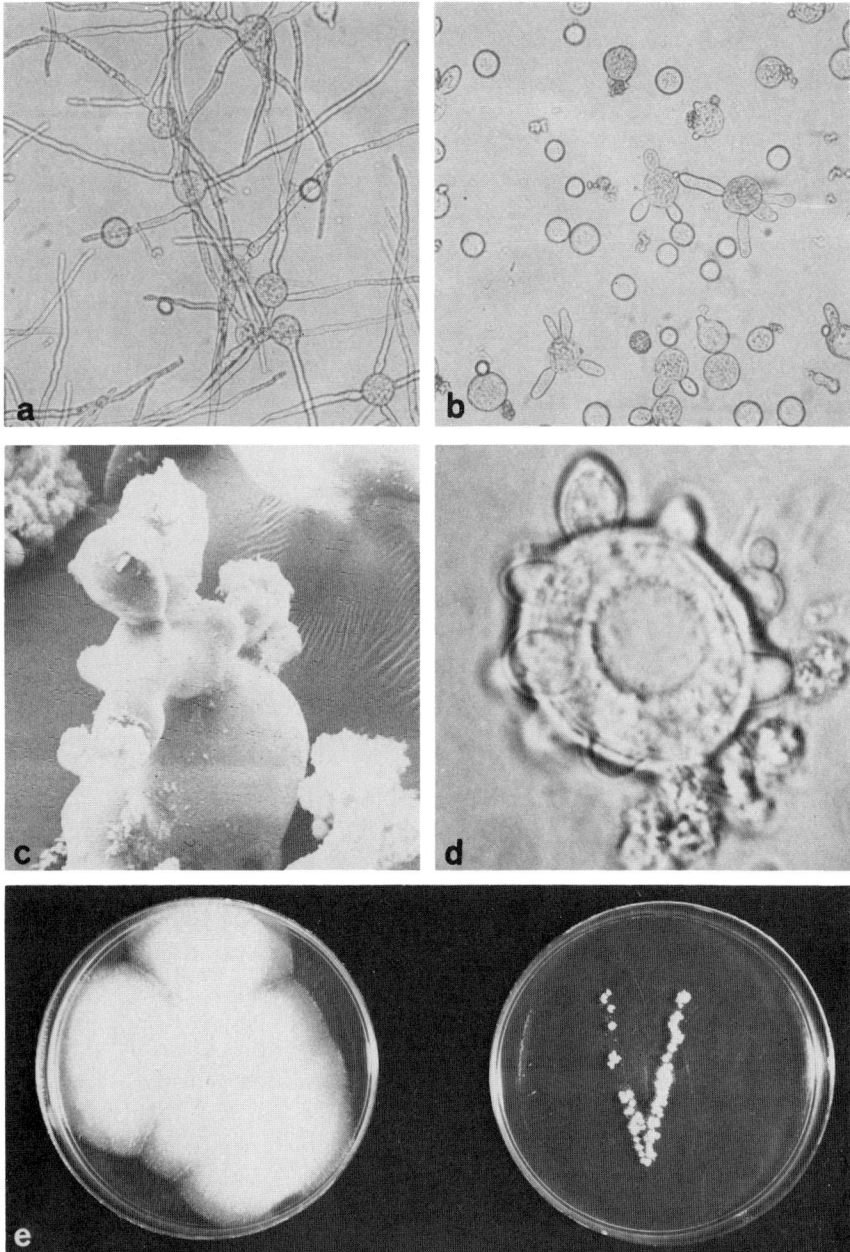

FIG. 2. (a) Sporangiospores of *Mucor hiemalis* germinating in growth medium. (b) Sporangiospores of *M. hiemalis* germinating in growth medium containing dicloran (2 μg/ml). (c) Scanning electron micrograph of budding yeastlike form of *M. pusillus* growing in dicloran (5 μg/ml). (d) Multipolar budding of yeastlike form of *M. pusillus* growing in dicloran (5 μg/ml). (e) *Mucor hiemalis* after 3 days' growth on agar: (i) mycelial growth on untreated agar; (ii) yeastlike growth on agar containing dicloran (0.2 μg/ml). a, ×150; b, ×180; c, ×2000; d, ×1700.

are representative of the organism as a whole, and the effects of toxicants must be clearly distinguished from the normal effects of aging.

Blank et al. (1960) were the first to study the action of toxicants on fungal fine structure. They treated mycelium of the dermatophyte *Trichophyton rubrum* with the antibiotic griseofulvin. Young, treated hyphae developed thickened and laminated cell walls, and the cell contents became grossly distorted. Two years later Matsui et al. (1962) reported the effect of immersion in an 0.1% solution of mercuric chloride on the fine structure of conidia of *Cochliobolus miyabeanus*. The treatment did not alter the structure of the organelles, but all membranes in treated cells had greatly increased contrast. The authors considered that these effects were due to preferential reaction of mercury with the SH and NH_2 groups of proteins within cell membranes.

Since these early studies there have been continuous advances in the techniques of electron microscopy, and new methods of fixation have revealed the structural complexity of fungal cells with greatly increased clarity. As a particular toxicant may act on several organelles, the effects of each compound will be considered separately. Fungicides will be considered first, followed by antibiotics and other toxicants.

A. Captan

Captan is a widely used protective fungicide which controls a number of important plant diseases. The compound reacts with intracellular thiols in conidia of *Neurospora crassa* and cells of *Saccharomyces pastorianus* (Richmond and Somers, 1968; Siegel, 1970). The effect of treatment with captan on the fine structure of resting conidia of *N. crassa* was studied by Richmond et al. (1967). Most organelles were unaffected by captan, but in permanganate-fixed sections the nuclear membrane had a more convoluted form than usual. As other membranes were unaltered, the change in nuclear membrane shape was probably due to alterations in the structure of the nucleus, not to an attack on the membrane itself. Conidia that had been treated with captan and then incubated in growth medium showed an almost complete loss of internal fine structure.

The mitochondria of untreated conidia which had been fixed in glutaraldehyde were stained heavily by uranyl acetate between the cristae. After treatment with captan the mitochondria could not be stained; this observation suggested that captan may have reacted with SH groups that control the swelling and contraction of mitochondria.

More recently, Nelson (1971) found that mitochondria from rat liver cells were much swollen after treatment with captan and contained no visible cristae. He suggested that captan had reacted with SH groups in the mitochondrial membranes to alter membrane permeability.

B. Benomyl

Richmond and Pring (1971a) studied the effect of benomyl on germinating conidia of *Botrytis fabae* using both chemically fixed thin sections and freeze-etched replicas. The organelles and endoplasmic reticulum of untreated young germ tubes were usually oriented toward the hyphal tip (Fig. 3a). After treatment with benomyl, this pattern was disorganized and the organelles appeared to be directed to several different points on the periphery of the germ tube (Fig. 3b). The form of the endoplasmic reticulum also changed from multiple strands surrounding the nuclei (Fig. 3a) to short broken fragments and, sometimes, a branched reticulate network. Some strands of endoplasmic reticulum lay parallel to the plasmalemma and others were directed to positions on the periphery of the cell, where they terminated in a series of small vesicles (Fig. 3b). This configuration was normally more characteristic of resting than of germinating conidia (Richmond and Pring, 1971b,c).

The effects of benomyl on the endoplasmic reticulum were confirmed by the examination of freeze-etched replicas. The endoplasmic reticulum of treated spores consisted of a fragmented series of platelets, tubules, and vesicles whereas in untreated conidia the endoplasmic reticulum formed multiple sheets surrounding the nuclei.

Benomyl-treated nuclei appeared deeply lobed and convoluted after freeze-etching, but no changes could be detected in chemically fixed nuclei. The thin sections were from material fixed in permanganate, a substance that destroys microtubules. To detect any effects of benomyl on nuclear division would have required the use of more delicate fixation procedures.

C. 2-Aminobutane

The *sec*-butylammonium cation (SBA, 2-aminobutane) is used to control decay of citrus fruits caused by *Penicillium digitatum* (Eckert and Kolbegen, 1964). The compound inhibits the active transport of amino acids into hyphae (Bartz and Eckert, 1972).

Zaki *et al.* (1973) showed that conidia of *P. digitatum* contained several dense spherical inclusions of phospholipid either in the cytoplasm or in vacuoles. The inclusions in vacuoles were associated with complex whorled unit membrane structures. When conidia germinated, these phospholipid inclusions were seen in the germ tubes, usually near the hyphal tips. Treatment of germinating conidia with SBA produced con-

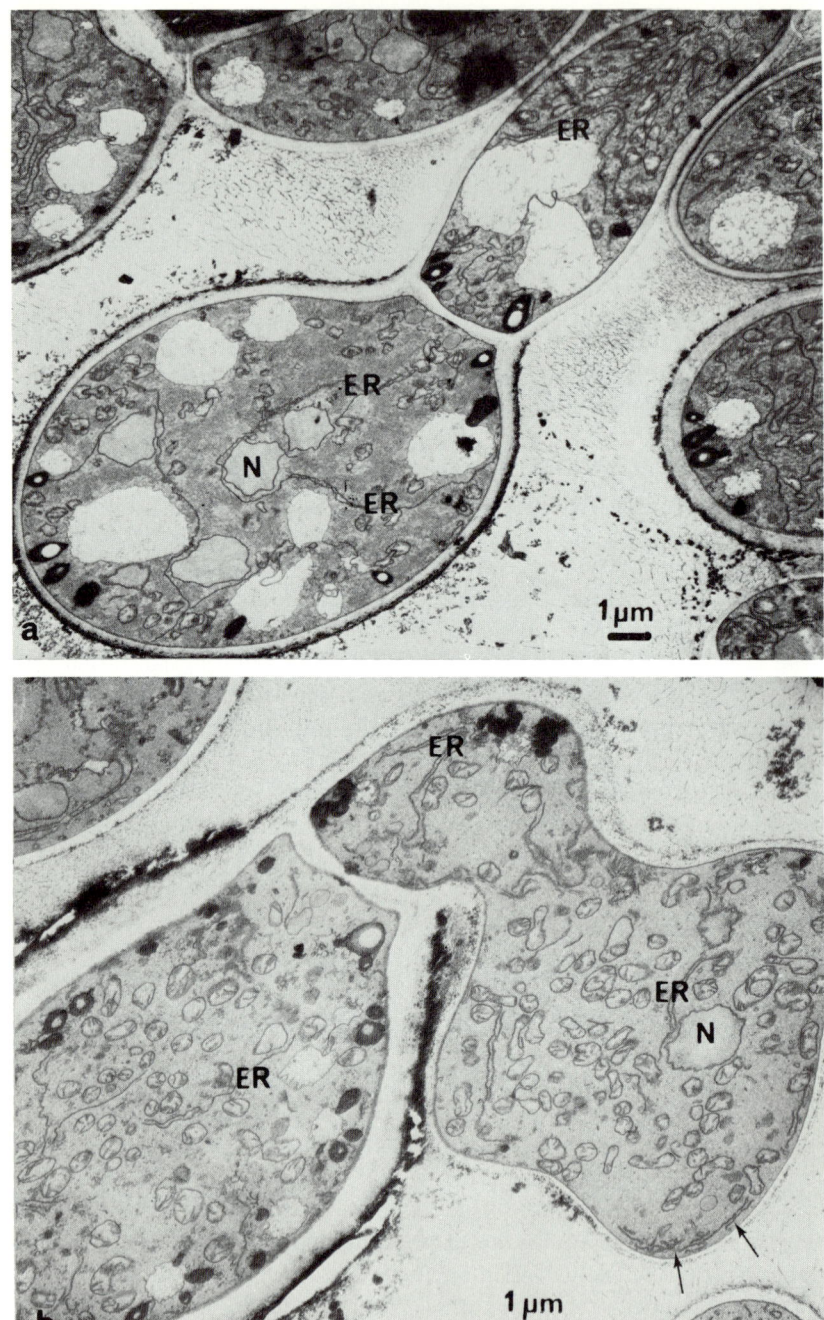

siderable changes in the inclusions within the vacuoles. The inclusions were broken down into smaller fragments and the whorled structures were replaced by numerous vesicles and membranes of various shapes. Other organelles were unaffected by SBA treatment. The authors suggested that the breakdown of lipoprotein inclusions and associated membranes was due to the greater activity of hydrolytic enzymes in treated cells. Treatment with SBA was considered to deprive conidia of the phospholipid required for the formation of developing organelles.

D. Oxycarboxin

Oxycarboxin and carboxin are effective systemic fungicides which control many diseases caused by Basidiomycetes. The fungicides inhibit mitochondrial respiration by blocking the oxidation of succinate (Mathre, 1971; White, 1971).

Lyr et al. (1972) have studied the effect of carboxin on the fine structure of *Rhodotorula mucilaginosa*. After treatment for 2 hours with carboxin (10 μM) the mitochondria became swollen with disorganized cristae, and the vacuolar membrane was damaged. At higher concentrations of carboxin (100 μM) the mitochondria were almost completely destroyed, but the endoplasmic reticulum and plasmalemma were unchanged. The cell wall had a layered appearance.

The concentration of carboxin required to inhibit the respiration of isolated mitochondria of sensitive fungi was similar to that required for intact cells. Lyr et al. (1972) confirmed that the primary site of action of carboxin was on succinic dehydrogenase. R. J. Pring and D. V. Richmond (unpublished observations, 1974) studied the effect of oxycarboxin on the fine structure of *Uromyces phaseolus* infecting leaves of French bean (*Phaseolus vulgaris* cv. The Prince). Oxycarboxin was effective in preventing rust infection when applied to the roots of French bean plants. The compound was also a good eradicant. Spores of *U. phaseoli* did, however, germinate and produce appressoria on leaves of bean plants treated with concentrations of oxycarboxin effective in preventing the appearance of disease symptoms. Oxycarboxin must therefore exert its preventive effect at some stage after appressorium formation.

Fig. 3. (a) Section of germinating conidium of *Botrytis fabae* showing strands of endoplasmic reticulum (ER) surrounding the nucleus (N). In the germ tube the organelles are oriented toward the hyphal tip. From Richmond and Pring (1971c). (b) Conidium of *B. fabae* germinating in the presence of benomyl (0.3 μg/ml) showing distorted multilobed germ tube. The endoplasmic reticulum (ER) is more fragmented than in untreated conidia. Some of the endoplasmic reticulum lies close to the plasmalemma and terminates in a series of vesicles (arrows). N, nucleus. From Richmond and Pring (1971a). a, ×5000; b, ×7000.

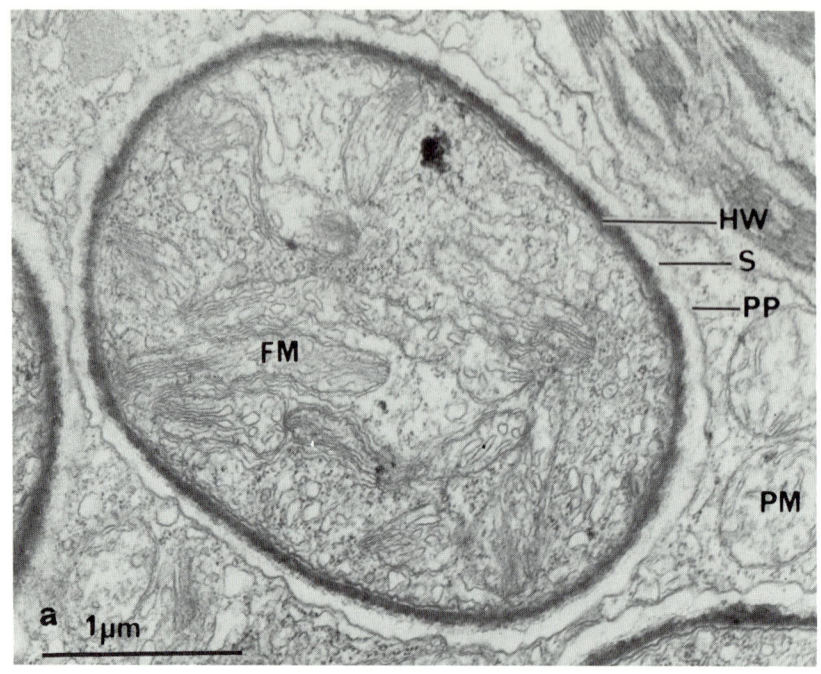
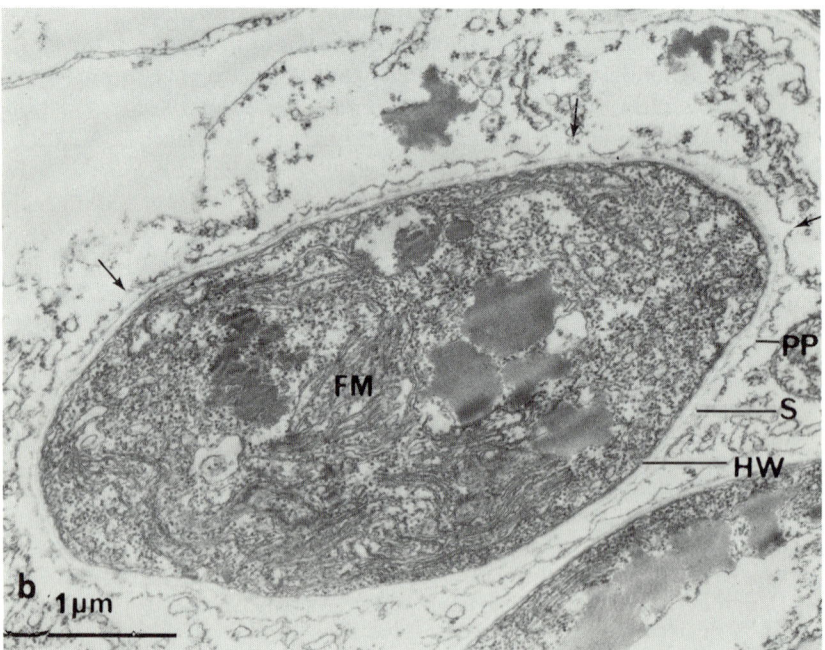

The eradicant action of oxycarboxin was studied by electron microscopy; plants were treated with oxycarboxin by soil application at the necrotic spot stage (about 5 days after inoculation). Changes in the shape and pattern of cristae in the haustorial mitochondria occurred 24 hours after treatment. After 2 days, further changes had occurred in the haustoria; lipid bodies accumulated, and the cytoplasm became disorganized. The host plasmalemma, which normally completely surrounds the haustorium (Fig. 4a), became broken and fragmented (Fig. 4b). The intercellular mycelium still appeared unaffected by the fungicidal treatment. After 6 days effects on haustoria were even more pronounced; more vacuoles appeared, and some haustoria became shrunken and filled with electron-dense material. Finally, the necrotic haustoria became completely surrounded by collars of pathogen-induced host wall apposition. After the haustoria were killed, changes appeared in the intercellular mycelium.

E. GRISEOFULVIN

Griseofulvin has fungistatic activity against many fungi but has no effect on bacteria or yeasts (Huber, 1967). The most striking visible effect of the antibiotic is its ability to produce abnormal curling in growing hyphae (Brian, 1949). Early work suggested that griseofulvin acted by inhibiting chitin synthesis, but it is now known that the primary action of the antibiotic is on nucleic acids (Dekker, 1969).

Treatment with griseofulvin increased the thickness of hyphal walls of *Botrytis allii* and *Aspergillus niger* but had no effect on walls of *Microsporum gypseum*. The fine structure of the walls and the orientation of the microfibrils was unaffected by the treatment (Bent and Moore, 1966; Evans and White, 1967). The antibiotic produced abnormal nuclei in *B. allii* and *A. niger*, suggesting that griseofulvin was affecting mitosis. Treated hyphae of both fungi had more vacuoles and lipid bodies than controls. Griseofulvin also appeared to increase the number of lomasomes in *B. allii* (Bent and Moore, 1966). Nuclei of *Aspergillus nidulans* in the later stages of mitosis become abnormally stretched in the presence

---

FIG. 4. (a) Haustorium of *Uromyces phaseoli* in French bean (*Phaseolus vulgaris*) 5 days after infection. The plant plasmalemma completely encircles the haustorium. (b) Haustorium of *U. phaseoli* treated with oxycarboxin (2.5 mg in 40 ml of water applied to roots) at necrotic spot stage (5 days after infection). Photograph shows the effect of fungicide 2 days after treatment. Several lipid bodies are present in the haustorium, and the plant plasmalemma is breaking up and is fragmented in several places (arrows). C, chloroplast; ER, endoplasmic reticulum; FM, fungal mitochondrion; HW, haustorial wall; L, lipid body; N, nucleus; PM, plant mitochondrion; PP, plant plasmalemma; S, sheath. a, ×27,000. b, ×24,000.

of griseofulvin owing to the breakdown of the spindle apparatus (Crackower, 1972). The effects of griseofulvin on fine structure closely resemble those produced by benomyl (Section IV,B); the two compounds may have similar modes of action.

F. Cycloheximide

Cycloheximide inhibits protein synthesis in many eukaryotic cells, but not in prokaryotes or mitochondria (Gale et al., 1972). Hyphae of *Aspergillus nidulans* continued to synthesize all the major cell wall components when cycloheximide was added to growing cultures. The hyphae did not elongate, but the walls were shown by electron microscopy to become thicker. Thus cycloheximide changed wall synthesis from apical growth to subapical thickening. The antibiotic may interfere in some way with the cytoplasmic transport system which directs wall precursors in vesicles toward the hyphal tip (Sternlicht et al., 1973).

G. Chloramphenicol

Chloramphenicol inhibits protein synthesis in bacteria and mitochondria, but not in eukaryotes (Gale et al., 1972). Smith and Marchant (1969) found that cells of *Rhodotorula glutinis* growing in high concentrations of chloramphenicol (500 μg/ml) developed extensive cell wall thickening. They attributed the wall thickening to the antibiotic restricting synthesis of all cell constituents other than wall material. Osumi and Kitsutani (1971) studied the effects of even higher concentrations of chloramphenicol (5 mg/ml) on *Rhodotorula mucilaginosa*. The mitochondria of cells growing in the presence of chloramphenicol were much smaller than those of control cells and their cristae were poorly developed. Clark-Walker and Linnane (1967) found that chloramphenicol (4 mg/ml) had similar effects on the mitochondria of *Saccharomyces cerevisiae*. They considered that chloramphenicol primarily inhibited the formation of mitochondrial cristae. Similar results have been reported for *Neurospora crassa* (Howell et al., 1971).

H. Ethidium Bromide

Wood and Luck (1971) found that a paracrystalline inclusion accumulated in hyphae of *Neurospora crassa* after growth in the presence of ethidium bromide or euflavine. After 20 hours of growth in the presence of the toxicant, needlelike cytoplasmic inclusions could be clearly seen by means of the light microscope. The electron microscope showed the inclusions to consist of bundles of parallel fibers. The authors suggested that the inclusion, which normally occurs in a soluble form in the cytoplasm, crystallizes as the result of a mitochondrial defect.

I. Other Compounds

Gale (1963a) studied the effect of an antifungal peptide (R02-7758) isolated from a streptomycete on the fine structure of *Mucor corymbifera*. The compound produced great increases in cell wall thickness.

The polyene antibiotics amphotericin B and filipin disorganized the fine structure of *Candida albicans* and produced a marked thinning of the cytoplasm (Gale, 1963b). The initial effect of thiobenzoate on *C. albicans* was to reduce the electron density of the nucleus; prolonged treatment caused complete loss of cell integrity (Gale and McLain, 1964).

Deshusses *et al.* (1970) examined the ultrastructural effects of isomytilitol, an anti-inositol on *Schizosaccharomyces pombe*. Cell separation was incomplete in the presence of isomytilitol, the cell septa developed in abnormal positions, and some parts of the cell wall were greatly thickened.

Maxwell and Spoerl (1972) examined the effects of iodoacetic acid on *Saccharomyces cerevisiae* using thin sections and freeze-etched replicas. Iodoacetic acid altered the structure of the vacuolar membranes and the mitochondria became extremely distorted.

Keyhani (1973) has studied the effect of high copper concentrations or copper deficiency on the structure of mitochondria of *Candida utilis*. Cells grown in the absence of copper had large mitochondria with disorganized cristae: cells grown in high copper media (112 μg/l) had giant mitochondria with poorly developed cristae. The author suggested that the giant mitochondria were formed by the fusion of smaller mitochondria.

The work described above demonstrates that fungicides and antibiotics can produce considerable and varied changes in fungal fine structure. Compounds acting on more than one biochemical pathway may produce several ultrastructural changes and hence the primary sites of action may be difficult to determine. Other more specific compounds may produce effects that are easier to interpret. Inhibitors of respiration produce changes in mitochondria; inhibitors of mitosis, changes in nuclei, and inhibitors of protein synthesis frequently produce abnormally thickened cell walls.

The diverse techniques of electron microscopy can clearly give valuable information on the modes of action of toxicants.

Acknowledgments

I would like to thank Dr. D. J. Fisher and Mr. R. J. Pring for allowing me to include some of their unpublished results in this review. The photomicrographs are by Mr. R. J. Pring.

References

Armitage, F. D., and Verdcourt, B. (1948). *Research* **1**, 236–237.
Baráth, Z., and Betina, V. (1972). *Biologia (Bratislava)* **27**, 485–490.
Baráthová, H., Betina, V., and Nemec, P. (1969). *Folia Microbiol. (Prague)* **14**, 475–483.
Baráthová, H., Baráth, Z., and Betina, V. (1972). *Biologia (Bratislava)* **27**, 469–478.
Bartels-Schooly, J., and MacNeill, B. H. (1971). *Phytopathology* **61**, 816–819.
Bartnicki-Garcia, S. (1963). *Bacteriol. Rev.* **27**, 293–304.
Bartnicki-Garcia, S. (1968a). *Annu. Rev. Microbiol.* **22**, 87–108.
Bartnicki-Garcia, S. (1968b). *J. Bacteriol.* **96**, 1586–1594.
Bartnicki-Garcia, S., and Lippman, E. (1972). *J. Gen. Microbiol.* **71**, 301–309.
Bartnicki-Garcia, S., and Nickerson, W. J. (1962a). *Biochim. Biophys. Acta* **58**, 102–119.
Bartnicki-Garcia, S., and Nickerson, W. J. (1962b). *J. Bacteriol.* **84**, 829–840.
Bartz, J. A., and Eckert, J. W. (1972). *Phytopathology* **62**, 239–246.
Bekker, Z. E., Suprun, T. P., and Nesterenko, E. I. (1971). *Mikol. Fitopatol.* **5**, 3–9.
Bent, K. J., and Moore, R. H. (1966). *Symp. Soc. Gen. Microbiol.* **16**, 82–110.
Betina, V., and Mičeková, D. (1972). *Z. Allg. Mikrobiol.* **12**, 355–364.
Betina, V., and Mičeková, D. (1973). *Z. Allg. Mikrobiol.* **13**, 287–298.
Betina, V., Betinová, M., and Kulková, M. (1966). *Arch. Mikrobiol.* **55**, 1–16.
Betina, V., Mičeková, D., and Nemec, P. (1972). *J. Gen. Microbiol.* **71**, 343–349.
Biehn, W. L. (1972). *Phytopathology* **62**, 493.
Blank, H., Taplin, D., and Roth, F. J., Jr. (1960). *AMA Arch. Dermatol. Syphilol.* **81**, 667–680.
Bollen, G. J. (1971). *Neth. J. Plant Pathol.* **77**, 187–193.
Bracker, C. E. (1967). *Annu. Rev. Phytopathol.* **5**, 343–374.
Brian, P. W. (1949). *Ann. Bot. (London)* [N.S.] **13**, 59–77.
Brian, P. W., Curtis, P. J., and Hemming, H. G. (1946). *Trans. Brit. Mycol. Soc.* **29**, 173–187.
Brody, S., and Allen, B. (1972). *J. Supramol. Struct.* **1**, 125–134.
Brown, I. F., and Hall, H. R. (1971). *Phytopathology* **61**, 886.
Buckley, P. M., Wyllie, T. D., and DeVay, J. E. (1969). *Mycologia* **61**, 240–250.
Burguillo, P. F., Nicolas, G., and Martin, J. F. (1972). *Trans. Brit. Mycol. Soc.* **59**, 512–516.
Cain, R. F. (1972). *Mycologia* **64**, 1–14.
Carter, S. B. (1967). *Nature (London)* **213**, 261–264.
Clark-Walker, G. D. (1973). *J. Bacteriol.* **116**, 972–980.
Clark-Walker, G. D., and Linnane, A. W. (1967). *J. Cell Biol.* **34**, 1–14.
Clemons, G. P., and Sisler, H. D. (1971). *Pestic. Biochem. Physiol.* **1**, 32–43.
Cochrane, V. W. (1958). "Physiology of Fungi." Wiley, New York.
Crackower, S. H. B. (1972). *Can. J. Microbiol.* **18**, 683–687.
Davidse, L. C. (1973). *Pestic. Biochem. Physiol.* **3**, 317–325.
Dekker, J. (1969). *In* "Fungicides" (D. C. Torgeson, ed.), Vol. 2, pp. 579–635. Academic Press, New York.

Deshusses, J., Oulevey, N., and Turian, G. (1970). *Protoplasma* **70**, 119–130.
De Terra, N., and Tatum, E. L. (1961). *Science* **134**, 1066–1068.
De Terra, N., and Tatum, E. L. (1963). *Amer. J. Bot.* **50**, 669–677.
Detroy, R. W., and Ciegler, A. (1971). *J. Gen. Microbiol.* **65**, 259–264.
Duran, R. (1962). *Plant Dis. Rep.* **46**, 115–118.
Eckert, J. W., and Kolbegen, M. J. (1964). *Phytopathology* **54**, 978–986.
Endo, A., Kakiki, K., and Misato, T. (1970). *J. Bacteriol.* **104**, 189–196.
Erwin, D. C. (1969). *World Rev. Pest Contr.* **8**, 6–21.
Esuruoso, O. F., Price, T. V. and Wood, R. K. S. (1968). *Trans. Brit. Mycol. Soc.* **51**, 405–410.
Evans, G., and White, N. H. (1967). *J. Exp. Bot.* **18**, 465–470.
Farkaš, V., Svoboda, A., and Bauer, Š. (1970). *Biochem. J.* **118**, 755–758.
Fries, N. (1973). *Trans. Brit. Mycol. Soc.* **60**, 1–21.
Gale, E. F., Cundliffe, E., Reynolds, P. E., Richmond, M. H., and Waring, M. J. (1972). "The Molecular Basis of Antibiotic Action." Wiley, New York.
Gale, G. R. (1963a). *J. Bacteriol.* **85**, 833–837.
Gale, G. R. (1963b). *J. Bacteriol.* **86**, 151–157.
Gale, G. R., and McLain, H. H. (1964). *J. Gen. Microbiol.* **36**, 297–301.
Garber, R. H., and Houston, B. R. (1966). *Phytopathology* **56**, 1121–1126.
Georgopoulos, S. G., and Zaracovitis, C. (1967). *Annu. Rev. Phytopathol.* **5**, 109–130.
Gottlieb, D., and Kumar, K. (1970). *Phytopathology* **60**, 1451–1455.
Gottlieb, D., and Shaw, P. D. (1970). *Annu. Rev. Phytopathol.* **8**, 371–402.
Gottlieb, D., Carter, H. E., Wu, L-C., and Sloneker, J. H. (1960). *Phytopathology* **50**, 594–603.
Griffin, M. J., and Coley-Smith, J. R. (1971). *J. Gen. Microbiol.* **69**, 117–134.
Gundersen, K., and Wadstein, T. (1962). *J. Gen. Microbiol.* **28**, 325–332.
Haidle, C. W., and Storck, R. (1966a). *J. Bacteriol.* **92**, 1236–1244.
Haidle, C. W., and Storck, R. (1966b). *Biochem. Biophys. Res. Commun.* **22**, 175–180.
Hall, M. J., and Kolankaya, N. (1973). *J. Gen. Microbiol.* **75**, XIV–XV.
Hammerschlag, R. S., and Sisler, H. D. (1973). *Pestic. Biochem. Physiol.* **3**, 42–54.
Hickey, K. D., and Drake, C. R. (1972). *Phytopathology* **62**, 669.
Higgons, D. J. (1961). SCI (*Soc. Chem. Ind., London*) *Monogr.* **15**, 132–141.
Hofsten, A. V. (1964). *Physiol. Plant.* **17**, 177–185.
Hofsten, A. V., and Hofsten, B. V. (1958). *Physiol. Plant* **11**, 106–117.
Horsfall, J. G. (1956). "Principles of Fungicidal Action." Chronica Botanica, Waltham, Massachusetts.
Houseworth, L. D., Brunton, E. W., and Tweedy, B. G. (1971). *Phytopathology* **61**, 896.
Howell, N., Zuiches, C. A., and Munkres, K. D. (1971). *J. Cell Biol.* **50**, 726–736.
Huber, F. M. (1967). *In* "Antibiotics" (D. Gottlieb and P. D. Shaw, eds.), Vol. 1, pp. 181–189. Springer-Verlag, Berlin and New York.
Hutchinson, S. A. (1973). *Annu. Rev. Phytopathol.* **11**, 223–246.
Hütter, R., Keller-Scherlein, W., Nüesch, J., and Zähner, H. (1965). *Arch. Mikrobiol.* **51**, 1–8.
Katz, D., and Rosenberger, R. F. (1971a). *J. Bacteriol.* **108**, 184–190.
Katz, D., and Rosenberger, R. F. (1971b). *Arch. Mikrobiol.* **80**, 284–292.
Keen, N. T., Wang, M. C., Long, M., and Erwin, D. C. (1971). *Phytopathology* **61**, 1266–1269.
Keyhani, E. (1973). *Exp. Cell Res.* **81**, 73–78.
Koch, W. (1971). *Pestic. Sci.* **2**, 207–210.

Koltin, Y., and Chorin-Kirsch, I. (1971). *J. Gen. Microbiol.* **66**, 145–151.
Koman, V., Baráth, Z., and Betina, V. (1972). *Biologia (Bratislava)* **27**, 479–483.
Lé John, H. B. (1971). *Nature (London)* **231**, 164–168.
Lingappa, B. T., Prasad, M., and Lingappa, J. (1969). *Science* **163**, 192–194.
Lingappa, B. T., Kaufman, A., and Lingappa, Y. (1971). *Bacteriol. Proc.* p. G235.
Links, J., Rombouts, J. E., and Keulen, P. (1957). *J. Gen. Microbiol.* **17**, 596–601.
Loveless, L. E., Spoerl, E., and Weisman, T. H. (1954). *J. Bacteriol.* **68**, 637–644.
Lyr, H., Ritter, G., and Casperson, G. (1972). *Z. Allg. Mikrobiol.* **12**, 271–280.
Macris, B., and Georgopoulos, S. G. (1973). *Z. Allg. Mikrobiol.* **13**, 415–423.
Malca, I., Erwin, D. C., Moje, W., and Jones, B. (1966). *Phytopathology* **56**, 401–406.
Malik, V. S. (1972). *Advan. Appl. Microbiol.* **15**, 297–336.
Marks, D. B., Keller, B. J., and Guarino, A. J. (1971). *J. Gen. Microbiol.* **69**, 253–259.
Mathre, D. E. (1971). *Pestic. Biochem. Physiol.* **1**, 216–224.
Matsui, C., Noyu, M., Kikumoto, T., and Maetsuura, M. (1962). *Phytopathology* **52**, 88–90.
Maxwell, W. A., and Spoerl, E. (1972). *Cytobiologie* **5**, 309–312.
Mirocha, C. J., and DeVay, J. E. (1971). *Can J. Microbiol.* **17**, 1373–1377.
Moore, R. T. (1971). *Proc. Int. Congr. Microbiol., 10th, 1970* pp. 49–64.
Moore-Landecker, E., and Stotzky, G. (1973). *Mycologia* **65**, 519–530.
Müller, W. E., Heicke, B., and Zahn, R. K. (1971). *Biochim. Biophys. Acta* **240**, 506–514.
Nečas, O., Svoboda, A., and Kopecká, M. (1968). *Exp. Cell Res.* **53**, 291–293.
Nelson, B. D. (1971). *Biochem. Pharmacol.* **20**, 749–758.
Norrman, J. (1968). *Arch. Mikrobiol.* **61**, 128–142.
Oliver, P. T. P. (1973). *Protoplasma* **76**, 279–281.
Osumi, M., and Kitsutani, S. (1971). *J. Electronmicrosc.* **20**, 23–31.
Pine, L. (1962). *In* "Fungi and Fungous Diseases" (G. Dalldorf, ed.), pp. 84–110. Thomas, Springfield, Illinois.
Pomerlau, R. (1970). *Can. J. Bot.* **48**, 2043–2057.
Priest, D., and Wood, R. K. S. (1961). *Ann. Appl. Biol.* **49**, 445–460.
Ragsdale, N. N., and Sisler, H. D. (1973). *Pestic. Biochem. Physiol.* **3**, 20–29.
Ramsey, G. B., Smith, M. A., and Heiberg, B. C. (1944). *Bot. Gaz. (Chicago)* **106**, 74–83.
Raper, J. R. (1968). *In* "The Fungi" (G. C. Ainsworth and A. S. Sussman, eds.), Vol. 3, pp. 677–693. Academic Press, New York.
Reavill, M. J. (1954). *Ann. Appl. Biol.* **41**, 448–460.
Reiss, J. (1971). *Z. Allg. Mikrobiol.* **11**, 637–638.
Reiss, J. (1972). *J. Gen. Microbiol.* **71**, 167–172.
Richmond, D. V., and Pring, R. J. (1971a). *J. Gen. Microbiol.* **66**, 79–94.
Richmond, D. V., and Pring, R. J. (1971b). *Ann. Bot. (London)* [N.S.] **35**, 175–182.
Richmond, D. V., and Pring, R. J. (1971c). *Ann. Bot. (London)* [N.S.] **35**, 493–500.
Richmond, D. V., and Somers, E. (1968). *Ann. Appl. Biol.* **62**, 35–43.
Richmond, D. V., Somers, E., and Millington, P. F. (1967). *Ann. Appl. Biol.* **59**, 233–237.
Robinson, P. M., and Park, D. (1966). *Nature (London)* **211**, 883–884.
Romano, A. H. (1966). *In* "The Fungi" (G. C. Ainsworth and A. S. Sussman, eds.), Vol. 2, pp. 181–209. Academic Press, New York.
Scherr, G. H., and Weaver, R. H. (1953). *Bacteriol. Rev.* **17**, 51–92.
Sherald, J. L., Ragsdale, N. N., and Sisler, H. D. (1973). *Pestic. Sci.* **4**, 719–727.
Siegel, M. R. (1970). *J. Agr. Food Chem.* **18**, 823–826.

Smith, D. G., and Marchant, R. (1969). *Antonie van Leeuwenhoek; J. Microbiol. Serol.* **35**, 113–119.
Solel, Z. (1970). *Phytopathology* **60**, 1186–1190.
Somers, E. (1962). *Sci. Progr. (London)* **50**, 218–234.
Sternlicht, E., Katz, D., and Rosenberger, R. F. (1973). *J. Bacteriol.* **114**, 819–823.
Stevenson, I. L. (1956). *J. Gen. Microbiol.* **15**, 372–380.
Storck, R., and Morrill, R. C. (1971). *Biochem. Genet.* **5**, 467–479.
Stover, R. H. (1969). *Plant Dis. Rep.* **53**, 830–833.
Strippoli, V., and Simonetti, N. (1973). *Mycopathol. Mycol. Appl.* **51**, 65–73.
Strzelczyk, A. B. (1968). *Can. J. Microbiol.* **14**, 901–906.
Terenzi, H. F., and Storck, R. (1968). *Biochem. Biophys. Res. Commun.* **30**, 447–452.
Terenzi, H. F., and Storck, R. (1969a). *Mycologia* **61**, 894–901.
Terenzi, H. F., and Storck, R. (1969b). *J. Bacteriol.* **97**, 1248–1261.
Threlfall, R. J. (1972). *J. Gen. Microbiol.* **71**, 173–180.
Tribe, H. T., and Mabadeje, S. A. (1972). *Trans. Brit. Mycol. Soc.* **58**, 127–137.
Turian, G., Peduzzi, R., and Lingappa, B. T. (1972). *Arch. Mikrobiol.* **85**, 333–340.
Verdcourt, B. (1952). *Mycologia* **44**, 377–386.
Vincent, J. M. (1947). *Nature (London)* **159**, 850.
Vogel, H. J., Thompson, J. S., and Shockman, G. D. (1970). *Symp. Soc. Gen. Microbiol.* **20**, 107–119.
Vonk, J. W., and Kaars Sijpesteijn, A. (1971). *Pestic. Sci.* **2**, 160–164.
Wang, M. C., and Bartnicki-Garcia, S. (1970). *J. Gen. Microbiol.* **64**, 41–54.
Weber, D. J., and Ogawa, J. M. (1965). *Phytopathology* **55**, 159–165.
Wessells, N. K., Spooner, B. S., Ash, J. F., Bradley, M. O., Luduena, M. A., Taylor, E. L., Wrenn, J. T., and Yamada, K. M. (1971). *Science* **171**, 135–143.
White, G. A. (1971). *Biochem. Biophys. Res. Commun.* **44**, 1212–1219.
Whittaker, R. H. (1969). *Science* **163**, 150–163.
Wildman, J. D. (1966). *J. Ass. Offic. Anal. Chem.* **49**, 562–566.
Wood, D. D., and Luck, D. J. L. (1971). *J. Cell Biol.* **51**, 249–264.
Wood-Baker, A. (1955). *Trans. Brit. Mycol. Soc.* **38**, 291–297.
Zaki, A. I., Eckert, J. W., and Endo, R. M. (1973). *Pestic. Biochem. Physiol.* **3**, 7–13.

SUBJECT INDEX

A

Actinomycins, 28, 36
  Micromonospora as source, 29, 34, 36
  Streptomyces as source, 29, 34, 36
AICAR (aminoimidazolecarboxylic acid riboside)
  converted to 5′-GMP, 227
  by fermentation, 219, 220, 223
American Type Culture Collection, handling of patent cultures, 2
Antibiotic production as culture characteristics, 25, 26
Anticapsin
  by *Streptomyces*, 29, 36
  by *Bacillus* sp., 29, 36

B

Benomyl, effects on fungi, 293–294, 309

C

Captan, effects on fungi, 308
CBS (Centraal Bureau voor Schimmelcultures), handling of patent cultures, 4
Cephem antibiotics
  by *Cephalosporium*, 29, 40
  by *Streptomyces*, 29, 40
Chloramphenicol, mode of action on fungi, 314
Citrinin
  by aspergilli, 30, 42
  by penicillia, 30, 42
  by *Crotalaria crispato*, 30, 42
  by *Clavariopsis aquatica*, 30, 42
CMI (Commonwealth Mycological Institute), handling of patent cultures, 4
Coenzyme A
  biosynthesis of intermediates, 240
  mechanisms of biosynthesis, 236–239
  production, 236
Composting of soil waste
  continuous composting, 138–141

C/N ratio, 141
  heat origin, 117–127
  moisture-aeration-temperature interrelationships, 143–148
  pH, 141–143
  role of actinomycetes, 130–132
  role of bacteria, 132–134
  role of fungi, 127–130
  variations, 116–117
Cyclic AMP, by fermentation, 215–216
Cycloheximide
  effects on fungi, 299–300
  mechanism of action, 314
Cycloserine
  by *Pseudomonas*, 29, 36
  by *Streptomyces*, 29, 36
Cytochalasins, effects on fungi, 298–299

D

Denitrification
  aeration, effects of, 161–164
  biochemistry of, 157–161
  C/N concentrations, 169–171
  pH, effects on, 164–166
  temperature, effects on, 164–166
5-Deoxynucleotides, by enzyme hydrolysis of RNA and DNA, 24–25
DNA:DNA association 60–63
DNA nucleotide composition of actinomycetes, 60–61
L-Dopa, enzymic synthesis, 278–285

E

Erythromycin A
  by *Arthrobacter*, 29, 38
  by *Streptomyces erythreus*, 29, 38

F

FAD (flavin adenine dinucleotide), from microbial cells, 234–236
Fermentation pilot plant
  assignments, 192–194

321

systemic fungicides, 292–295
  benomyl, 293–294
  triarmol and triforine, 295
Transribosylation of nucleosides, 227–230
β-Tyrosinase
  cofactor requirements, 254–255
  catalytic properties, 25
    α,β elimination, 255–256, 258–264
    racemization, 257–264
    β replacement, 256–264, 270–273
    synthetic reactions, 257–258, 264–266
  immobilization on Sepharose, 273–274
  microbial production, 275–278
  microbial sources, 274–275
  molecular weight, 253–254
  purification procedures, 250–253
  stability, 253
L-Tyrosine, enzymic synthesis, 278–285

X

5'-XMP (5'-xanthinemonophosphate) by fermentation, 218, 220–221, 223

CONTENTS OF PREVIOUS VOLUMES

Volume 1

Protected Fermentation
  Miloš Herold and Jan Nečásek

The Mechanism of Penicillin Biosynthesis
  Arnold L. Demain

Preservation of Foods and Drugs by Ionizing Radiations
  W. Dexter Bellamy

The State of Antibiotics in Plant Disease Control
  David Pramer

Microbial Synthesis of Cobamides
  D. Perlman

Factors Affecting the Antimicrobial Activity of Phenols
  E. O. Bennett

Germfree Animal Techniques and Their Applications
  Arthur W. Phillips and James E. Smith

Insect Microbiology
  S. R. Dutky

The Production of Amino Acids by Fermentation Processes
  Shukuo Kinoshita

Continuous Industrial Fermentations
  Philip Gerhardt and M. C. Bartlett

The Large-Scale Growth of Higher Fungi
  Radcliffe F. Robinson and R. S. Davidson

AUTHOR INDEX–SUBJECT INDEX

Volume 2

Newer Aspects of Waste Treatment
  Nandor Porges

Aerosol Samplers
  Harold W. Batchelor

A Commentary on Microbiological Assaying
  F. Kavanagh

Application of Membrane Filters
  Richard Ehrlich

Microbial Control Methods in the Brewery
  Gerhard J. Hass

Newer Development in Vinegar Manufactures
  Rudolph J. Allgeier and Frank M. Hildebrandt

The Microbiological Transformation of Steroids
  T. H. Stoudt

Biological Transformation of Solar Energy
  William J. Oswald and Clarence G. Golueke

SYMPOSIUM ON ENGINEERING ADVANCES IN FERMENTATION PRACTICE

Rheological Properties of Fermentation Broths
  Fred H. Deindoerfer and John M. West

Fluid Mixing in Fermentation Processes
  J. Y. Oldshue

Scale-up of Submerged Fermentations
  W. H. Bartholemew

Air Sterilization
  Arthur E. Humphrey

Sterilization of Media for Biochemical Processes
  Lloyd L. Kempe

Fermentation Kinetics and Model Processes
  Fred H. Deindoerfer

Continuous Fermentation
W. D. Maxon

Control Applications in Fermentation
George J. Fuld

AUTHOR INDEX–SUBJECT INDEX

Volume 3

Preservation of Bacteria by Lyophilization
Robert J. Heckly

Sphaerotilus, Its Nature and Economic Significance
Norman C. Dondero

Large-Scale Use of Animal Cell Cultures
Donald J. Merchant and C. Richard Eidam

Protection Against Infection in the Microbiological Laboratory: Devices and Procedures
Mark A. Chatigny

Oxidation of Aromatic Compounds by Bacteria
Martin H. Rogoff

Screening for and Biological Characterizations of Antitumor Agents Using Microorganisms
Frank M. Schabel, Jr., and Robert F. Pittillo

The Classification of Actinomycetes in Relation to Their Antibiotic Activity
Elio Baldacci

The Metabolism of Cardiac Lactones by Microorganisms
Elwood Titus

Intermediary Metabolism and Antibiotic Synthesis
J. D. Bu'Lock

Methods for the Determination of Organic Acids
A. C. Hulme

AUTHOR INDEX–SUBJECT INDEX

Volume 4

Induced Mutagenesis in the Selection of Microorganisms
S. I. Alikhanian

The Importance of Bacterial Viruses in Industrial Processes, Especially in the Dairy Industry
F. J. Babel

Applied Microbiology in Animal Nutrition
Harlow H. Hall

Biological Aspects of Continuous Cultivation of Microorganisms
T. Holme

Maintenance and Loss in Tissue Culture of Specific Cell Characteristics
Charles C. Morris

Submerged Growth of Plant Cells
L. G. Nickell

AUTHOR INDEX–SUBJECT INDEX

Volume 5

Correlations between Microbiological Morphology and the Chemistry of Biocides
Adrien Albert

Generation of Electricity by Microbial Action
J. B. Davis

Microorganisms and the Molecular Biology of Cancer
G. F. Gause

Rapid Microbiological Determinations with Radioisotopes
Gilbert V. Levin

The Present Status of the 2,3-Butylene Glycol Fermentation
Sterling K. Long and Roger Patrick

Aeration in the Laboratory
W. R. Lockhart and R. W. Squires

Stability and Degeneration of Microbial Cultures on Repeated Transfer
Fritz Reusser

Microbiology of Paint Films
Richard T. Ross

The Actinomycetes and Their Antibiotics
Selman A. Waksman

Fusel Oil
A. Dinsmoor Webb and John L. Ingraham

AUTHOR INDEX–SUBJECT INDEX

Volume 6

Global Impacts of Applied Microbiology: An Appraisal
Carl-Göran Hedén and Mortimer P. Starr

Microbial Processes for Preparation of Radioactive Compounds
D. Perlman, Aris P. Bayon, and Nancy A. Giuffre

Secondary Factors in Fermentation Processes
P. Margalith

Nonmedical Uses of Antibiotics
Herbert S. Goldberg

Microbial Aspects of Water Pollution Control
K. Wuhrmann

Microbial Formation and Degradation of Minerals
Melvin P. Silverman and Henry L. Ehrlich

Enzymes and Their Applications
Irwin W. Sizer

A Discussion of the Training of Applied Microbiologists
B. W. Koft and Wayne W. Umbreit

AUTHOR INDEX–SUBJECT INDEX

Volume 7

Microbial Carotenogenesis
Alex Ciegler

Biodegradation: Problems of Molecular Recalcitrance and Microbial Fallibility
M. Alexander

Cold Sterilization Techniques
John B. Opfell and Curtis E. Miller

Microbial Production of Metal–Organic Compounds and Complexes
D. Perlman

Development of Coding Schemes for Microbial Taxonomy
S. T. Cowan

Effects of Microbes on Germfree Animals
Thomas D. Luckey

Uses and Products of Yeasts and Yeastlike Fungi
Walter J. Nickerson and Robert G. Brown

Microbial Amylases
Walter W. Windish and Nagesh S. Mhatre

The Microbiology of Freeze-Dried Foods
Gerald J. Silverman and Samuel A. Goldblith

Low-Temperature Microbiology
Judith Farrell and A. H. Rose

AUTHOR INDEX–SUBJECT INDEX

Volume 8

Industrial Fermentations and Their Relations to Regulatory Mechanisms
Arnold L. Demain

Genetics in Applied Microbiology
S. G. Bradley

Microbial Ecology and Applied Microbiology
Thomas D. Brock

The Ecological Approach to the Study of Activated Sludge
Wesley O. Pipes

Control of Bacteria in Nondomestic Water Supplies
Cecil W. Chambers and Norman A. Clarke

The Presence of Human Enteric Viruses in Sewage and Their Removal by Conventional Sewage Treatment Methods
Stephen Alan Kollins

Oral Microbiology
Heiner Hoffman

Media and Methods for Isolation and Enumeration of the Enterococci
Paul A. Hartman, George W. Reinbold, and Devi S. Saraswat

Crystal-Forming Bacteria as Insect Pathogens
Martin H. Rogoff

Mycotoxins in Feeds and Foods
Emanuel Borker, Nino F. Insalata, Colette P. Levi, and John S. Witzeman

AUTHOR INDEX–SUBJECT INDEX

Volume 9

The Inclusion of Antimicrobial Agents in Pharmaceutical Products
A. D. Russell, June Jenkins, and I. H. Harrison

Antiserum Production in Experimental Animals
Richard M. Hyde

Microbial Models of Tumor Metabolism
G. F. Gause

Cellulose and Cellulolysis
Brigitta Norkrans

Microbiological Aspects of the Formation and Degradation of Cellulosic Fibers
L. Jurášek, J. Ross Colvin, and D. R. Whitaker

The Biotransformation of Lignin to Humus—Facts and Postulates
R. T. Oglesby, R. F. Christman, and C. H. Driver

Bulking of Activated Sludge
Wesley O. Pipes

Malo-lactic Fermentation
Ralph E. Kunkee

AUTHOR INDEX–SUBJECT INDEX

Volume 10

Detection of Life in Soil on Earth and Other Planets. Introductory Remarks
Robert L. Starkey

For What Shall We Search?
Allan H. Brown

Relevance of Soil Microbiology to Search for Life on Other Planets
G. Stotzky

Experiments and Instrumentation for Extraterrestrial Life Detection
Gilbert V. Levin

Halophilic Bacteria
D. J. Kushner

Applied Significance of Polyvalent Bacteriophages
S. G. Bradley

Proteins and Enzymes as Taxonomic Tools
Edward D. Garber and John W. Rippon

Mycotoxins
Alex Ciegler and Eivind B. Lillehoj

Transformation of Organic Compounds by Fungal Spores
Claude Vézina, S. N. Sehgal, and Kartar Singh

Microbial Interactions in Continuous Culture
Henry R. Bungay, III and Mary Lou Bungay

Chemical Sterilizers (Chemosterilizers)
Paul M. Borick

Antibiotics in the Control of Plant Pathogens
M. J. Thirumalachar

AUTHOR INDEX–SUBJECT INDEX

CUMULATIVE AUTHOR INDEX–CUMULATIVE TITLE INDEX

Volume 11

Successes and Failures in the Search for Antibiotics
Selman A. Waksman

Structure–Activity Relationships of Semisynthetic Penicillins
K. E. Price

Resistance to Antimicrobial Agents
J. S. Kiser, G. O. Gale, and G. A. Kemp

Micromonospora Taxonomy
George Luedemann

Dental Caries and Periodontal Disease Considered as Infectious Diseases
William Gold

The Recovery and Purification of Biochemicals
Victor H. Edwards

Ergot Alkaloid Fermentations
William J. Kelleher

The Microbiology of the Hen's Egg
R. G. Board

Training for the Biochemical Industries
I. L. Hepner

AUTHOR INDEX–SUBJECT INDEX

Volume 12

History of the Development of a School of Biochemistry in the Faculty of Technology, University of Manchester
Thomas Kennedy Walker

Fermentation Processes Employed in Vitamin C Synthesis
Miloš Kulhánek

Flavor and Microorganisms
P. Margalith and Y. Schwartz

Mechanisms of Thermal Injury in Nonsporulating Bacteria
M. C. Allwood and A. D. Russell

Collection of Microbial Cells
Daniel I. C. Wang and Anthony J. Sinskey

Fermentor Design
R. Steel and T. L. Miller

The Occurrence, Chemistry, and Toxicology of the Microbial Peptide-Lactones
A. Taylor

Microbial Metabolites as Potentially Useful Pharmacologically Active Agents

D. Perlman and G. P. Peruzzotti

AUTHOR INDEX–SUBJECT INDEX

Volume 13

Chemotaxonomic Relationships Among the Basidiomycetes
Robert G. Benedict

Proton Magnetic Resonance Spectroscopy—An Aid in Identification and Chemotaxonomy of Yeasts
P. A. J. Gorin and J. F. T. Spencer

Large-Scale Cultivation of Mammalian Cells
R. C. Telling and P. J. Radlett

Large-Scale Bacteriophage Production
K. Sargeant

Microorganisms as Potential Sources of Food
Jnanendra K. Bhattacharjee

Structure–Activity Relationships Among Semisynthetic Cephalosporins
M. L. Sassiver and Arthur Lewis

Structure–Activity Relationships in the Tetracycline Series
Robert K. Blackwood and Arthur R. English

Microbial Production of Phenazines
J. M. Ingram and A. C. Blackwood

The Gibberellin Fermentation
E. G. Jeffreys

Metabolism of Acylanilide Herbicides
Richard Bartha and David Pramer

Therapeutic Dentifrices
J. K. Peterson

Some Contributions of the U.S. Department of Agriculture to the Fermentation Industry
George E. Ward

Microbiological Patents in International Litigation
John V. Whittenburg

Industrial Applications of Continuous Culture: Pharmaceutical Products and Other Products and Processes
R. C. Righelato and R. Elsworth

Mathematical Models for Fermentation Processes
A. G. Frederickson, R. D. Megee, III, and H. M. Tsuchija

AUTHOR INDEX–SUBJECT INDEX

Volume 14

Development of the Fermentation Industries in Great Britain
John J. H. Hastings

Chemical Composition as a Criterion in the Classification of Actinomycetes
H. A. Lechevalier, Mary P. Lechevalier, and Nancy N. Gerber

Prevalence and Distribution of Antibiotic-Producing Actinomycetes
John N. Porter

Biochemical Activities of Nocardia
R. L. Raymond and V. W. Jamison

Microbial Transformations of Antibiotics
Oldrich K. Sebek and D. Perlman

In Vivo Evaluation of Antibacterial Chemotherapeutic Substances
A. Kathrine Miller

Modification of Lincomycin
Barney J. Magerlein

Fermentation Equipment
G. L. Solomons

The Extracellular Accumulation of Metabolic Products by Hydrocarbon-Degrading Microorganisms
Bernard J. Abbott and William E. Gledhill

AUTHOR INDEX–SUBJECT INDEX

Volume 15

Medical Applications of Microbial Enzymes
Irwin W. Sizer

Immobilized Enzymes
K. L. Smiley and G. W. Strandberg

Microbial Rennets
Joseph L. Sardinas

Volatile Aroma Components of Wines and Other Fermented Beverages
A. Dinsmoor Webb and Carlos J. Muller

Correlative Microbiological Assays
Ladislav J. Haňka

Insect Tissue Culture
W. F. Hink

Metabolites from Animal and Plant Cell Culture
Irving S. Johnson and George B. Boder

Structure–Activity Relationships in Coumermycins
John C. Godfrey and Keeneth E. Price

Chloramphenicol
Vedpal E. Malik

Microbial Utilization of Methanol
Charles L. Cooney and David W. Levine

Modeling of Growth Processes with Two Liquid Phases: A Review of Drop Phenomena, Mixing, and Growth
P. S. Shah, L. T. Fan, I. C. Kao, and L. E. Erickson

Microbiology and Fermentations in the Prairie Regional Laboratory of the National Research Council of Canada 1946–1971
R. H. Haskins

AUTHOR INDEX–SUBJECT INDEX

Volume 16

Public Health Significance of Feeding Low Levels of Antibiotics to Animals
Thomas H. Jukes

Intestinal Microbial Flora of the Pig
R. Kenworthy

Antimycin A, a Piscicidal Antibiotic
Robert E. Lennon and Claude Vézina

Ochratoxins
Kenneth L. Applegate and John R. Chipley

Cultivation of Animal Cells in Chemically Defined Media, A Review
Kiyoshi Higuchi

Genetic and Phenetic Classification of Bacteria
R. R. Colwell

Mutation and the Production of Secondary Metabolites
Arnold L. Demain

Structure–Activity Relationships in the Actinomycins
Johannes Meienhofer and Eric Atherton

Development of Applied Microbiology at the University of Wisconsin
William B. Sarles

AUTHOR INDEX–SUBJECT INDEX

Volume 17

Education and Training in Applied Microbiology
Wayne W. Umbreit

Antimetabolites from Microorganisms
David L. Pruess and James P. Scannell

Lipid Composition as a Guide to the Classification of Bacteria
Norman Shaw

Fungal Sterols and the Mode of Action of the Polyene Antibiotics
J. M. T. Hamilton-Miller

Methods of Numerical Taxonomy for Various Genera of Yeasts
I. Campbell

Microbiology and Biochemistry of Soy Sauce Fermentation
F. M. Yong and B. J. B. Wood

Contemporary Thoughts on Aspects of Applied Microbiology
P. S. S. Dawson and K. L. Phillips

Some Thoughts on the Microbiological Aspects of Brewing and Other Industries Utilizing Yeast
G. G. Stewart

Linear Alkylbenzene Sulfonate: Biodegradation and Aquatic Interactions
William E. Gledhill

The Story of the American Type Culture Collection—Its History and Development (1899–1973)
William A. Clark and Dorothy H. Geary

Microbial Penicillin Acylases
E. J. Vandamme and J. P. Voets

SUBJECT INDEX

Volume 18

Microbial Formation of Environmental Pollutants
Martin Alexander

Microbial Transformation of Pesticides
Jean-Marc Bollag

Taxonomic Criteria for Mycobacteria and Nocardiae
S. G. Bradley and J. S. Bond

Effect of Structural Modifications on the Biological Properties of Aminoglycoside Antibiotics Containing 2-Deoxystreptamine
Kenneth E. Price, John C. Godfrey, and Hiroshi Kawaguchi

Recent Developments of Antibiotic Research and Classification of Antibiotics According to Chemical Structure
János Bérdy

SUBJECT INDEX